The Handbook of Environmental Chemistry

Volume 93

Founding Editor: Otto Hutzinger

Series Editors: Damià Barceló • Andrey G. Kostianoy

In over three decades, *The Handbook of Environmental Chemistry* has established itself as the premier reference source, providing sound and solid knowledge about environmental topics from a chemical perspective. Written by leading experts with practical experience in the field, the series continues to be essential reading for environmental scientists as well as for environmental managers and decision-makers in industry, government, agencies and public-interest groups.

Two distinguished Series Editors, internationally renowned volume editors as well as a prestigious Editorial Board safeguard publication of volumes according to high scientific standards.

Presenting a wide spectrum of viewpoints and approaches in topical volumes, the scope of the series covers topics such as

- local and global changes of natural environment and climate
- anthropogenic impact on the environment
- water, air and soil pollution
- remediation and waste characterization
- environmental contaminants
- biogeochemistry and geoecology
- chemical reactions and processes
- chemical and biological transformations as well as physical transport of chemicals in the environment
- environmental modeling

A particular focus of the series lies on methodological advances in environmental analytical chemistry.

The Handbook of Envir onmental Chemistry is available both in print and online via http://link.springer.com/bookseries/698. Articles are published online as soon as they have been reviewed and approved for publication.

Meeting the needs of the scientific community, publication of volumes in subseries has been discontinued to achieve a broader scope for the series as a whole.

The Rivers of Montenegro

Volume Editors: Vladimir Pešić · Momir Paunović ·
Andrey G. Kostianoy

With contributions by

N. Blagojević · Z. Bulić · B. Csányi · A. Divac Rankov · J. Đorđević ·
A. Farnleitner · P. Glöer · M. Grabowski · S. Hadžiablahović · B. Ilić ·
M. Ilić · S. Ixenmaier · S. Jarić · D. Joksimović · S. Jokanović ·
J. Jovanović · M. Jovanović · A. K. T. Kirschner · S. Kolarević ·
A. G. Kostianoy · E. A. Kostianaia · J. Kostić-Vuković ·
M. Kračun-Kolarević · B. Karadžić · R. Linke · D. Marić ·
R. Martinović · M. Mitrović · M. Paunović · A. Pavićević ·
P. Pavlović · V. Pešić · N. Popović · M. M. Radulović · M. Raković ·
G. Reischer · A. Savić · D. Savio · G. Sekulić · D. M. Soloviev ·
K. Sunjog · N. Tomić · J. Tomović · V. Vukašinović-Pešić ·
B. Vuković-Gačić

Springer

Editors
Vladimir Pešić
Department of Biology
University of Montenegro
Podgorica, Montenegro

Momir Paunović
Institute for Biological Research "Siniša Stanković" – National Institute of the Republic of Serbia
University of Belgrade
Belgrade, Serbia

Andrey G. Kostianoy
Shirshov Institute of Oceanology
Russian Academy of Sciences
S.Yu. Witte Moscow University
Moscow, Russia

ISSN 1867-979X ISSN 1616-864X (electronic)
The Handbook of Environmental Chemistry
ISBN 978-3-030-55711-9 ISBN 978-3-030-55712-6 (eBook)
https://doi.org/10.1007/978-3-030-55712-6

This Springer imprint is published by the registered company Springer Nature Switzerland AG.
The registered company address is: Gewerbestrasse 11, 6330 Cham, Switzerland

Series Preface

With remarkable vision, Prof. Otto Hutzinger initiated *The Handbook of Environmental Chemistry* in 1980 and became the founding Editor-in-Chief. At that time, environmental chemistry was an emerging field, aiming at a complete description of the Earth's environment, encompassing the physical, chemical, biological, and geological transformations of chemical substances occurring on a local as well as a global scale. Environmental chemistry was intended to provide an account of the impact of man's activities on the natural environment by describing observed changes.

While a considerable amount of knowledge has been accumulated over the last four decades, as reflected in the more than 150 volumes of *The Handbook of Environmental Chemistry*, there are still many scientific and policy challenges ahead due to the complexity and interdisciplinary nature of the field. The series will therefore continue to provide compilations of current knowledge. Contributions are written by leading experts with practical experience in their fields. *The Handbook of Environmental Chemistry* grows with the increases in our scientific understanding, and provides a valuable source not only for scientists but also for environmental managers and decision-makers. Today, the series covers a broad range of environmental topics from a chemical perspective, including methodological advances in environmental analytical chemistry.

In recent years, there has been a growing tendency to include subject matter of societal relevance in the broad view of environmental chemistry. Topics include life cycle analysis, environmental management, sustainable development, and socio-economic, legal and even political problems, among others. While these topics are of great importance for the development and acceptance of *The Handbook of Environmental Chemistry*, the publisher and Editors-in-Chief have decided to keep the handbook essentially a source of information on "hard sciences" with a particular emphasis on chemistry, but also covering biology, geology, hydrology and engineering as applied to environmental sciences.

The volumes of the series are written at an advanced level, addressing the needs of both researchers and graduate students, as well as of people outside the field of "pure" chemistry, including those in industry, business, government, research establishments, and public interest groups. It would be very satisfying to see these volumes used as a basis for graduate courses in environmental chemistry. With its high standards of scientific quality and clarity, *The Handbook of Environmental Chemistry* provides a solid basis from which scientists can share their knowledge on the different aspects of environmental problems, presenting a wide spectrum of viewpoints and approaches.

The Handbook of Environmental Chemistry is available both in print and online via www.springerlink.com/content/110354/. Articles are published online as soon as they have been approved for publication. Authors, Volume Editors and Editors-in-Chief are rewarded by the broad acceptance of *The Handbook of Environmental Chemistry* by the scientific community, from whom suggestions for new topics to the Editors-in-Chief are always very welcome.

Damià Barceló
Andrey G. Kostianoy
Series Editors

Contents

The Rivers of Montenegro: Introductory Remarks

Vladimir Pešić, Momir Paunović, and Andrey G. Kostianoy

Contents

Abstract The river network in Montenegro is primarily affected by the geological processes in the karst which dominates the Montenegrin landscape. All the surface and groundwater from the territory of Montenegro flow into the Black and Adriatic Seas, with approximately 45.4% (6,268 km^2) of its surface belonging to the Adriatic and 7,545 km^2 (about 54.6%) to the Black Sea (Danube) basin.

As one of the last remaining strongholds of unvalorized resources, the Montenegrin rivers have become a point of possible conflict as they become the object of the social, political, ecological, and economic changes through which the country has passed, most especially in recent decades. As a result of the poor management of aquatic resources and the absence of an established set of equitable sharing principles among all the parties involved, the number of conflicts over the exploitation of the water resources has increased, not only in its role as an economic and biodiversity issues but also with the active involvement and participation of the local community. This book aims to provide an extensive overview of the various

V. Pešić (✉)
Faculty of Sciences, University of Montenegro, Podgorica, Montenegro
e-mail: vladopesic@gmail.com

M. Paunović
Institute for Biological Research "Siniša Stanković" - National Institute of Republic of Serbia, University of Belgrade, Belgrade, Serbia
e-mail: mpaunovi@ibiss.bg.ac.rs

A. G. Kostianoy
P.P. Shirshov Institute of Oceanology, Russian Academy of Sciences, Moscow, Russia

S.Yu. Witte Moscow University, Moscow, Russia
e-mail: kostianoy@gmail.com

Vladimir Pešić, Momir Paunović, and Andrey G. Kostianoy (eds.), *The Rivers of Montenegro*, Hdb Env Chem (2020) 93: 1–12, DOI 10.1007/698_2019_416,

limnological and hydrobiological aspects of the Montenegrin river basins, so as to stimulate a science-based management approach to the use of water resources.

Keywords Adriatic basin, Adriatic Sea, Black Sea Basin, Karst springs, Montenegro, River basin, Rivers, Skadar Lake

1 Rivers as Hydrological Objects

"It's not the *river* that flows – it's the water.
It's not the time that passes – it's us!"
Ivo Andrić, Yugoslav novelist, the Nobel Prize winner in literature (1961)

The network of rivers in Montenegro (Fig. 1) is strongly affected by the latitudinal arrangements of the landforms. The mountain chain of the Dinarides (the Dinaric Alps) that extends in a southeast-northwest direction, parallel to the current shore of the Adriatic Sea coast, dominates in the relief of Montenegro [2, 3]. The direction of the mountain chains in Montenegro and a strong altitudinal gradient in a southwest to northeast direction are the major barriers to surface runoff that modify the direction of the flows. In the Adriatic basin, groundwater flow, which generally mirrors the surface water flow, flows to the south and southwest, while the groundwater regime in the Danube basin flows to the north and northwest [4].

In the landscape of Montenegro, three main geomorphological units can be distinguished, (1) the Mediterranean coastal area (Coastal Montenegro), (2) the sub-Mediterranean central area (Central Montenegro), and (3) the mountainous northern – northeastern area (Northern Montenegro) [5]. The main rock type in the Dinarides is karst, which dominates the majority of the Montenegrin landscape. Dinaric karst is distinguished by a complex tectonic set of diverse formations from the Paleozoic, Mesozoic, and Cenozoic ages, dominated by limestones and dolomite sedimentary rocks in which, due to tectonic activity, folding, faulting, and overthrusting, the process of karstification has come to full expression [2]. Typical Dinaric karst forms a diverse array of landforms of which karren fields, sinkholes (dolines), uvalas, fields, and caves are particularly well developed in Montenegro [3].

According to the geological map of Montenegro [6], four geotectonic units exist in Montenegro: (1) the *Durmitor* tectonic unit, which is in the northeast, (2) the *Visoki Krš* tectonic zone that includes two tectonic units: *Starocrnogorska Kraljušt* and *Kučka Kraljušt*, (3) toward the southwest we find the *Budva-Cukali* zone, and (4) the *Paraautohton* (*Adriatic-Ionian*) zone extends along the Adriatic coast. The chain of high coastal mountains (NW–SE oriented) composed of Mts. Orjen, Lovćen, Sutorman, and Rumija run toward the northeast and encounters the *High Karst* zone. This zone, characterized by diverse underground karst forms, at its southeastern limit runs into the Zeta-Skadar depression, while toward the northeast,

Fig. 1 Map of Montenegro with its main river basins. Modified from [1]

it descends into Nikšić polje (field) and the valley of the River Zeta (the Bjelopavlići Valley) [7]. From this area, karst terrains rapidly expand to the central part of Montenegro forming a large number of high mountains (e.g., Njegoš, Žijevo, Prokletije, Komovi, Visitor, Pivske Planine, Sinjajevina, Durmitor, and others), which are bisected by river valleys and impressive canyons such as those of the rivers Morača, Piva (Fig. 2), Tara (Fig. 3), Čehotina, Lim, and Ibar [7].

Geological processes in the karst have had a decisive influence on hydrology and the formation of the hydrographic network [8]. From the territory of Montenegro,

Fig. 2 The confluence of the river Tara and the river Piva at Šćepan Polje at the border between Montenegro and Bosnia and Herzegovina. The Montenegrin rivers (the Piva, the Tara, the Lim, and the Ćehotina) represent about one third of the entire basin of the River Drina, but they provide more than 52.9% of the mean annual river flow [4]. Photo by D. Marić

surface and groundwater flow into both the Black and Adriatic Seas, with approximately 45.4% (about 6,268 km^2) of its surface belong the Adriatic basin and around 54.6% (about 7,545 km^2) to the Black Sea (Danube) basin (Fig. 1) [9].

The terrain along the drainage boundaries is mainly formed by the Mesozoic carbonate sediment that builds the famous karst terrains of the Dinarides, which leads that the hydrological drainage boundaries between the Adriatic and Black Sea's basins are mostly underground [7]. This makes it difficult to define the drainage boundaries between individual catchments, especially between smaller basins in the holokarstic terrain (e.g., between Trebišnjica and Boka Kotorska Bay on the one hand and the basin of the Nikšić polje on the other) [7].

The Adriatic basin generally experiences a "Mediterranean" type climate, with a predominantly hot and dry summer with average temperature of >22°C in the warmest month, and much lower temperatures in the winter, although they still remain above freezing point [10]. A moderately warm climate with a dry summer but without a pronounced dry period over the year is present in the northwestern, western, and southwestern part of the Adriatic basin at altitudes ≥650 m [7]. On the other hand, the Black Sea basin experiences a "continental" type climate without a dry period over the year, with relatively cool and humid summers and long and harsh winters.

Fig. 3 The canyon of the River Tara, 78 km long and 1,350 m deep, is the deepest canyon in Europe and the second deepest in the world after the Colorado River Canyon [7]. In Montenegro the canyon is protected as a part of the Durmitor National Park, and since 1977 it has been a UNESCO World Heritage Site. Photo by S. Popović

Fig. 4 Most rivers found in coastal Montenegro such as the river Ljuta (between the towns of Kotor and Perast) are short and do not have floodplain habitats; usually they rise from strong karstic springs and descend abruptly into the sea after short stretches. Photo by V. Pešić

The amount of precipitation rapidly increases from the southwest toward the northeast [7]. The greatest annual amount of precipitation of 8,063 mm was registered at Crkvine on Mt. Orjen in 1938, which is still the European record [7]. The average annual amount of precipitation as well as the mean annual number of days with precipitation strongly varies both between the river basins and across the year. November and December are the rainiest months, while July and August are the driest months [7].

Montenegrin rivers are characterized by a complex flow regime that depends on the climatic conditions and morphogenetic zones from where these rivers flow. Most of the rivers of the Black Sea basin are characterized by large flows in the spring (April and May) and late autumn (November and December) with their minimum flows coming between August and September [7]. The rivers in the Adriatic basin, which are closer to the sea, exhibit their highest flows in November and December. On the upper part of the Morača and the Zeta, the greatest water flow was recorded in the spring [7].

There is considerable variability in the flow between different basins. The River Ćehotina has the most stable average annual flow, while the Ibar exhibits the greatest variability [7]. On the temporal scale, the variability of monthly flows is greatest in the autumn [7]. Moreover, a considerable difference in the river density between the basins was found, in that the river network is denser in the Black Sea basin and less dense in the Adriatic one (Fig. 4).

The contribution of surface runoff to the river basin discharge varies both spatially and temporally. Underground water drainage is the main factor that affects the stability of the water in the karstic terrain. Most rivers and streams in the Adriatic basin dry up in the summer months as a result of the fact that field evaporation exceeds precipitation in this period of the year. In general, it can be noticed that there is a tendency toward decreasing precipitation moving from the northwest toward the southeast, in the direction of Cetinje-Ostros, as well in toward Nikšić-Podgorica [7], consequently leading to a denser network of intermittent rivers and ephemeral streams (IRES) along those gradients. Intermittent rivers that have been often neglected in limnological studies dominate in the river network of South Montenegro, and they are analyzed here in a special chapter in this book.

2 The Rivers of Montenegro as Scientific Objects: Challenges and Perspectives

The first systematic limnological and hydrobiological research of Montenegro began in the second half of the nineteenth century and was mostly performed by foreign researchers (e.g., for hydrogeology, see [3]; for ichthyology and fisheries, see [11]; for limnological research in the Lake Skadar basin, see [12]).

The period after World War II was characterized by the formation of institutions involved in the fundamental and applied research of the aquatic resources of Montenegro. The Geological Survey of Montenegro, the main public institution responsible for regional hydrogeological investigations, was founded in 1945. The Hydrometeorological Institute (Montenegrin: Hidrometeorološki zavod Crne Gore or RHMZ), the main national hydrological and the meteorological service in the country involved in the monitoring of meteorological, hydrological, and ecological parameters, began working in 1947. The first hydrobiological institution, the Fishing Station of the People's Republic of Montenegro, was founded in 1952 and worked until 1965 when it was transformed into the biological station, which worked within the Institute for Biological and Medical Research (from 1973) as part of the state University in Titograd (the modern Podgorica) [13].

Regardless of the large amount of disciplinary-specific knowledge acquired, most of the data that are crucial for understanding the history of the Montenegrin river basins and their living world [14] remain highly fragmented, and a comprehensive meta-analysis is long overdue. An analysis of the comprehensive bibliography on the Lake Shkodra/Skadar basin [12] revealed the difference in the amount of data of discipline-specific knowledge and variation of the number of publications over the time [15]. For example, in most disciplines, relatively few studies were reported prior to the 1940s, with most publications occurring from 2000 up to the present (e.g., see [15] and references cited therein).

The largest number of limnological and hydrobiological studies done so far are related to the Lake Skadar basin which is paleo- and phylogeographically the best

Fig. 5 The canyon of the river Cemi/Cijevna that flows through Albania and Montenegro, situated at the fifth kilometer from Podgorica; in places, it is less than 1 m wide. Photo by S. Popović

and most widely studied hydrological area of Montenegro (see [14] for an overview). Phylogeographic studies have confirmed that the lake itself cannot be referred to as an ancient lake but is rather a very recent water body that some 1,200 years ago took over a former freshwater marshlands [16]. However, regardless of the recent formation of this lacustrine system, the Lake Skadar basin with its large system of karst springs is definitely ancient, having originated more than 2.5 million years ago and being isolated for most of its history, with a high number of crenal and fluvial endemics, at both the morphospecies level and potentially also at the level of lineage (cryptic or pseudocryptic species) [14].

While employing an integrative methodology based on comparative phylogeographic studies on the biota of Montenegrin rivers, this also needs to be complemented by geological evidence with the aim of building up reliable knowledge on the changes throughout which the river basins (Fig. 5) of Montenegro and their living world pass [14]. This would help to more accurately forecast even those changes that are likely to occur, which is particularly important because, for many rhitrobiontic species, there is enough evidence that they are being altered by ongoing climate change.

Climate change, which has markedly increased since the early 1990s, has had a particularly significant impact on the frequency, duration, and intensity of extreme hydrologic events in Montenegro [10]. For example, in the last decade, Montenegro has experienced severe droughts (in 2000, 2003, 2007, and 2012) and has suffered losses from damaging floods, most notably in 2010. The increased frequency and duration of drought period affects the river ecosystems by increasing the share of IRES in the regional hydrological network, by reducing the flow in the lotic and increasing the water temperature in the lentic parts of the perennial rivers. On the other hand, the higher frequency and abundance of floods accelerate bank erosion, increasing suspended sediment load into the rivers, which leads to reduced diversity among aquatic biota and their communities. In addition, the changes in local hydrology as the result of regional warming may also reduce the hydropower capacity of exposed rivers.

Fig. 6 The small hydropower plant at Bistrica river in Bijelo Polje Municipality The loss of the small streams in a rural regions, the loss of a fish and biodiversity, as well as the degradation of other environmental services such as using the water for irrigation but also for recreation have been recognized by the local community as the main threats caused by building small hydro power plants in the rural regions of Montenegro. Photos by S.Višnjić

The rivers in Montenegro are not only exposed to climate change but also to various types of anthropogenic activities. The impacts of land-use changes, urbanization, deforestation, agriculture, industry, sand and gravel extraction, fisheries, mining, hydropower development, and hazardous weather events are recognized as the main pressures threatening the integrity of the rivers and their basins in Montenegro. Some of the above-mentioned factors have had a serious impact on the economy, especially in the agricultural and livestock production sectors, due to their dependence on the water conditions.

In the more rural parts of Montenegro, especially in its northern regions, more and more pristine rivers have become dammed, which significantly affects their functionality and biodiversity. The construction of dams and reservoirs for hydroelectricity has transformed some rivers into standing waters (e.g., the accumulation on the river Piva), changing the pattern of the water flows, with a direct influence on the depth and the substrate contents consequently leading to changes in fish and macrozoobenthic communities and their interaction with environmental parameters [17]. In the last decade, the building of small hydropower plants (<1 MW) (Fig. 6) has significantly increased, causing a number of conflicts with local community and their interests.

The main problem that appears here is that on the largest number of small rivers and streams in Montenegro where small hydro power plants were built, the watercourses do not have enough water in the summer months to meet the prescribed minimum for them to function (and each power plant without its own accumulations cannot provide such a flow). The result is that the streams which have installed small hydropower plants completely or partially dry up in the summer months, destroying the habitats for aquatic biota and preventing the local community from exploiting the water. Poor management of aquatic resources and the absence of an established set of equitable sharing principles among the contending parties have led to an increase in the number of conflicts over these facilities.

In urban areas, rivers are often used as a sewage recipient, degrading water quality and its capacity for self-purification. The increase in industrial and transportation activities in the major urban centers, such as Podgorica and Nikšić, has caused increasingly degraded water quality in the lower part of the rivers Morača and Zeta, where the impact of sewage effluents is often accompanied by reduced biotic diversity, changes in the trophic status, and the greater toxicity of the sediment (see [15] and references therein). In the agricultural landscapes of the Zeta-Skadar valley, the watercourses have become threatened by the accelerated process of eutrophication as a result of human activities, most especially through fertilizers that are washed out from the soil.

Montenegro is rich in water resources, but as with many other countries, it needs to comply with national and international targets for the protection and sustainable management of those water resources. Due to increased demand for water resources and in the face of ongoing climate change (for example, variation in rainfall), it can be expected that tensions will become more frequent and intensive in the future. The conflicts that mostly occur between biodiversity conservation on the one hand and general human activity on the other are in most cases complex and can severely affect both socioeconomic and biological parameters [18].

As one of the last remaining strongholds of largely unvalorized resources, the rivers have become a point of possible conflict as they become the focus of social, ecological, political, and economic changes. Recent practices in protecting water resources in Montenegro (e.g., the hydroelectric dam controversy on the River Tara), as well as in other countries in the Balkans, reveal that only ensuring political commitment can provide the institutional and legal basis for the sustainable development of water resources in this region.

The central role of researchers and scientific institutions in the conservation of water resources has often been emphasized by governmental and nongovernmental institutions in Montenegro, but, in reality, they have been rarely involved in the management of those resources. The absence of comprehensive in-depth meta-analysis between and/or within river basins impedes the estimation of conservation needs and the planning of the proper conservation strategy for the area. A scientifically based approach should ensure the development of methods that need to provide mechanisms to identify the threats to the river ecosystem and propose measures to ensure the protection of these valuable ecosystems. Collaboration between all the contending parties, from scientists to decision-makers and resource managers, and defining the obligations and responsibilities of the key stakeholders involved in the implementation of the measures planned for the protection of the riverine environment, is seen as the main factor that will prevent potential conflict and guarantee the sustainable use of water resources. The usefulness of such a comprehensive approach has recently been demonstrated by the publishing of a holistic study, "The Skadar/Shkodra Lake Environment," [19] devoted to the environment of the unique but threatened Lake Skadar which is the largest freshwater hydrological object in both Montenegro and the Balkans. The purpose of the current book is to provide an extensive overview of the various aspects of the riverine environment that should stimulate a science-based management approach in the use of water resources in Montenegro.

Acknowledgments We thank Saša Popović (Podgorica), Siniša Višnjić (Bijelo Polje) and Drago Marić (Podgorica) for their permission to use their photos.

References

1. Sekulić G, Radulović M (2020) The hydrology and hydrogeology of Montenegro. In: Pešić V, Paunović M, Kostianoy A (eds) The rivers of Montenegro. The handbook of environmental chemistry. Springer, Cham
2. Barović G, Spalević V, Pešić V, Vujačić D (2018) The physical and geographical characteristics of the Lake Skadar Basin. In: Pešić V, Karaman G, Kostianoy A (eds) The Skadar/Shkodra Lake environment. The handbook of environmental chemistry, vol 80. Springer, Cham, pp 11–23
3. Radulović MM (2018) Hydrogeology of the Skadar Lake Basin. In: Pešić V, Karaman GS, Kostianoy AG (eds) The Skadar/Shkodra Lake environment. The handbook of environmental chemistry, vol 80. Springer, Cham, pp 25–45
4. IWRM (2016) Support to water resources management in the Drina River Basin. Montenegro – IWRM Study and Plan – Background Paper. World Bank. http://www.wb-drinaproject.com/pdf/Roof%20Report%20-%20English%20-%20Final.pdf. Accessed 5 Aug 2019
5. Lenaerts T, Nyssen J, Spalević V, Frankl A (2013) Regional geomorphological mapping of Montenegro: landform genesis and present processes. In: Agrosym 2013, pp 974–981
6. Mirković M, Žvaljević M, Đokić V, Perović Z, Kalezić M, Pajović M (1985) Geological map of Montenegro 1:200,000. The Republic Self Managing Community of Interest for Geological Exploration of Montenegro, Titograd
7. Hrvačević S (2004) Resources of surface water in Montenegro (in Montenegrin). EPCG AD, Podgorica
8. Bešić Z (1969) Geology of Montenegro–Karst of Montenegro (in Serbian). Geological Survey of Montenegro, Podgorica
9. IJC (Institute Jaroslav Cerni) (2001) Water master plan of Montenegro (in Serbian). Government of Montenegro, Podgorica
10. Kostianoy AG, Serykh IV, Kostianaia EA (2018) Climate change in the Lake Skadar/Shkodra region. In: Pešić V, Karaman G, Kostianoy A (eds) The Skadar/Shkodra Lake environment. The handbook of environmental chemistry, vol 80. Springer, Cham, pp 63–88
11. Marić D (2010) Bibliografija o ribarsko-ihtiološkim istraživanjima u Crnoj Gori (150 godina istraživanja) (Bibliography on fishery-ichthyological investigations in Montenegro (150 years of research)). JU "Prirodnjački muzej Crne Gore", Posebno izdanje No 5, Podgorica, 140 pp
12. Kashta L, Dhora D, Sokoli F, Rakaj M, Hoti M, Beka I, Bekteshi A, Ziu T, Pulević, V, Hadžiablahović S, Kasom G, Rakočević-Nedović J, Nikčević, S, Pešić V, Ražnatović A, Ćirović R, Saveljić D, Bušković V, Micev B, Filipović S, Fuštić B, Radulović M, Žižić-Velimirović O (2001) Bibliography on Shkodra/Skadar Lake. Checklist of species. The Regional Environmental Center for Central and Eastern Europe, Activity 1.2.1

13. Pešić V, Karaman GS, Kostianoy AG (2018) Introduction. In: Pešić V, Karaman G, Kostianoy A (eds) The Skadar/Shkodra Lake environment. The handbook of environmental chemistry, vol 80. Springer, Cham, pp 1–10
14. Grabowski M, Jabłońska A, Wysocka A, Pešić V (2018) The obscure history of the Lake Skadar and its biota: a perspective for future research. In: Pešić V, Karaman G, Kostianoy A (eds) The Skadar/Shkodra Lake environment. The handbook of environmental chemistry, vol 80. Springer, Cham, pp 47–61
15. Pešić V, Karaman GS, Kostianoy AG, Vukašinović-Pešić V (2018) Conclusions: recent advances and the future prospects of the Lake Skadar/Shkodra environment. In: Pešić V, Karaman G, Kostianoy A (eds) The Skadar/Shkodra Lake environment. The handbook of environmental chemistry, vol 80. Springer, Cham, pp 481–500
16. Mazzini I, Gliozzi E, Koci R et al (2015) Historical evolution and middle to late Holocene environmental changes in Lake Shkodra (Albania): new evidences from micropaleontological analysis. Palaeogeogr Palaeoclimatol Palaeoecol 419:47–59
17. Fu X, Tang T, Jiang W, Li F, Zhou S, Wu N, Cai Q (2008) Impacts of small hydropower plants on macroinvertebrate communities. Acta Ecol Sin 28:45–52
18. Young J, Watt A, Nowicki P, Alard D, Clitherow J, Henle K, Johnson R, Laczko E, McCracken D, Matouch S, Niemela J, Richards C (2005) Towards sustainable land use: identifying and managing conflicts between human activities and biodiversity conservation in Europe. Biodivers Conserv 14:1641–1661
19. Pešić V, Karaman G, Kostianoy A (2018) The Skadar/Shkodra Lake environment. The handbook of environmental chemistry, vol 80. Springer, Cham, p 508. https://doi.org/10.1007/978-3-319-99250-1

The Hydrology and Hydrogeology of Montenegro

Goran Sekulić and Milan M. Radulović

Contents

Abstract The waters of Montenegro flow through the Adriatic Sea Basin and Danube River Basin. The main rivers within the Adriatic Sea Basin are the Morača, the Zeta, the Cijevna and the Bojana, and within the Danube River Basin, they are the Piva, the Tara, the Lim, the Ibar and the Ćehotina. The largest onshore surface water body is Lake Skadar, which is shared between Montenegro and Albania (around 60% of the lake surface belongs to Montenegro). The Adriatic Sea Basin, although richer in water, has a poorer river network, primarily because of the widespread distribution of karst (terrain made of highly permeable carbonate rocks). The paradox is evident in the fact that a surface river network is almost entirely absent in the western part of Montenegro, i.e. in the area with the highest precipitation in Europe (the mean annual precipitation at Mount Orjen is around 5,000 mm). Atmospheric water from this area easily infiltrates into the permeable

G. Sekulić (✉) and M. M. Radulović
Faculty of Civil Engineering, University of Montenegro, Podgorica, Montenegro
e-mail: sgoran2000@gmail.com; radulovicmilan33@yahoo.com

Vladimir Pešić, Momir Paunović, and Andrey G. Kostianoy (eds.), *The Rivers of Montenegro*, Hdb Env Chem (2020) 93: 13–42, DOI 10.1007/698_2019_413,

ground and flows through a subsurface network of karst channels. It only appears in springs distributed along the Adriatic coast and along the edges of large karst depressions. The total mean annual runoff from the territory of Montenegro is around 595 m^3/s (the average specific runoff is 43 $l/s/km^2$). When taking into account these mean annual values, it can be concluded that Montenegro is extremely rich in water. However, the karst regime of flow causes a significantly uneven time distribution of water resources throughout the year. In the dry season, discharges are more than a thousand times lower than discharges in the rainy period; even at the end of summer, numerous springs and riverbeds are completely dry.

Keywords Adriatic Basin, Danube Basin, Groundwater, Montenegro, Surface water

1 Introduction

The natural beauty of Montenegro is especially reflected in the beauty of its coast, mountains, rivers and lakes. Rivers with deep canyons that cut through limestone massifs represent one of the most attractive part of the landscape. In addition to their beauty, the rivers are characterized by great riches of water for most of the year.

However, Montenegro has a specific hydrology, which is, among other factors, a consequence of the great distribution of karst (very permeable carbonate rocks such as limestone and dolomite).

Generally, in karstic terrains there is a significantly uneven time distribution of water resources throughout the year. The discharges of rivers and karstic springs are often more than a thousand times greater in the rainy period than during the summer.

Surface water is also unevenly distributed in spatial terms. There is a paradoxical situation in relation to this issue because a river network is at its most dense in the area with the lowest precipitation (the north-eastern part of Montenegro where the precipitation rate is 800 mm/year), and it is almost completely absent in the area with the greatest precipitation (the western part of the country which could be called "the Montenegrin desert", at least in terms of water availability, despite having a precipitation rate of 5,000 mm/year).

The territory of Montenegro is almost equally divided between the Adriatic Sea Basin (45.4%) and the Danube River (Black Sea) basin (54.6%) (Fig. 1).

The main rivers of the Adriatic Sea Basin are the Morača, the Zeta, the Cijevna and the Bojana. The largest onshore surface water body is Lake Skadar. It receives water from the Morača river and several small rivers, as well as from many sublacustrine springs. Lake Skadar is drained by the river Bojana which flows into the Adriatic Sea near Ulcinj.

The main rivers of the Danube River Basin are the Piva, the Tara, the Lim, the Ibar and the Ćehotina. The largest artificial lake in Montenegro is Lake Piva, formed after the construction of a 220-m-high dam on the Piva river (the Mratinje Dam).

Fig. 1 Map of the main river basins and subbasins in Montenegro

One of the reasons for writing this paper is that there are not enough publications in English that describe the hydrology of the entire Montenegrin territory. This work presents a short overview of the geological, hydrogeological and hydrological characteristics of Montenegro as a whole.

2 Geology

The first systematic geological and hydrogeological research in Montenegro began at the end of the nineteenth century. These early studies were mostly performed by foreign researchers [1–7]. After World War II, when the first generation of local geologists began to work, the level of research was increased significantly [8–12].

Carbonate rocks (limestone and dolomite) have the largest distribution on the territory of Montenegro. The geological map of Montenegro [12] presents these rocks as formations from the Mesozoic Age (the Triassic, the Jurassic and the Cretaceous). The thickness of the carbonate rock can be more than 5 km.

Triassic, Cretaceous and Eocene flysch sediments and Palaeozoic foliated metamorphic rocks are also widely distributed, especially in the southeastern part of Montenegro [12]. The topographic depressions are filled with clastic material of different origins (including alluvial, glacial, fluvioglacial, lacustrine, eluvial and diluvial sediments).

Within the territory of Montenegro, four separate tectonic zones have been distinguished [12, 13] (Fig. 2):

1. The Paraautochton (Adriatic–Ionian) zone – it extends along the Adriatic coast (this area is mainly formed of limestone and flysch).
2. The Budva–Cukali zone – it extends along a belt which is parallel to the Adriatic coast (this area is represented by a complex of rocks: claystone, marlstone, sandstone, limestone, chert and volcanic rock).
3. The Visoki Krš zone – it is found in the central part of Montenegro (this area is mainly formed of limestone and dolomite).
4. The Durmitor zone – it covers the northern part of Montenegro (this area consists of carbonate rock but also of flysch and foliated metamorphic rock).

The tectonic zones are separated by reverse faults which take the general direction northwest–southeast, so that the Durmitor tectonic zone is pushed up over the Visoki Krš zone, the Visoki Krš over the Budva–Cukali zone and the Budva–Cukali zone over the Paraautochton zone [12, 13]. Folds and faults generally extend along the Dinaric direction, i.e. along the northwest–southeast direction.

3 Hydrogeology

The hydrogeology of Montenegro has been studied by numerous researchers [4, 9, 11, 13–30].

Montenegro has an interesting hydrogeology, primarily due to the extensive distribution of karst (over 60% of the territory is formed of highly permeable limestone and dolomite). Low-permeable rocks (shale, marl, sandstone) are found only in its north-eastern region, where the river network is much better developed.

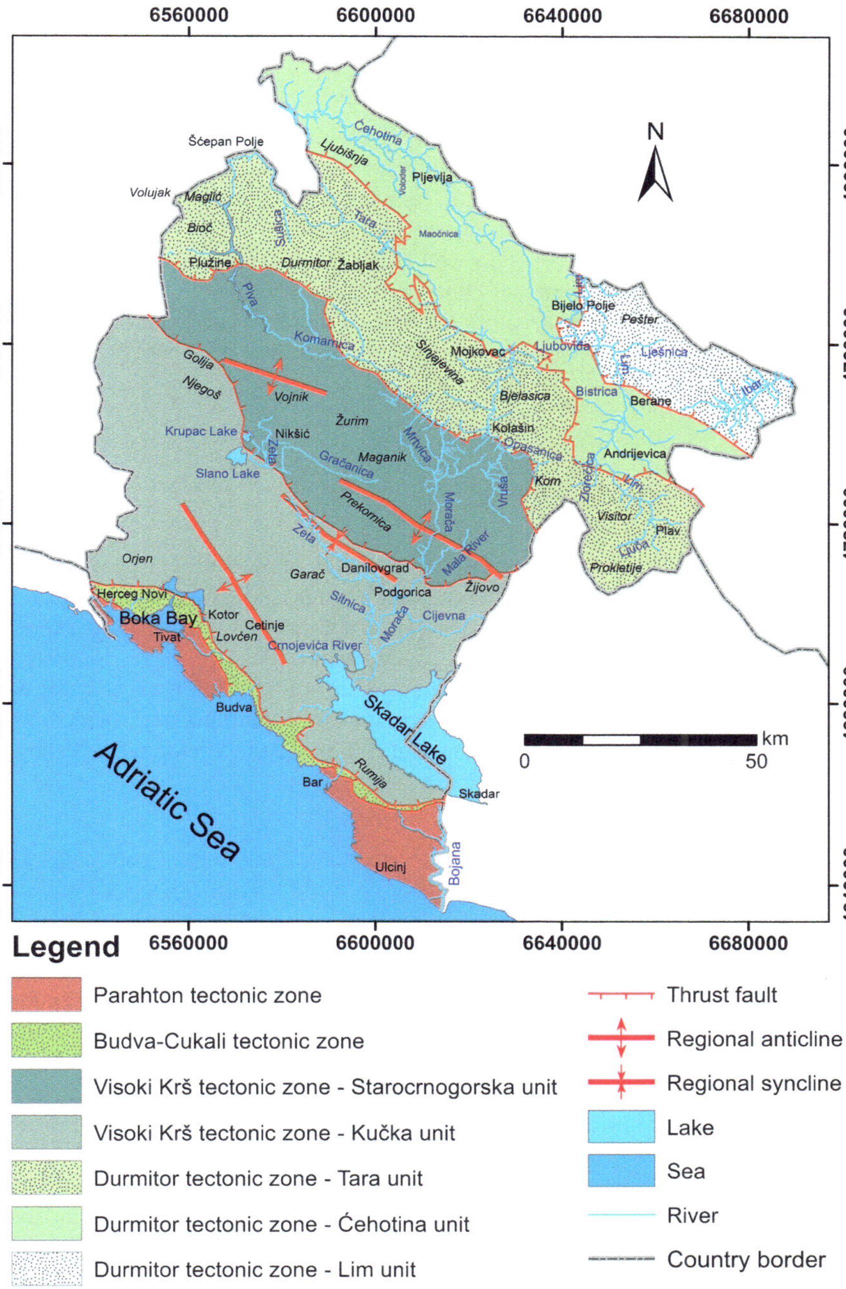

Fig. 2 Tectonic map of Montenegro (after [12, 13])

The boundary between the External and Internal Dinarides crosses the northern and north-eastern part of Montenegro. This boundary is mostly represented by the contact between the Visoki Krš and Durmitor tectonic units. The External Dinarides are mostly characterized by bare, highly karstified terrain (*holokarst*) with poor river networks. Permeable carbonate rocks are present to a significant degree even in the area of the Internal Dinarides, but their distribution is often intersected by low-permeable rocks. Generally, the carbonate rocks of the Internal Dinarides have a lower degree of karstification (*merokarst*) [4].

Bearing in mind all the above, as well as the fact that the majority of the Adriatic Sea Basin belongs to the External Dinarides, and the majority of the Danube River Basin within Montenegrin territory relates to the Internal Dinarides, it can be concluded that both the hydrogeology and the hydrology of these two basins are significantly different.

3.1 The Adriatic Sea Basin

A karst aquifer, which is characterized by very high permeability, is dominant in the Adriatic Sea Basin (Fig. 3). An intergranular aquifer is mostly found in the areas of the Zeta Valley, Niksic and Ulcinj. A fissured aquifer is widespread only along the Adriatic coast and in the northern part of the basin [26].

Low-permeable and impermeable rocks are most frequently present in the north-eastern part of the basin and along the Adriatic coast. Furthermore, these rocks are distributed along a narrow zone from Kuči, through the Bjelopavlića valley and Niksic, to Mount Golija.

3.1.1 The Karst Aquifer

Recharging of the karst aquifer usually occurs by the infiltration of atmospheric water in areas of high karst plateaus (at around 800–1,200 m asl). Villages that are located on the karst plateaus often do not have access to drinking water, even though these areas have the highest precipitation in Europe (e.g. the mean annual precipitation in the hinterland of Boka Bay, at Mount Orjen, is from 3,000 to 5,000 mm/year). Because of the high permeability of the rocks, almost the entire amount of the rainwater infiltrates into the ground and flows out only at sea level or along the edges of deep karst depressions. Even the deepest wells in these villages are left dry during the summer, so it is necessary to drill over 300 m to reach the water table. There is virtually no water anywhere on the surface of karst plateaus, and even though streams do sometimes appear, they usually sink through shallow holes after a very short distance.

The groundwater passes from the recharge zone to the discharge points of the aquifer (in springs), in a relatively short time, often in only a few days. This is because the network of karst channels, which play the role of the conductor of water

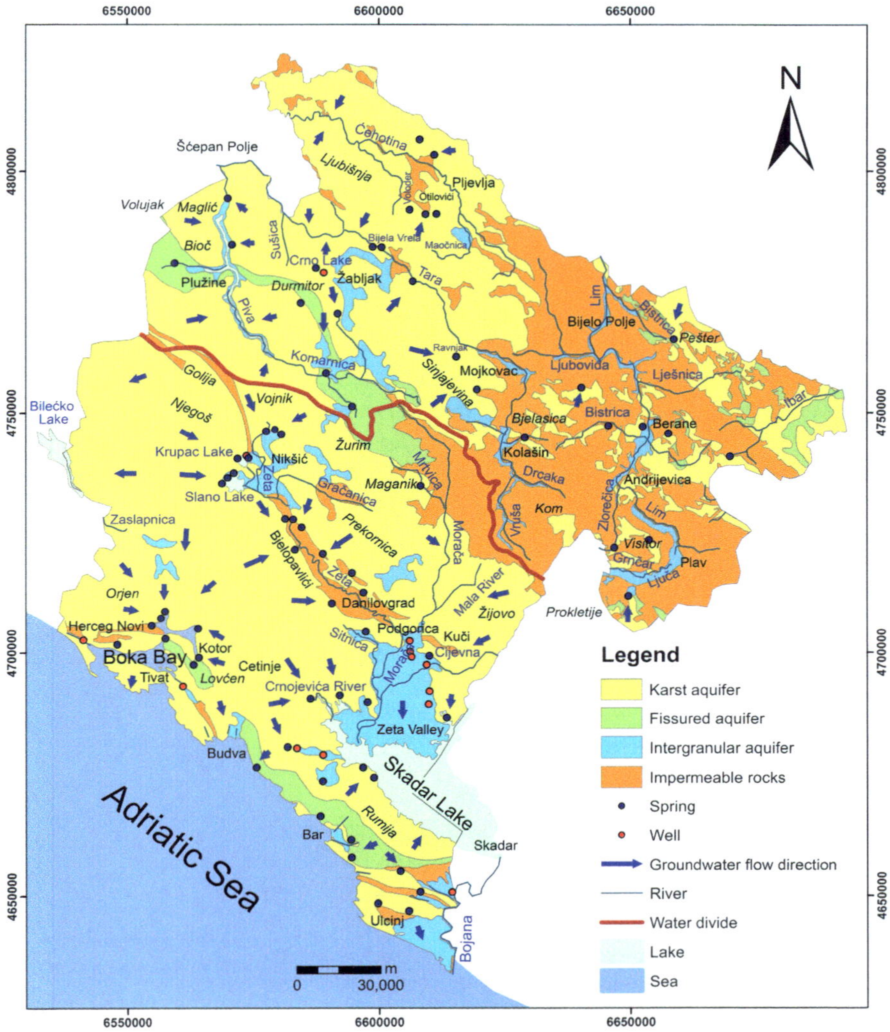

Fig. 3 Hydrogeological map of Montenegro

towards the springs, is very well developed. The average velocity of groundwater flow determined by performing artificial tracer tests is around 2.65 km/day [28].

The aquifer discharge occurs through major karst springs (Table 1; Fig. 3) which are generally distributed along:

- The Adriatic coast (Fig. 4)
- The edges of large karst depressions (Nikšić Polje, the Bjelopavlića valley, the Zeta valley, Lake Skadar and Bilećko Lake)

Table 1 List of the main karstic springs in the Montenegrin part of the Adriatic Sea Basin (after [28])

Name	Area	Coordinates		Estimated minimum discharge (l/s)
		X	Y	
Gač	Ulcinj	6,605,865	4,646,923	30
Salč	Ulcinj	6,599,759	4,648,493	2
Klezna	Ulcinj	6,608,244	4,650,315	15
Kaliman	Bar	6,604,208	4,655,135	5
Zaljevo	Bar	6,594,559	4,657,924	16
Kajnak	Bar	6,594,468	4,661,497	40
Brca	Bar	6,588,425	4,666,345	40
Reževica Spring	Budva	6,575,651	4,676,461	75
Vrmac	Kotor	6,562,965	4,696,140	20
Škurda	Kotor	6,564,153	4,698,364	200
Ljuta	Kotor	6,563,533	4,704,989	100
Plavda	Tivat	6,556,620	4,702,282	20
Risan Cave	Risan	6,557,613	4,708,540	4
Sopot Springs	Risan	6,556,386	4,707,957	250
Morinj	Risan	6,553,804	4,704,915	550
Opačica	Herceg Novi	6,548,028	4,701,672	80
Zaslapnica	Zaslap	6,549,391	4,726,426	54
Krnjice	Lake Skadar	6,598,023	4,674,361	10
Raduš	Lake Skadar	6,596,169	4,676,281	60
Velje Oko	Lake Skadar	6,587,168	4,674,899	50
Podgor	Lake Skadar	6,581,876	4,680,527	230
Crnojevica Spring	Lake Skadar	6,583,367	4,690,257	240
Karuč Springs	Lake Skadar	6,591,526	4,691,071	2,500
Bolje Sestre	Lake Skadar	6,597,562	4,690,132	2,300
Vitoja Springs	Lake Skadar	6,612,722	4,687,720	50
Ribnica	Podgorica	6,607,254	4,699,998	10
Mareza	Podgorica	6,597,779	4,704,531	2,000
Oraška Jama	Bjelopavlici	6,590,158	4,710,224	150
Žarica Jama	Bjelopavlici	6,596,476	4,711,932	50
Slatina Springs	Bjelopavlici	6,594,693	4,716,585	15
Viš Springs	Bjelopavlici	6,587,643	4,718,881	10
Milojevica Springs	Bjelopavlici	6,583,354	4,721,384	50
Dobro Polje	Bjelopavlici	6,585,104	4,721,248	1,000
Glava Zete	Bjelopavlici	6,582,117	4,726,152	2,000
Obošnica	Bjelopavlici	6,581,587	4,725,182	200
Slano Springs	Niksic	6,570,311	4,737,249	450
Krupačko Oko	Niksic	6,572,487	4,741,348	130
Poklonci	Niksic	6,573,723	4,741,053	200
Gornjepoljski Vir	Niksic	6,575,404	4,745,079	400

(continued)

Table 1 (continued)

Name	Area	Coordinates		Estimated minimum discharge (l/s)
		X	Y	
Vidrovan	Niksic	6,577,611	4,746,138	350
Bijeli Nerini	Mrtvica	6,611,402	4,731,760	500

Coordinates are indicated in the MGI/Balkans coordinate system (Zone 6)

Fig. 4 Sopot Springs in Boka Bay after a period of intense rainfall (groundwater flows out through the cave and submarine spring) (photo by Dejan Stojanović; interpretation by Milan Radulović)

- Deep river canyons (including those of the Morača, the Mrvica, the Mala Rijeka, the Cijevna and the Zaslapnica).

In some cases, aquifer discharge occurs below the level of the sea or lake, as happens in Boka Bay and Lake Skadar, where there are numerous submarine/ sublacustrine springs.

The quality of the groundwater from the karst aquifer is relatively good [30]. Generally, it has low TDS (around 300 mg/l), with the highest concentration of HCO_3^{2-} and Ca^{2+} ions. Periodically, there is an increase in turbidity and bacteria in the spring water, but that happens only after extreme rainfall events [13]. The groundwater quality of the coastal aquifer is disturbed during the summer period because of sea water intrusion (in that salinity, i.e. the concentration of Na^{+}, Cl^{-} and SO_4^{2-} ions, exceeds the threshold values).

3.1.2 The Intergranular Aquifer

The intergranular aquifers are widely distributed in the Nikšić area, the Zeta valley and the Ulcinj valley. In addition to the infiltration of atmospheric water, intergranular aquifers are also recharged by the infiltration of surface water, as well as by subterranean inflow from the bottom karst aquifer. The general direction of the groundwater flow within the three mentioned intergranular aquifers is from north to south.

The most important intergranular aquifer is located in the Zeta valley. It is very rich in water, and it has a relatively high hydraulic conductivity. The discharge of this aquifer occurs through diffuse springs along the southern edge of the valley. The artificial discharge of the aquifer cannot be neglected when assessing the water budget because large quantities of groundwater are abstracted through wells. This groundwater is used for the water supply of one part of Podgorica (the capital of Montenegro), as well as for irrigation and industry. The groundwater quality of the Zeta valley aquifer is poor in areas distributed downstream from large pollution sources, such as the Aluminium Plant Podgorica and other large agriculture areas [31].

3.2 The Danube River Basin

A considerable part of the Danube River Basin within the territory of Montenegro is formed of low-permeable and impermeable rocks that are mostly found in the eastern part of the basin (Fig. 3).

Limestone and dolomite (karst aquifer) are also present to a significant degree. They mostly occur in mountain areas. The intergranular aquifer is represented by the alluvial sediments of the main rivers (the Piva, Tara, Lim, Ćehotina and Ibar rivers). A fissured aquifer is found only to a limited degree, mainly within the Piva and Ibar River Basins [12, 26].

3.2.1 The Karst Aquifer

The karst aquifer recharge within the Danube River Basin generally occurs through the direct infiltration of atmospheric water, but concentrated infiltration is also common along mountain streams. The amount of precipitation in the Danube River Basin is significantly lower than in Adriatic Sea Basin, which affects the conditions of groundwater recharge.

The aquifer recharge zones are represented by mountain areas and high karst plateaus (a typical example is the Sinjajevina plateau). Generally, groundwater flow directions are from the recharge zones towards the erosion bases, i.e. towards the nearest river valleys along which karst springs are distributed (Figs. 3 and 5). The

Fig. 5 Ravnjak Spring (photo by Milan Radulović)

groundwater flow directions (Fig. 3) are assessed by the analysis of the hydrogeological characteristics [32], as well as on the basis of the results of various artificial tracer tests. The obtained results were often surprising, which is not unusual in karst terrains which have a well-developed subsurface network of conduits and caves, as is the case in Montenegro.

The results of the tracer test performed in the area of Lake Crno represent one such example [32]. The lake water fills two depressions that are unequal in diameter, and so it is subdivided into Malo and Veliko Crno. By injecting a tracer into the shallow holes of Malo Crno Lake, their hydraulic connection with the Dubrovska Springs in the Komarnica Canyon was demonstrated (some 19 km south of the lake). In the second test, the tracer which was injected into the shallow hole of a stream that flows from Veliko Crno Lake appeared in an entirely different side, in the Tara Canyon (10 km north-east of the lake). What is surprising in the results of the second test is that the tracer did not appear in those springs located on the left (closer) bank of the Tara but at Bijela Spring which is located on the right bank of the Tara. This means that a part of the groundwater from the Durmitor area actually flows under the riverbed and appears under pressure on the more distant bank of the Tara river. Furthermore, this example shows that in karst terrains, it is possible that one lake can actually represent a dividing point between two river basins [32].

Almost all of the water supply systems in the northern part of Montenegro (within the Danube River Basin) are based on karst springs (Table 2), which are usually of a good quality throughout the year [16, 33]. These springs have temporary problems only in terms of increased turbidity and higher bacteria content, which is something that usually happens after intense precipitation, especially in the area around Pljevlja.

Table 2 List of the main karstic springs in the Montenegrin part of Danube River Basin (after [28])

Name	Area	Coordinates		Estimated minimum discharge (l/s)
		X	Y	
Oko Bijele	Savnik	6,594,763	4,751,299	30
Šavnička Glava	Savnik	6,589,593	4,758,139	100
Bukovica Springs	Durmitor	6,590,955	4,768,445	200
Dubrovska Vrela	Savnik	6,577,641	4,762,720	500
Sutulija	Piva	6,679,877	4,740,273	50
Medjedjak	Piva	6,570,064	4,784,914	500
Kaludjerovo Vrelo	Piva	6,569,553	4,794,748	100
Oko (Zminje Lake)	Durmitor	6,587,606	4,779,854	40
Mušovica Springs	Kolasin	6,629,062	4,744,807	80
Plašnica Springs	Kolasin	6,614,758	4,749,601	100
Ravnjak	Tara	6,615,387	4,761,525	1,150
Ljutica	Tara	6,606,237	4,777,825	2,000
Bijela Vrela	Tara	6,604,979	4,779,973	100
Alipaša Springs	Gusinje	665,024	4,712,952	2,500
Krkor	Andrijevica	6,646,806	4,721,802	100
Merića Vrelo	Berane	6,645,673	4,747,310	200
Manastirsko Vrelo	Berane	6,652,533	4,747,101	80
Dapsića Vrelo	Berane	6,657,487	4,746,199	45
Vrelo Ibra	Ibar	6,671,409	4,741,341	100
Vrelo Bistrice	Ljubovidja	6,640,229	4,755,702	500
Djalovića Cave	Pester	6,656,792	4,771,681	300
Bezarska Vrela	Pljevlja	6,611,508	4,791,102	5
Zmajevac	Pljevlja	6,609,331	4,791,046	45
Mandovac	Pljevlja	6,606,135	4,791,982	10
Breznica	Pljevlja	6,611,071	4,803,350	50
Jugoštica	Pljevlja	6,608,175	4,806,522	5

Coordinates are indicated in the MGI/Balkans coordinate system (Zone 6)

3.2.2 The Intergranular Aquifer

The intergranular aquifer is mostly distributed along the main rivers: the Piva, the Tara, the Lim, the Ćehotina and the Ibar. These are relatively narrow, but elongated alluvial aquifers, with heterogeneous characteristics in terms of their hydraulic conductivity.

In the rainy season, when surface water levels are high, each intergranular aquifer is recharged by river water, and groundwater flow directions are from each river towards the edges of their alluvial plains. During the dry season, the opposite is true, i.e. the groundwater from the intergranular aquifer flows towards the relevant river and feeds its bed through diffuse springs.

These alluvial aquifers are locally very rich in groundwater. In the alluvium of the Tara river there is a potential water source (located upstream from Kolašin) which consists of four wells with an individual depth of 20–25 m. The wells are drilled through sandy-gravel sediments of relatively high hydraulic conductivity. During the pumping of groundwater from the wells at a pumping rate of 89 l/s in total, the drawdown was only around 0.3 m [34]. The results of physical, chemical and microbiological analyses indicated the high quality of the groundwater.

4 Hydrology

4.1 General Features of the Area and Its Hydrographical Network

The spatial arrangement of the surface water of Montenegro (its rivers, lakes and sea) is primarily determined by the morphological and hydrogeological characteristics of the area. The basic feature of Montenegrin hydrography is the existence of two approximately equal drainage basins: Danube River (the Black Sea) and Adriatic Sea (Fig. 1). The Adriatic Sea Basin covers 45.4% and the Black Sea Basin 54.6% of the total territory of Montenegro. Generally speaking, both of these basins are rich in water, even by global standards.

The total area covered by the Black Sea Basin is about 7,545 km^2. The Ibar, the Lim, the Ćehotina, the Tara and the Piva are the main river within this basin. The total area covered by the Adriatic Sea Basin is 6,268 km^2 [35]. The Morača river, with its tributaries the Zeta and the Cijevna, and Rijeka Crnojevića and the Orahovištica form part of this basin. All these rivers first flow into Lake Skadar, and then the water moves along the river Bojana into the Adriatic Sea. Besides the river Bojana, several streams also flow into the Adriatic; however, here there are no constant observations nor measurements of the overall hydrological regime [35].

A significant number of surface water bodies in Montenegro show flooding features. They are grouped into flood systems according to their distinctive geography headings: the Adriatic coast, Lake Skadar, Boka Bay, Niksic, Cetinje, Podgorica, Pljevlja, the Lim river and others.

There are many natural lakes in Montenegro, with the largest of them are located on valleys of southern part of the country. Lake Skadar, formed in a spacious depression, is the largest lake in the Balkans. Three fifths of the lake is controlled by Montenegro. During its highest water level of about 9.85 m asl, this lake occupies an area of 525 km^2. Lake Šas is the second largest lake in Montenegro and is situated between Lake Skadar Lake, the river Bojana and the Adriatic Sea.

Mountain lakes – specifically Lake Crno, Lake Plavsko and Lake Biogradsko – are also natural reserves, as typical examples of glacier lakes. Besides these lakes, there are numerous smaller lakes, which are either glacier or karstic in their origin.

Montenegrin hydrography has been made significantly richer due to various artificial lakes created on the Piva, the Ćehotina, the Zeta (in the region of Niksic) and the stream at Grahovo. Part of the territory of Montenegro is sunken due to the creation of the artificial Bilećko Lake, which forms part of the hydropower plant system of the Trebišnjica river [28, 36].

The Adriatic Sea, between Montenegro and Italy, is only around 200 km wide and here forms a part of the South Adriatic Basin, where the deepest point of the Adriatic Sea is located (about 1,400 m). The total length of the seacoast of Montenegro is around 300 km. About 80% of coast is made out of rocks, with great sea depths just by the coast, while the rest of coast is shallow with a sand pebble bottom. The largest beach is Long Beach in Ulcinj. There are numerous places (on shore and by the sea) that are suitable for tourism and recreation.

The approximate amplitude of the high and low tides is about 23 cm [35]. This is a relatively warm sea. The predominant direction of the sea currents is parallel to the coastline towards the northwest. The salinity of the water in the Southern Adriatic is slightly below the Mediterranean average (being 38.6‰ as opposed to the overall average of 39‰) [28].

4.2 Hydrological Data

Both Montenegrin basins have an overall total of only 24 hydrometrical stations from which a long time series could be obtained for the reconstruction of monthly and annual flows. This number of stations is insufficient for an area the size of Montenegro, especially given the fact that numerous rivers cannot be measured, most especially in the upper parts of the basins. The consequence of this limitation is that numerous streams in the upper parts of basins which have significant water potential cannot be hydrologically analysed.

The longest time series of measured data exist from 1948 only for a few of the available stations, but a time series of longer than 50 years can be followed at numerous hydrographic stations. Tables 3 and 4 and Fig. 1 show the current arrangement of stations and the year when measurements first started there.

4.3 Specific Runoff

The obvious conclusion that the area around Montenegro is very rich in water is further confirmed by various measurements that have been performed. The Morača river basin contains most of the water, where the specific runoff is approximately equal and has values between 56.6 and 68.5 l/s/km^2 [35]. This is followed by the Piva and the Tara, with specific runoffs of between 40 and 50 l/s/km^2. The Lim river basin is extraordinarily changeable along its course, showing runoffs of between 29.9 and 53.0 l/s/km^2 as the specific runoff as it exits Montenegro actually falls below 30 l/s/km^2 [35]. The Ibar has a specific runoff of between 13.9 and 17 l/s/km^2, and the Ćehotina shows values between 15 and 20 l/s/km^2 [28, 35]. The highest specific runoff is for the Crnojevića catchment area (82.3 l/s/km^2) [28].

Table 3 Network of hydrometrical stations in the Black Sea Basin [37]

Station	River	Basin	Beginning (year)	Elevation of “0” (m asl)	Basin area (km^2)	Water level (cm)		Runoff (m^3/s)	
						Min	Max	Min	Max
Andrijevica	Zlorecica	Lim	1985	742	169.1	14	250	0.31	198
Ravna Rijeka	Ljubovidja	Lim	1948	582	351	8	278	0.98	115
Bijelo Polje	Lim	Drina	1948	559.7	2,182.7	−16	455	8.2	1,077
Gubavač	B. Bistrica	Lim	1949	545	484	−2	245	0.36	65.6
Dobrakovo	Lim	Drina	1961	531.6	2,875	0	612	6.46	1,518
Crna Poljana	Tara	Drina	1957	965.8	247	10	344	0.85	368
Trebaljevo	Tara	Drina	1959	894.8	506	2	497	1.82	701
Pljevlja	Ćehotina	Drina	1948	755	361	10	262	0.6	145
Gradac	Ćehotina	Drina	1963	665	809.8	30	572	2.1	414
Rožaje	Ibar	Morava	1968	1,035.1	98.2	4	210	0.04	102
Vusanje	Grlja	Lim	1959	960	48.5	2	194	0	84.4
Timar	Bukovica	Piva	2006	1,065	70.1	−2	194	–	–
Gornja Bijela	Bijela	Piva	2006	1,050	24.9	11	186	–	–

Table 4 Network of hydrometrical stations in the Adriatic Sea Basin [37]

Station	River/lake	Basin	Beginning (year)	Elevation “0” (m asl)	Basin area (km^2)	Water level (cm)		Runoff (m^3/s)	
						Min	Max	Min	Max
Podgorica	Morača	Lake Skadar	1948	24.6	31.40	18	1,226	7.93	2,073
Zlatica	Morača	Lake Skadar	1983	53.0	98.5	0	1,004	0	1,320
Pernica	Morača	Lake Skadar	1948	178.72	440.9	20	534	1.14	874
Međurječje	Mrtvica	Morača	1948	187.5	207.7	−6	413	0.48	4.6
Danilovgrad	Zeta	Morača	1948	33.3	1,215.8	60	1,295	4.68	577
Duklov Most	Zeta	Morača	1955	615.2	342.2	4	243	0.07	286
R. Crnojevića	R. Crnojevića	Lake Skadar	–	–	–	–	288	–	–
Plavnica	Lake Skadar	Lake Skadar	1948	4.56	–	−2	588	–	–
Vranjina	Lake Skadar	Lake Skadar	1965	5.01	–	−23	553	–	–
Ckla	Lake Skadar	Lake Skadar	1950	4.86	–	0	569	–	–
Fraskanjel	Bojana	Adriatic Sea	1960	−0.07	1,652	–	636	2	–

The average specific runoff of the remaining streams in Montenegro has not been sufficiently examined so far, which is a consequence of insufficient funding for hydrological observation and measurement, the level of the reliability of such data and the lack of precision in defining catchment areas.

By analysing the known measured data, the following conclusion can be drawn. Net rainfall ranges from about 800 mm/year (in the Ibar and Ćehotina basins) to over 3,000 mm/year in the Boka Bay Basin. The runoff coefficients give values of 0.45 in the Ibar River Basin, while for most profiles of the Morača and Zeta basins, these coefficients show values of above 0.80 [28]. Evapotranspiration is mostly between 400 and 600 mm annually [35]. Bearing in mind that evaporation in karst basins is the most complex component of the hydrological balance, in most cases it is defined indirectly, and it needs to be considered as a rough approximation, mainly because of unknown hydrogeological watersheds. The low values of evaporation in the Morača river basin can be partially explained by the hydrogeological and pedological features of terrain, as well as by its poor vegetation [35].

4.4 *Hydrological Balance*

Based on applied hydrological and climate processing and analysis, the approximate water balance of Montenegro was calculated (Table 5). From these observations, it is evident that around 595 m^3/s of Montenegro's own waters run off from its territory. This is equivalent to a runoff layer of 1,360 mm/year or a volume of about 18.7 billion m^3/year. The approximate specific runoffs are about 43 $l/s/km^2$ [35]. Transit waters make only an insignificant addition of 29 m^3/s to the total [35], so that domicile waters make up 95% of the total runoff from Montenegro [28]. Considering basins, the Adriatic Basin (6,268 km^2), although smaller than the Black Sea Basin (7,545 km^2), produces about 11.35 billion m^3/year, which is about 50% more in terms of runoff than the Black Sea runoff (7.4 billion m^3/year). The approximate specific runoff in the Adriatic Basin is about 59.5 $l/s/km^2$, while it is about 31 $l/s/km^2$ in the Black Sea Basin [35]. Another various elements that need to be considered is that around one third of the Adriatic Basin is usually drained by subterranean routes, which has not yet been reliably explored or measured, meaning that the accuracy of their estimation is lower than it is for the estimation for surface water [35].

4.5 *Mean Monthly and Annual Discharges*

The mean annual discharge (flow) represents the feature of the water regime which points to the total water potential of a certain basin. Mean discharges are defined for considered hydrometrical stations (profiles) based on a time series of observed and recorded data (Table 6). It should be emphasized that the reliability of results is

Table 5 Water balance of Montenegrin rivers [35]

River	A_{vl} (km^2)	A_{tr} (km^2)	V_{vl} (10^6 m^3)	V_{tr} (10^6 m^3)	V_{ul} (10^6 m^3)	W_{vl} (10^6 m^3)	W_{tr} (10^6 m^3)	W_{ot} (10^6 m^3)	W_{isp} (10^6 m^3)	W_{uk} (10^6 m^3)
Ibar	433	0	423	0	423	188	0	188	235	423
Lim	2,291	270	3,001	272	3,274	1,910	272	2,182	1,091	3,274
Ćehotina	1,113	16	1,043	7	1,050	538	7	546	505	1,050
Tara	1,915	125	3,261	95	3,356	2,450	95	2,545	811	3,356
Piva	1,731	53	2,976	79	3,055	2,292	79	2,372	683	3,055
Rest	62	0	55	0	55	22	0	22	33	55
Danube River (Black Sea) basin	7,545	464	10,759	453	11,213	7,401	451	7,855	3,358	11,213
Morača and Zeta	2,649	0	6,194	0	6,194	5,096	0	5,096	1,097	6,194
Skadar Lake	1,624	225	3,653	461	4,114	2,671	461	3,133	981	4,114
Ulcinj - Budva	656	0	1,191	0	1,191	883	0	883	308	1,191
Boka Bay	728	0	2,258	0	2,258	1,892	0	1,892	365	2,258
Trebišnjica, Gatačko Polje (based on estimations)	541	0	936	0	936	755	0	755	181	936
Rest	70	0	84	0	84	55	0	55	29	84
Adriatic Sea Basin	6,268	225	14,315	461	14,776	11,353	461	11,814	2,962	14,776
Total (Montenegro)	13,813	689	25,074	915	25,989	18,754	915	19,669	6,320	25,989

A_{vl} area of own basin, A_{tr} area of the basin outside Montenegrin borders, V_{vl} precipitation volume on the basin inside Montenegrin territory, V_{tr} precipitation volume on the basin outside Montenegrin territory, V_{ul} total volume of precipitation, W_{vl} runoff from Montenegrin territory, W_{tr} runoff from areas outside Montenegrin territory, W_{ot} total runoff volume, W_{isp} evaporation volume, W_{uk} total water balance

Table 6 Mean monthly and annual discharges (m^3/s) (after [35])

Station (river)	Jan	Feb	Mar	Apr	May	Jun	Jul	Aug	Sep	Oct	Nov	Dec	Mean annual
Rožaje (Ibar)	1.62	2.29	3.04	5.06	5.39	2.41	1.55	1.13	0.91	1.41	2.03	2.65	2.46
Bač (Ibar)	3.97	5.31	7.30	11.11	11.46	5.92	3.71	2.40	2.17	3.25	4.82	5.93	5.61
Plav (Lim)	15.09	14.40	15.47	29.94	41.14	28.89	14.41	7.56	7.20	13.63	22.15	21.59	19.29
Andrijevica (Lim)	24.89	24.25	27.54	50.11	66.96	40.80	19.29	10.3	10.7	19.39	31.37	32.56	29.85
Berane (Lim)	41.97	44.40	49.84	81.55	95.42	52.84	25.44	14.66	16.31	28.12	46.92	56.60	46.17
Zaton (Lim)	52.25	54.33	62.19	97.88	112.5	62.98	29.13	16.28	18.55	31.89	55.97	67.05	55.10
Bijelo Polje (Lim)	60.28	63.98	73.56	116.3	129.1	75.34	35.73	20.13	22.1	39.06	67.90	80.46	65.43
Pljevlja (Ćehotina)	6.76	8.84	10.88	11.56	7.16	5.29	4.01	2.85	3.33	5.05	6.16	8.85	6.73
Gradac (Ćehotina)	14.50	17.88	21.51	21.45	14.76	10.91	8.06	5.50	6.03	8.86	13.08	17.96	13.37
Crna Poljana (Tara)	12.11	12.38	13.73	25.18	22.81	9.15	3.44	2.33	3.28	7.43	15.32	18.10	12.11
Trebaljevo (Tara)	24.96	26.48	29.56	48.64	44.77	20.07	7.91	4.58	6.76	16.28	32.58	37.23	24.99
Bistrica (Tara)	35.36	35.55	38.90	64.80	62.67	27.55	11.37	7.39	9.55	22.09	43.54	51.66	34.20
Dj. Tara (Tara)	56.32	55.37	62.03	103.5	109.9	62.02	31.89	18.91	20.31	35.32	68.76	79.78	58.66
Sćepan Polje (Tara)	77.11	75.06	85.51	139.4	149.5	85.05	42.10	24.61	26.97	49.87	92.50	104.25	79.36
Sćepan Polje (Piva)	73.46	71.16	80.30	125.0	131.5	75.82	33.19	21.00	25.31	52.18	102.06	107.57	74.88
Duški Most (Morača)	13.20	13.24	16.26	27.48	27.50	12.66	4.50	3.41	4.48	11.08	23.64	23.26	15.06
Pernica (Morača)	27.42	27.83	30.46	49.34	57.51	29.27	9.74	6.59	10.44	25.14	46.58	41.63	30.16
Zlatica (Morača)	79.69	77.03	73.60	90.69	78.41	36.14	9.90	4.61	14.07	43.67	94.88	99.89	58.55
Podgorica (Morača)	214.94	213.0	203.7	236.1	200.7	103.2	40.86	27.13	50.64	124.19	253.93	274.06	161.90
Duklov Most (Zeta)	22.44	23.45	24.84	33.13	23.87	9.60	2.26	1.22	4.09	14.38	30.72	31.95	18.50
Danilovgrad (Zeta)	113.34	110.41	104.77	109.25	79.27	43.28	21.98	15.04	26.35	1.25	122.26	134.69	78.49
Trgaj (Cijevna)	26.94	28.10	26.80	39.52	41.61	22.93	7.85	4.56	8.15	19.06	36.42	36.34	24.86
Brodska Njiva (R. Crnojevića)	9.96	8.6	6.84	9.21	3.38	1.63	0.91	0.99	2.83	7.05	13.73	14.08	6.59
Orahovo (Orahovštica)	5.65	5.37	5.01	4.29	2.55	1.06	0.37	0.23	0.73	2.12	5.32	6.12	3.23

relatively low for stations where original observations are short, so these results can be considered only as approximate indicators of the water regime [34].

Analysing the trends in mean annual discharges shows that there is a decreasing trend at almost all of the hydrometric stations [33, 38, 39].

4.6 Low Flows

Knowing about the low flows of rivers is especially significant for water quality protection. Therefore, it is essential to define the relevant minimum flows in order to assess the possibility of using the water for water supply but also to provide favourable conditions for the survival of plants and animals in critical periods [35]. The analysis of low flows was performed based on time series of different lengths, covering minimum annual flows and minimum average monthly flows (Table 7).

Table 7 Minimum average monthly flows in terms of their characteristic probabilities of occurrence (m^3/s) [35]

Station	50%	80%	90%	95%
Rožaje	0.50	0.32	0.25	0.22
Bač	1.05	0.58	0.43	0.35
Plav	4.82	3.55	3.11	2.85
Andrijevica	8.09	5.63	4.47	3.56
Berane	11.22	8.25	7.06	6.25
Zaton	12.39	9.36	8.23	7.50
Bijelo Polje	15.05	11.68	10.47	9.71
Pljevlja	1.59	1.03	0.78	0.60
Gradac	3.33	2.36	2.04	1.85
Crna Poljana	1.68	1.19	1.02	0.90
Trebaljevo	3.17	1.73	1.15	0.75
Bistrica	5.69	4.09	3.42	2.96
Djurdjevića Tara	12.93	9.53	8.12	7.13
Sćepan Polje (Tara)	17.28	12.81	11.20	10.19
Sćepan Polje (Piva)	15.66	11.89	10.32	9.22
Duški Most	2.13	1.43	1.15	0.96
Pernica	4.55	3.14	2.65	2.36
Podgorica	20.65	14.33	12.07	10.64
Duklov Most	0.46	0.22	0.15	0.11
Danilovgrad	11.87	8.04	6.71	5.91
Trgaj	3.28	1.91	1.36	0.98
Brodska Njiva	0.84	0.65	0.57	0.52

4.7 High Flows

High flows are features of the water regime, and their assessment is important for the purposes of building flood protection facilities and the dimensioning of hydro-technical facilities [35]. Table 8 shows the adopted values of maximum high flows (Q_{max}) for each profile in terms of their characteristic probabilities, as well as flows for the upper limit of the 95% confidence interval (Q_a).

It can be concluded that the results of the calculations of high flows are relatively reliable for return periods of less than 100 years [35]. The results of the calculations of high flows, and the probability of exceeding 0.1% and 0.01% drawn from these, do not, however, have the required level of reliability in calculated high flows for designing hydro-technical facilities, and especially not for final stages of design. Rather, it is necessary to observe high flows more precisely based on the comprehensive analysis of the conditions that preceded the formation of each flood event [35].

These results of the analyses of characteristic flows can only be used as preliminary results when considering both flows and water regimes. In all cases of the preparation of technical documentation for certain hydro-technical facilities, the results should be tested through new (special) hydrological and hydrogeological measurements.

4.8 Water Levels in Lake Skadar and the River Bojana

Lake Skadar is the most important water body of Montenegro in a hydrological, ecological, managerial and tourist sense [40]. The lake basin occupies a large area of Montenegrin territory (Figs. 6 and 7), so its significance for water management is undoubtedly huge [33, 41–44].

The river Drim, which mostly flows through Albania, is one of the factors that affects the lake water level. For the last few centuries, the Drim has constantly changing its streaming route in the Skadar Valley (there were times when the route flowed towards the river Bojana, in the direction of Zadrimlje and Lješ). Whenever it had a flowing route towards the Bojana, a disorder in the Lake Skadar level would appear, which was mirrored in both the rivers Bojana and Morača; in other words, catastrophic floods would occur [45]. The process of forming and formulating the system of Lake Skadar, the river Bojana and the river Drim has been intensified since the great floods of 1848 and 1858, up until November 1896, when the river Drim began to flow entirely into the river Bojana. This remains the situation today [45, 46].

In the 1970s, with the construction of reservoirs on the Drim, the situation became somewhat more favourable. The river is regulated by the reservoirs at Globočica and Špilje in Macedonia, as well as at Vaudeis, Fierza and Komana in Albania, with a usable volume of about 2.8 billion m^3, which is significantly more than the usable

Table 8 Absolute maximum flows in terms of their characteristic probabilities of occurrence (m^3/s) (after [35])

Station (river)	Type	0.01%	0.1%	1%	2%	5%	10%	20%	50%
Rožaje (Ibar)	Q_{max}	218	161	106	89	67	52	37	20
	Q_a	–	–	–	–	–	–	–	–
Bač (Ibar)	Q_{max}	545	313	215	173	130	100	72	39
	Q_a	–	–	–	–	–	–	–	–
Plav (Lim)	Q_{max}	511	401	298	269	229	199	168	122
	Q_a	688	516	365	323	269	229	189	128
Andrijevica (Lim)	Q_{max}	982	744	539	482	410	355	301	221
	Q_a	1,372	983	668	585	482	410	336	232
Berane (Lim)	Q_{max}	1,496	1,143	835	749	639	556	473	350
	Q_a	2,043	1,481	1,020	897	744	633	525	367
Zaton (Lim)	Q_{max}	1,560	1,206	893	805	691	606	518	389
	Q_a	2,068	1,525	1,071	949	794	682	571	407
Bijelo Polje (Lim)	Q_{max}	2,124	1,578	1,128	1,007	855	744	634	478
	Q_a	2,942	2,058	1,376	1,202	991	842	699	499
Pljevlja (Ćehotina)	Q_{max}	167	131	99.4	–	78.5	69.4	–	–
	Q_a	301	214	144	–	104	88.6	–	–
Gradac (Ćehotina)	Q_{max}	973	730	493	423	334	270	210	142
	Q_a	1,192	885	586	499	387	305	226	155
Crna Poljana (Tara)	Q_{max}	827	631	454	404	339	290	240	167
	Q_a	1,224	880	592	514	416	346	278	178
Trebaljevo (Tara)	Q_{max}	1,970	1,324	844	726	583	485	391	268
	Q_a	3,236	1,983	1,144	951	730	584	454	287
Bistrica (Tara)	Q_{max}	1,988	1,432	984	865	718	612	508	363
	Q_a	3,089	2,058	1,296	1,108	882	728	583	386
Djurdjevića Tara (Tara)	Q_{max}	2,344	1,704	1,175	1,034	856	727	599	420
	Q_a	3,388	2,315	1,488	1,278	1,024	846	678	443
Sćepan Polje (Tara)	Q_{max}	3,076	2,171	1,465	1,283	1,060	901	747	537
	Q_a	4,642	3,038	1,886	1,608	1,278	1,054	846	567
Sćepan Polje (Piva)	Q_{max}	2,452	1,788	1,246	1,101	920	788	659	477
	Q_a	3,832	2,582	1,645	1,412	1,132	939	757	506
Duški Most (Morača)	Q_{max}	2,736	1,425	719	579	430	338	262	176
	Q_a	4,930	2,247	986	761	533	401	295	186
Pernica (Morača)	Q_{max}	1,601	1,248	934	845	729	641	552	418
	Q_a	2,207	1,630	1,147	1,016	852	732	614	439
Zlatica (Morača)	Q_{max}	1,905	1,699	1,458	1,375	1,252	1,146	1,021	791
	Q_a	2,136	1,891	1,606	1,509	1,366	1,243	1,100	827
Podgorica (Morača)	Q_{max}	3,184	2,725	2,244	2,091	1,879	1,706	1,516	1,203
	Q_a	3,888	3,236	2,578	2,374	2,098	1,880	1,647	1,248
Duklov Most (Zeta)	Q_{max}	411	357	300	281	255	234	210	171
	Q_a	506	427	346	321	286	259	229	177
Danilovgrad (Zeta)	Q_{max}	693	623	547	523	488	459	426	369
	Q_a	802	705	604	571	526	490	450	378

(continued)

Table 8 (continued)

Station (river)	Type	0.01%	0.1%	1%	2%	5%	10%	20%	50%
Trgaj (Cijevna)	Q_{max}	875	732	590	546	486	439	387	305
	Q_a	1,151	924	709	645	561	496	429	318
Brodska Njiva (R. Crnojevića)	Q_{max}	213	208	200	196	189	181	171	148
	Q_a	285	277	260	252	238	225	208	160
Orahovo (Orahovštica)	Q_{max}	70	68	65	64	62	60	57	50
	Q_a	77	74	71	69	67	64	61	52

reservoir of Lake Skadar (which is around 1.57 billion m^3) (Bošković et al. 2004). The reservoirs practically enable the full levelling of the river Drim, which is of great importance from the standpoint of flooding [45, 47].

Lake Skadar receives its largest amount of water from the Morača river (67.5%), so the variations in the lake water level are directly dependant on its inflow (Table 9). Aside from the Morača, some other rivers flows into Lake Skadar, including the Crmnica, the Orahovštica, the Poseljanska, the Crnojevića river, the stream of Bazagurska and the Karatuna (Fig. 6). Considerable amounts of water come through the short streams of the northern lakeshore (including those of the Plavnica, the Zetica, the Gostiljska, the Mala Mrka, the Velika Mrka, the Raičevića Žalica, etc.) which are formed by the discharge of the Zeta Valley intergranular aquifer. In addition, the lake receives groundwater from various sublacustrine springs (known as "vrulje" or "okas") which usually appear along karstic lakeshores [38]. The total mean annual inflow of groundwater into Lake Skadar amounts to around 55 m^3/s [13].

The discharge of the lake takes place through the river Bojana. As mentioned above, since 1896, the Bojana has received the total discharge of the river Drim. The mouth of the Drim is around 2 km downstream from the lake. The main consequence of this change from the nineteenth century is the increased level of both the Bojana and the lake as a whole. Today, during high water levels, the river Drim slows the lake drainage, and during periods of extremely high flows of the water from the Drim, it even partially flows towards the lake (and on occasion within the lake there is a hydraulic barrier and reverse flow) [17].

What needs to be emphasized is that the level of the examination and the number of continuous hydrological measurements are relatively low as compared to the significance of the lake, especially bearing in mind the potential that it has. Therefore, despite the demonstrable significance of the river Bojana, measurements have not been performed on any profile of its 25-km-long flow along the border with Albania. The approximated values of discharge, which were taken from existing reports [35, 38], only represent rough assumptions. Therefore, it can be stated that the collection of basic hydrological data and the level of hydrological examination that has taken place so far are not in accordance with the complexity and the intensity of the issues along the river Bojana.

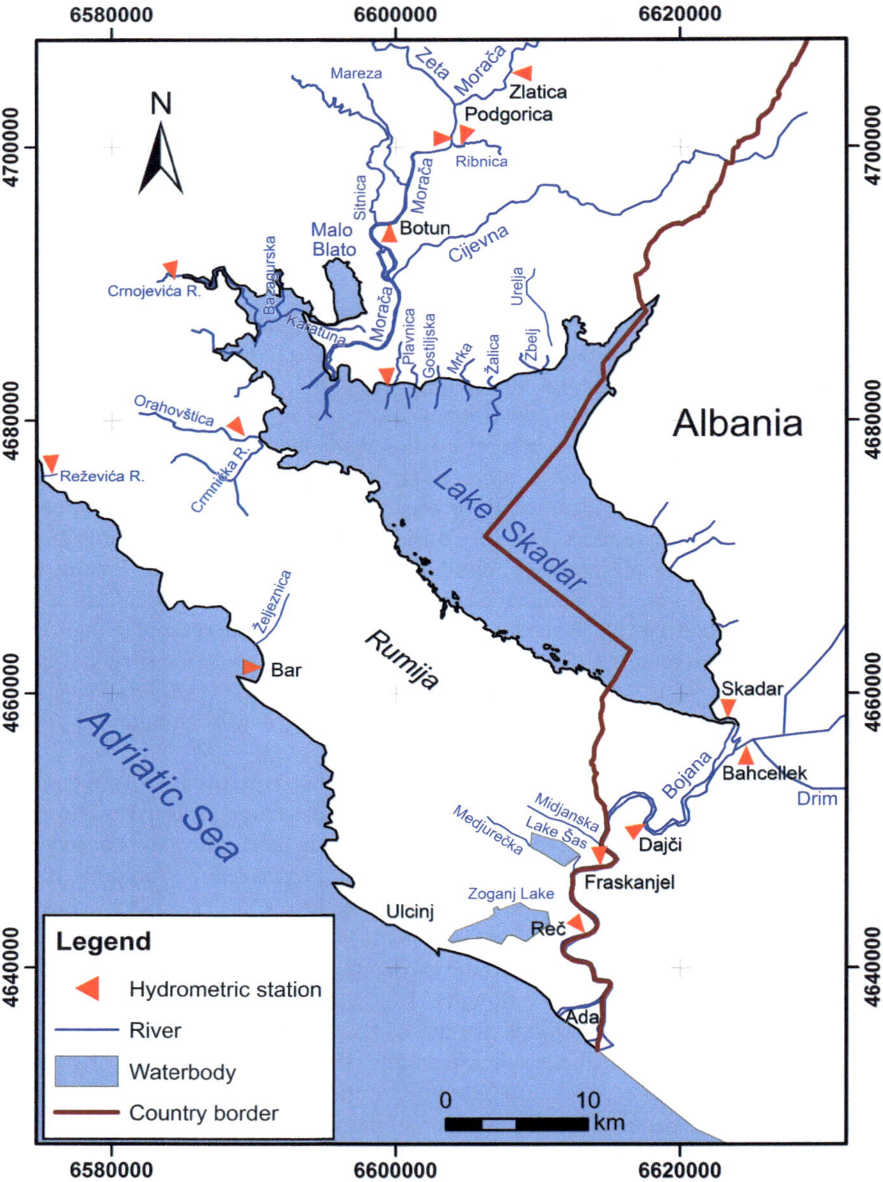

Fig. 6 Locations of hydrometric stations in the area of Lake Skadar

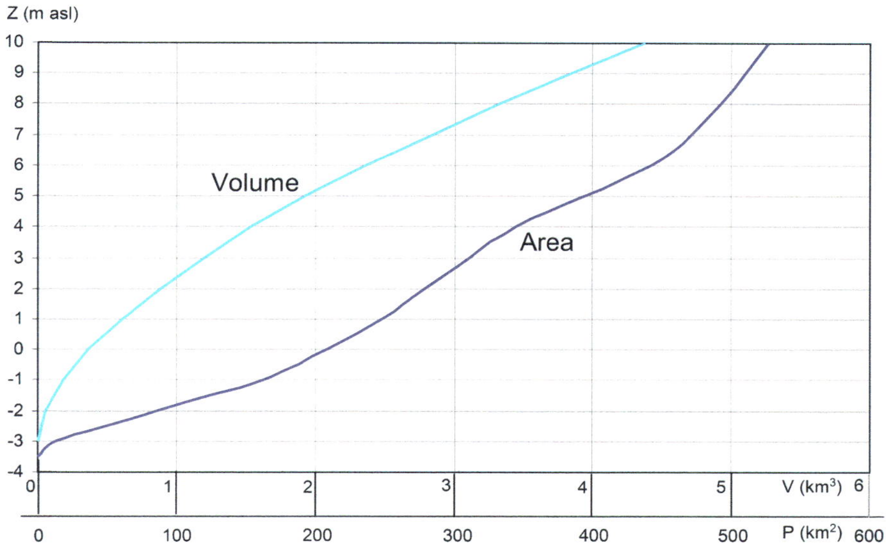

Fig. 7 The curves of volume and area of Lake Skadar based on a bathymetric survey conducted in 2009 [38]

Table 9 General water balance of Lake Skadar [38]

Type of inflow/outflow	Inflow/outflow (m^3/s)
The Morača river (HS Botun)[a]	170
The Sitnica river (profile Komanski bridge)	7
The River Cijevna (HS Trgaj)	25
Rijeka Crnojevića (HS Brodska Njiva)	6
The Orahovštica river (HS Orahovo)	3
Uncontrolled surface water inflow	10
Precipitation over the surface of the lake	20
Total surface water inflow from the Montenegrin side	241
Discharge from the lake (the river Bojana, HS Skadar)	310
Evaporation from the lake surface	16
Total outflow	326
Unmeasured inflow (groundwater inflow, surface inflow from the Albanian side, etc.)	85

[a]The measured mean annual discharge at the upstream station (HS Podgorica) which has a much longer time series is 160 m^3/s

4.9 Water Levels of Other Significant Natural Lakes in Montenegro

So far, aside from the extensive data about Lake Skadar, there are insufficient measured data about water level fluctuations for the other natural lakes in Montenegro (35 in total). For their particular beauty and tourism significance, Lake Biogradsko and Lake Plavsko are especially important, as is Lake Crno on Mount Durmitor.

The results of the analysis of the water levels of Lake Biogradsko and Lake Plavsko are given in Table 10.

Based on the observed results, the following can be concluded:

- The levels of Lake Biogradsko vary by about 4.5 m on average, while at a probability of 1%, the variations total 7.5 m [35]. A precise bathymetric survey of the lake floor morphology has not been performed yet. This lake loses certain amounts of water through swallow holes. Indeed, during periods when there is no continual flow from the River Jezerštica (which flows out from the lake) and when there is a significant inflow from the River Biogradska, the lake water level still decreases, which clearly indicates the sinking of water through fractured volcanic rocks and glacial sediments.
- The water level of Lake Plavsko varies by about 1.5 m on average, while at a probability of 1%, the variation is about 2.5 m [35]. During recent years, some hydrological analysis and lake water balances have been performed, as well as the hydrodynamic analysis of lake water currents. Furthermore, an analysis of the erosion production of slit and slit transport towards the lake have been conducted. Based on these examinations, it can be concluded that the usable volume in lake is decreasing, i.e. a filling has appeared in the shallow parts of the lake, which, in places, is up to 3 m in depth, and which cannot be neglected, since its expansion rate is 5–10 cm annually.

Lake Crno on Mount Durmitor is a unique water body, whose floor consists of two attached depressions divided by a submerged ridge between them. During the summer, lowering water levels means this ridge appears on the surface, dividing the lake into two separate water bodies, Malo Crno and Veliko Crno.

Table 10 Water levels of Lake Biogradsko and Lake Plavsko [35]

			Water level (m asl) for different probabilities of occurrence				
Station	Water level (m asl)		1%	2%	5%	10%	50%
Lake Biogradsko	Min	1,095.84	1,093.1	1,093.4	1,093.9	1,094.3	1,095.8
	Average	1,098.61	1,099.4	1,099.3	1,099.2	1,099.2	1,098.7
	Max	1,100.25	1,100.7	1,100.6	1,100.6	1,100.5	1,100.3
Lake Plavsko	Min	906.98	906.9	906.9	906.9	906.9	907.0
	Average	907.30	907.5	907.4	907.4	907.4	907.3
	Max	908.44	909.3	909.2	909.0	908.9	908.4

The water level in Veliko Crno, compared to the average level (1,418.72 m asl), oscillates from a hydrological maximum of 6.35 m above the average to a hydrological minimum of 1.32 m below it. At its lowest level (1,398 m asl), the water volume in Veliko Crno is 4,190 m^3 and the size of the water surface is 1.7 ha. At the average water level (1,418.72 m asl), the water volume is about 3.18 million m^3, and the size of the water surface is about 26.7 ha. For its maximum water level (1,423.07 m asl), the water volume is 4.4 million m^3, and the size of the water surface is about 35.8 ha [28].

5 Conclusion

Using the results of the calculations and analysis that was performed [35], it can be stated that the hydrological stations monitor over 10,200 km^2 or about 74% of Montenegrin territory. The whole territory of Montenegro cannot be hydrometrically monitored because a significant part of the groundwater flows out directly below the sea or lake levels, or else it flows through subterranean routes to the territories of neighbouring countries. The measured mean annual runoff amounts to around 440 m^3/s (13.87 billion m^3/year), but the assessed total mean annual runoff from the entire territory of Montenegro is around 595 m^3/s (18.76 billion m^3/year).

Measured by the mean annual specific runoff, which is around 43 l/s/km^2, Montenegro is thus one of the richest countries in terms of water resources. However, the uneven regime of flow causes water scarcity during the summer, when many springs and riverbeds dry up, which pushes Montenegro to extremes.

The huge water potential on the one hand and the uneven runoff regime and sea water intrusion on the other produce a significant contradiction between opportunities and the problem of insufficient quantities of drinking water during the hydrological minimum.

Most of the rivers in Montenegro are formed within its territory, so it is the people of Montenegro who are primarily responsible for their present and future status. This is an additional reason why we have to pay particular attention to the protection and preservation of our waters as one of the most precious natural resources that Montenegro has.

Acknowledgements We are extremely grateful to photographer Dejan Stojanović (Falco Nero Ltd.), who sent us the wonderful aerial photo of the Sopot Springs.

References

1. Tietze E (1884) Geologische Uebersicht von Montenegro. Hölder, Vienna
2. Baldacci L (1886) Ricognizionegeologico–minerariadel Montenegro. Boll R Comit Geol Ital Boll 17:9–10

3. Hassert K (1895) Beiträgezurphysischen Geographie von Montenegro. Petermanns Mitteilungen Ergänzungsheft, Gotha
4. Cvijić J (1899) Glacial and morphological studies of the mountains of Bosnia, Herzegovina and Montenegro (in Serbian). Voice Serb Roy Acad Sci 57:1–196
5. Nopcsa F (1916) Begleitwortzur geol. Karten von Nordalbanien, Rasien und Ostmontenegro. Földtani Közlöny 46:7–12
6. Bourcart J (1926) Nouvelles observations sur la structure des Dinarides adriatiques. In: Actasdel XIV Congreso Geológico Internacional, Madrid, pp 1899–1941
7. Waisse JG (1948) Les bauxites de l'Europecentrale (Province dinariqueet Hongrie). Dissertation, University of Lausanne, Lausanne
8. Milovanović B (1965) Epeirogenic and orogenic dynamics in the area of External Dinarides and the problems of paleokarstification and geological evolution of holokarst (in Serbian). J Geol Surv Serbia 4(5):5–44
9. Bešić Z (1969) Geology of Montenegro–Karst of Montenegro (in Serbian). Geological Survey of Montenegro, Podgorica
10. Grubić A (1975) Tectonics of Yugoslavia. Acta Geol 41:365–384
11. Vlahović V (1975) Karst of Niksic polje and its hydrogeology (in Serbian). The Society of Sciences and Arts of Montenegro, Titograd
12. Mirković M, Živaljević M, Đokić V, Perović Z, Kalezić M, Pajović M (1985) Geological map of Montenegro 1:200,000. The Republic Self Managing Community of Interest for Geological Exploration of Montenegro, Titograd
13. Radulović MM, Radulović M, Stevanović Z, Sekulić G, Radulović V, Burić M, Novaković D, Vako E, Blagojević M, Dević N, Radojević D (2015) Hydrogeology of the Skadar Lake basin (Southeast Dinarides) with an assessment of considerable subterranean inflow. Environ Earth Sci 74:71–82. https://doi.org/10.1007/s12665-015-4090-7
14. Torbarov K, Radulović V (1966) Regional hydrogeological research of Montenegro and Eastern Herzegovina (in Serbian). Geological Survey of Montenegro, Titograd
15. Radulović V (1989) Hydrogeology of Skadar Lake watershed (in Serbian). Geological Survey of Montenegro, Titograd
16. Radulović M (2000) Karst hydrogeology of Montenegro (in Serbian). Geological Survey of Montenegro, Podgorica
17. Radulović MM (2012) Multi-parameter analysis of groundwater recharge in karstic aquifers–case examples from Skadar Lake basin (in Serbian). Dissertation, University of Belgrade
18. Radulović M, Popović Z, Vujisić M, Novaković D (1989) Basic hydrogeological map of SRY 1:100,000, "Bar" sheet (in Serbian). Geological Survey of Montenegro, Titograd
19. Radulović M, Popović Z, Vujisić M, Novaković D (1998) Guide book for the Basic hydrogeological map of SRY 1:100,000, "Bar" sheet (in Serbian). Geological Survey of Montenegro, Titograd
20. Radulović MM, Novaković D, Sekulić G (2013) Geological and hydrogeological characteristics of the Montenegrin part of the Skadar Lake catchment area (in Serbian). In: Sekulić G, Bushati S (eds) Development of hydrogeological and hydraulic study of regulation of Skadar Lake and Bojana river water regime (IPA Project)–Volume 1. Montenegrin Academy of Science and Arts, Podgorica, pp 9–115
21. Burić M (1993) Water supply of tourist settlements of the Montenegrin coast (in Serbian). Dissertation, University of Belgrade
22. Burić M (2010) Atlas of waters of Montenegro (in Montenegrin). Montenegrin Academy of Sciences and Arts, Podgorica
23. Hrvačević S (1999) Hydrogeological characteristics of Piva and Tara river basins from the aspect of utilization and protection of water potential (in Montenegrin). Master thesis, University of Belgrade
24. Hrvačević S (2004) Resources of surface water in Montenegro (in Montenegrin). EPCG AD, Podgorica

25. Radulović V, Radulović M (1999) Groundwater bodies (in Serbian). In: Gomilanović M (ed) Mineral resources and mine production in Montenegro. Ministry of Industry, Energy and Mining of Montenegro, Podgorica, pp 595–760
26. Radulović M, Radulović V (2004) Hydrogeological map of Montenegro 1:200,000 (in Serbian). Geological Survey of Montenegro, Podgorica
27. Lješević M (2004) Karst of Piva (in Serbian). Montenegrin Academy of Sciences and Arts, Podgorica
28. Djordjević B, Sekulić G, Radulović M, Šaranović M (2010) Water potentials of Montenegro (in Serbian). Montenegrin Academy of Sciences and Arts, Podgorica
29. Radojević D, Nedić D, Blečić V, Zečević M (2011) Basic hydrogeological map of Montenegro 1:100,000, sheets: "Gusinje", "Ivangrad" and "Šavnik". Geological Survey of Montenegro, Titograd
30. Stevanović Z, Kukurić N, Pekašš Ž, Jolović B, Pambuku A, Radojević D (2016) Dinaric karst aquifer – one of the world's largest transboundary systems and an ideal location for applying innovative and integrated water management. In: Stevanović Z, Krešić N, Kukurić N (eds) Karst without boundaries. Taylor & Francis Group, London, pp 3–25
31. Radulović MM (2018) Hydrogeology of the Skadar Lake Basin. In: Pešić V, Karaman G, Kostianoy A (eds) Skadar/Shkodra lake environment. Springer International Publishing AG, Cham, Switzerland, pp 25–45
32. Radulović M, Danilović T, Radulovic MM (2005) Specifities of hydrogeologic watersheds and directions of movement of aquifer waters in karstic terrains of Montenegro. In: Stevanović Z, Milanović P (eds) Water resources and environmental problems in karst. Proceedings of the International conference and field seminars, Belgrade & Kotor, 13–19 September 2005, pp 437–442
33. MARD (2017) Strategy for water management in Montenegro (in Montenegrin). In: Government of Montenegro. Ministry of agriculture and rural development, Podgorica
34. Matović et al (2018) Final report about results of detailed hydrogeological investigations of the aluvion of Tara River, downstream from Mateševo, for the purposes of water supply of facilities along the highway Bar-Boljare, section Smokovac-Uvač-Mateševo. Geoprojekt Ltd, Podgorica
35. IJC (Institute Jaroslav Cerni) (2001) Water master plan of Montenegro (in Serbian). Government of Montenegro, Podgorica
36. Djordjević B, Šaranović M (2007) Hydropower potential of Montenegro (in Serbian). Montenegrin Academy of Science and Arts, Podgorica
37. IHSM (Institute of Hydrometeorology and Seismology of Montenegro) (2006) The web page "Hydrology". Accessed 2006
38. Sekulić G, Radulović MM, Novaković D, Popović M, Alilović N, Abazi E et al (2013) First year report (Volume I) of the IPA project "Development of hydrogeological and hydraulic study of regulation of Skadar Lake and Bojana river water regime" (in Serbian). Montenegrin Academy of Science and Arts, Podgorica
39. Sekulić G, Ivković M, Rosić N (2014) Development of hydrogeological and hydraulic study of regulation of Skadar Lake and Bojana river water regime (IPA Project)–Volume 2. Montenegrin Academy of Science and Arts, Podgorica
40. Pesić V, Karaman G, Kostianoy A (2018) The Skadar/Shkodra lake environment. Springer International Publishing AG, Cham, Switzerland, p 508
41. SPM (2008) Spatial plan of Montenegro until. Government of Montenegro, Ministry of Economic Development, Podgorica, p 2020
42. MSPE (2010) Detailed spatial plan for the area of multi-purpose reservoirs on the River Morača–Draft (in Montenegrin). Government of Montenegro, Ministry for Spatial Planning and Environment, Podgorica
43. ME (2011) Map of resources that includes natural and cultural characteristics, geographical position, human and other resources with an assessment of optimal directions of specialization of the region of Montenegro (in Montenegrin). In: Government of Montenegro. Ministry of Economy, Podgorica

44. MARD (2018) Report on the implementation of the strategy for water management (in Montenegrin). In: Government of Montenegro. Ministry of Agriculture and Rural Development, Podgorica
45. Bošković M, Popović M, Alilović N (2004) Supplement to Skadar Lake geogenesis, its inflow and outlet components and background on its regulation activities. In: Proceedings of the conference in water observation and information system for decision support, BALWOIS, Ohrid, 25–29 May 2004
46. Milošević M (1983) Background of previous activities on regulation of Skadar Lake, Drim and Bojana. In: Ivanović B (ed) Skadar Lake, book 9 of scientific meetings. Montenegrin Academy of Sciences and Arts, Titograd
47. Dragovic D (1997) Some natural potentials of Skadar Lake region and possibility of their usage in accordance with the ecological requirements (in Serbian). In: Vojvodić V, Mijušković M (eds) Natural values and protection of Skadar Lake. Scientific meeting on Natural values and protection of Skadar Lake, Podgorica, November 1995. Montenegrin Academy of Sciences and Arts, Podgorica, p 561

The Utilization of the Hydropower Potential of Rivers in Montenegro

Goran Sekulić

Contents

Abstract Previous hydrological researches of the surface watercourse system in Montenegro imply quite high watercourse water capacity compared to relatively small territory of Montenegro. This paper systematizes theoretical and remaining technically feasible hydropower potential of the major surface watercourses and their tributaries, in Montenegro. Analyses are based on the results of the most recent analysis of the theoretical hydropower potential of the specialized studies and projects for development of strategic basis and urban plan of Montenegro. Montenegro has such a lot of hydropower potential at its disposal that it could be among the top hydropower producers in the world, while retaining control over the ecological and social environments. From about 9.8 TWh/annually of linear potential on larger water streams, about 5.5–6.1 TWh/annually is already categorized (or soon will be) as economically usable potential.

An overview of existing built hydropower plants with their basic characteristics is given. Moreover, we analyze the state of development of small hydropower plants and possibilities for their further construction. It also gives an overview of the possibility of constructing new hydroelectric power plants under conditions that

G. Sekulić (✉)
Faculty of Civil Engineering, University of Montenegro, Podgorica, Montenegro
e-mail: sgoran2000@gmail.com

Vladimir Pešić, Momir Paunović, and Andrey G. Kostianoy (eds.), *The Rivers of Montenegro*, Hdb Env Chem (2020) 93: 43–68, DOI 10.1007/698_2019_412,

have changed significantly since the time when the previous plans for their construction were adopted.

Keywords Ecological and social environments, Hydropower plant, Hydropower potential, Montenegro

1 Introduction

Hydropower is the most significant, but also the most desirable, use of a watercourse, because it does not decrease the water balance within the framework of the closed annual hydrological cycle. Watercourse use for hydropower purposes not only has zero negative effects on water quality but can also – with proper protective measures – offer benefits in terms of actually improving water quality [1]. In the case of Montenegrin watercourses (Fig. 1), both components of improving water quality are highly desirable, because the watercourse flows are not equal during the various seasons, with extremely small flows during periods of low waters, when river and coastal ecosystems[1] are endangered [2].

Based on a highly accurate analysis of the hydrological balance, it is estimated that the total sum of usable hydropower potential in Montenegro is within the range of 5.6–6.1 TWh/annually, depending on the way the water streams are used, with small hydropower plants left out of this calculations, which would account for a further potential of about 0.45 TWh/annually [3]. The existing facilities create 1.6 TWh/annually, and the installed power creation is currently 658 MW, which are only 25–27% of the real potential. The approximate requirements of the electricity supply in Montenegro vary depending on processing, although it is estimated that they will total 6.9 TWh/annually in 2021 [4]. This means that the coverage of electric energy consumption by electricity produced from its own hydropower plants could total over 85%. If such an economic and ecological healthy national developing strategy for the more extensive use of its own watercourse potential was agreed – within the range of integral systems for arranging, using, and protecting watercourses – Montenegro as a country could be counted among those privileged countries with solid grounds to satisfy most of their domestic need for electricity supplying from one of the most renewable, ecologically pure energy sources, watercourses [4]. Only Norway has greater potential in this regard [5].

[1]A typical example is the River Moraca, having an average flow at Zlatica of 52.5 m^3/s, which also dries in low water periods, with all the painful ecological consequences of such an extravagantly variable water regime. In order to observe the extreme unevenness of the water regime better, it needs to be emphasized that flood flow on the Moraca according to that profile, having a probability value of 1% $Q_{1\%}$, is ~1,456 m^3/s.

Fig. 1 Hydrographical map of Montenegro

The orographic, hydrographic, and hydrological circumstances in Montenegro are very favorable for the realization of integral systems regarding watercourses. With a total runoff of domicile waters totalling 18.75×10^9 m^3, and an average perennial specific module of runoff of 39.2 $L/s/km^2$ [6], Montenegro can be considered as (a) one of the European countries that is richest in watercourses, observing the average values of domestic watercourses, and (b) one of the countries that have not used much of this valuable renewable resource, resulting in the following: instead of exporting valuable peak power from large reservoirs of hydropower plants, the country imports electrical energy, which is a significant burden and hindrance for its further development [4, 7]. From the total of available watercourses within Montenegrin territory, the majority is considered to be water resource, which has the potential to be transferable into hydro-energy production.

The total volume of the runoff of rivers belonging to the Adriatic basin is 11.814×10^6 m^3 (of that, domestic water streams total 11.353×10^6 m^3), while the total runoff volume of the Black Sea Basin is 7.855×10^6 m^3 (of which domestic water streams are 7.401×10^6 m^3) [6]. The total amount of rainfall and transit waters from Montenegrin territory is $W_{\text{tot}.n} = 26 \times 10^9$ m^3, of which $W_{\text{evp}} = 6.32 \times 10^9$ m^3 evaporates [4]. What follows, then, is that the total runoff from Montenegrin territory has a volume of $W_t = 19.7 \times 10^9$ m^3, while if we only consider domestic waters, the volume is $W_{\text{CG}} = 18.75 \times 10^9$ m^3 or approximately 595 m^3/s [6]. That flow is the result of runoffs from all the water streams of the Black Sea Basin, with the runoffs in the Adriatic basin for the water streams of the Moraca, the Zeta, and the Cijevna, plus other river streams, including those groundwater streams which flow towards the River Trebisnjica, the Adriatic Sea, and the Lake Skadar through karst formations.

If we separate and calculate all the flows which have absolutely no physical potential for hydropower within Montenegrin territory, the estimation of that water potential with resource features is implied, because it can be transformed into partial hydropower potential. This value is $V_p \approx 13.34 \times 10^9$ m^3.

Results of the analysis of water potential in Montenegro (Fig. 2) [4, 7], with a summary delimitation of two key categories of potential, show next:

1. The total amount of domestic water in Montenegro "*present* in the basins" is 18.754×10^6 m^3.
2. The sum of the flow within Montenegrin territory which have the feature of being a "water resource" is 13.342×10^6 m^3, because they might be transformed into partial potential – whether for hydropower or other uses, within the range of integral waterpower engineering, as well as the arrangement of that category of potential along all the major rivers.

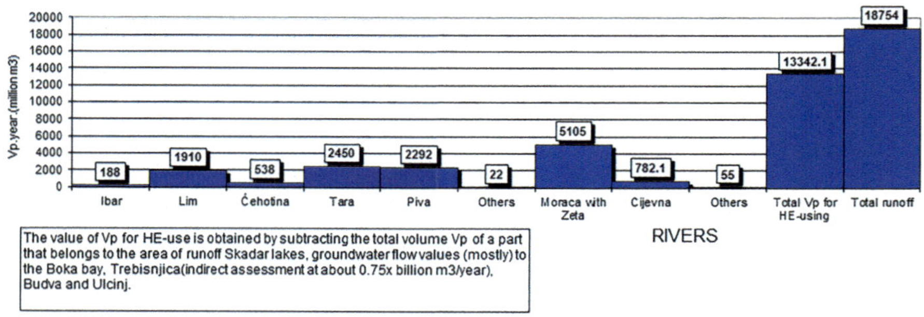

Fig. 2 Total water potential (million m^3) of the water streams and summary for Montenegro. Modified after [2, 4]

The water volume, quantified as a "water resource" with a total amount within Montenegrin territory, totals $V_p = 13.34 \times 10^9$ m^3 and matches an average flow of $Q = 423$ m^3/s, which is 71.1% of the domestic surface waters of Montenegro [2]. That value, V_p, is the sum of two other values: (a) the runoff of the rivers in the Black Sea river basin and (b) the runoff towards the Adriatic basin for the Rivers Moraca and Zeta and their tributaries. The volume of the part of the runoff towards Boka Bay and Trebisnjica and the immediate area and Lake Skadar groundwater runoff, which is 5.41×10^9 m^3 in total, are not included in this calculation of potential. The conclusion is that a significant part of the water streams in the basin cannot be used due to the absence of the right conditions for their use [7].

Analyzing the hydropower potential for planning requires some categories of gross potential, which are very important in terms of identifying the technically usable hydropower potential [4].

2 The Gross Potential of Montenegrin Rivers

Gross hydropower potential, calculated by the water availability in a so-called average hydrological year, for Q_{av}, has been measured multiple times, and the last time, it was within the study *The Waterpower Engineering Basis of Montenegro* (2001) [7]. Analysis was done for nine major rivers, and the gross potential for each was determined over line potential (calculated on 5 km of river flow) [7]. The final results are shown in Table 1, and their graphical representation is provided in Fig. 3. Average power, indicated in MW, represents the power which can be realized for perennial average flows (Q_{av}), which are defined for the hydrological series during the period 1947–2012, which was chosen as the most hegemonic and relevant for this analysis [1].

Table 1 Hydropower potential along the major rivers of Montenegro [2, 4]

River	Approximate power for Q_{sr} (MW)	Energy (GWh/annually)
Piva	155	1,361
Tara	257	2,255
Cehotina	53	463
Lim	164	1,438
Ibar	14	118
Moraca to the confluence with the Zeta river	168	1,469
Zeta	229	2,007
Mala River	52	452
Cijevna	32	283
Total	1,124	9,846

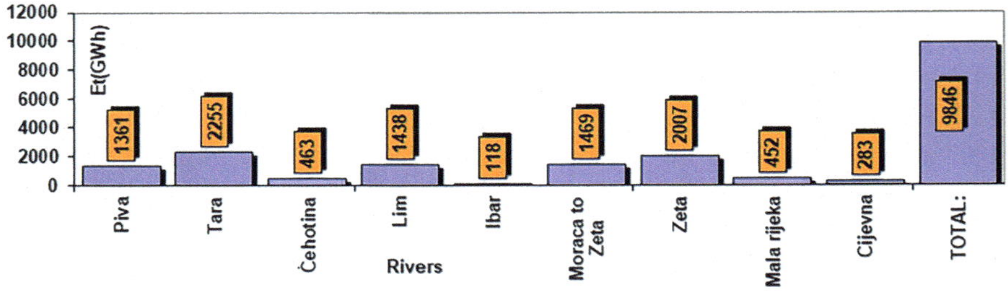

Fig. 3 Overview of gross potential (GWh/annually) of the major water streams in Montenegro. Modified after [2, 4]

The implication is that Montenegro has a potential of 9.846 GWh/annually on the nine major rivers at its disposal, while gross energy potential on minor water streams can be estimated at around 800–1,000 GWh/annually [1, 2, 12]. The greatest hydropotential is shown by the River Tara (2,225 GWh/annually), followed by the Zeta (2,007 GWh/annually), the Moraca (1,469 GWh/annually), the Lim (1,438 GWh/annually), and the Piva (1,361 GWh/annually), while the potential of the other rivers is lower [2, 3, 7].

It is very interesting to analyze the linear potential on the major rivers of Montenegro [2, 3, 7]. The analysis was performed through the discretization of each 5 km of river flow. What can be assumed is that the highest concentration of linear potential in Montenegro is shown on the Piva and the Tara (components of the River Drina), in the wider zone of "Scepan Polje." The highest concentration of potential was registered on the River Piva between kilometer 10 and kilometer 15 (about 37 MW), and this part is located exactly in the zone of the realization of the hydroelectric power plant, HPP, "Piva." An extremely high concentration of hydropotential on the River Zeta can be noticed in the Niksic downstream zone, which was successfully used for the derivational plant of HPP Perucica [2, 3, 7].

A significant concentration of potential is also evident on the River Tara, in the Mojkovac downstream zone. The Lim contains a considerable and relatively equal potential concentration along its whole flow (about 1.6–2.5 MW/km) [2, 3, 7], which proves the expedience of building a continuous cascade system on this river. Large concentrations of linear potential are also visible on the upper and middle Moraca streams, exactly in the zone predicted for the realization of cascades of hydropower plants on this river. The linear potential on the Cehotina and the Ibar is relatively modest, but these water streams have certain other advantage – including the possibility of creating large volume reservoirs for the purpose of arranging the water regimes by annual regulation [2, 3, 7].

The highly unequal disposition of potential in the river basins can be observed from the specific surface potential of certain rivers. The specific potential of individual basins, calculated only for the major water streams, is the Piva, 0.78 kWh/m^2; the Tara, 1.1 kWh/m^2; the Cehotina, 0.41 kWh/m^2; the Lim, 0.58 kWh/m^2; the Moraca and the Zeta (basin as a whole), 1.3 kWh/m^2; and the Ibar, 0.27 kWh/m^2 [2, 3, 7].

It can be concluded that the Moraca basin has the highest concentration of specific potential per unit of basin area (1.3 kWh/m^2), which places this basin on the list of the most interesting European basins (regarding potential power output) [2, 3, 8]. The next highest potential is shown in the Tara basin at 1.1 kWh/m^2. The least specific potential is found in the Ibar basin, but it has a large advantage – the possibility of creating large reservoirs by annual regulation [2, 3, 7].

3 Existing Hydropower Facilities

Several separate periods can be distinguished in the process of developing energy in Montenegro. The first covers the period after the first use of power (fuels) for industrial purposes and the building of the first power plants up until World War II.

The first time electrical energy was used in Montenegro was in 1906 in Bar, when the Railway Society, while building the Bar-Virpazar railway, established a diesel electric power plant for construction site energy requirements, which was also partially used for electric street lightning in Bar. However, the first public diesel power plant was built in Cetinje in 1910, having a power output of 100 kW, with city electrification (by cable networking and substations) involving 2100/150 V. Afterwards, diesel power plants were built in Niksic in 1912, in Podgorica in 1924, and later on in other Montenegrin cities. The Montenegrin domestic creation of electric power plants was achieved after building a small thermo-power plant in Pljevlja, with a power output of 70 kW in 1937 [3, 9], by using coal from the superficial coal seams of the Pljevlja region. The first hydropower plant at Podgor had a power output of 300 kW. It was built in 1939, and its purpose was to supply the pumps for the water supply system between Podgor and Cetinje with electricity.

The production of electrical power in diesel power plants, specifically TPP Pljevlja and HPP Podgor, was only 932 MWh in 1939, while electrification extended to only 2% of the towns in Montenegro [9].

After World War II, and up until 1955, the reconstruction and renovation of diesel power plants destroyed in the war were carried out, and small hydropower plants were built. In this way, the first plan for the electrification of Montenegro was accomplished. In the creation of small hydropower plants – at Rijeka Crnojevica, Savnik, and Lijeva Rijeka, which all had communal significance – three more significant small hydropower plants were built. In this period, a small thermo-power plant at Pljevlja was built, having a power output of 2 MW, and in 1950, a small thermo-power plant at Kotor was also built, having a power output of 1 MW. The small hydropower plant at Musovica Rijeka near Kolasin with an installed capacity of 15 MW was built in 1950, while HPP Slap Zete was built in 1955 (with an installed capacity of 5.0 MW) [3, 9]. The third period, after 1955, started with the development of significant metallurgical, metal industry, and textile industry capacities. The construction of HPP Perucica started and so too did the formation of the basis of the contemporary electro-power system in Montenegro.

The use of water streams for electrical power supply, or, to be more precise, hydropower potential, started in Montenegro shortly after the building of the first hydropower plants in the world. As is widely known, the first hydropower plant for commercial use was built in Wisconsin, USA, in 1882 [9]. The first important HPP was built by Nikola Tesla on one of the falls of the Niagara. In the area of the former Yugoslavia, the first such plant was built in 1908 at Velika Gubavica on Cetina by using waterfalls. The building of this facility encouraged the examination of the water stream capacities of Montenegro (during 1909 and 1910). In 1911, the Kingdom of Montenegro introduced a law on concessions related to the building of HPP Tara – Moraca by delivering River Tara water into the River Moraca and using the slope between the Tara and the Moraca of about 700 m and another power plant on the Moraca near Donja Gorica in Podgorica [3, 9]. The First World War interrupted the realization of this project. In 1939, this project was renewed, but it was interrupted by World War II.

The first renewed idea about a project to create HPP Tara – Moraca appeared in 1956. Following the concepts of that period, this was quite a radical project, which could affect urban planning and social issues. Simultaneously, a project for using the River Zeta for that purpose was executed. During 1922–1928, the so-called Skadar section was pursued, which had the task of regulating Lake Skadar, the River Bojana, the Niksic karst valley, and the valley of Bjelopavlici. In order to supply pumps for the amelioration of the Bjelopavlici Valley, the building of two hydropower plants was planned: HPP Slap Zete and HPP Glava Zete. The Second World War prevented the completion of this project. After World War II, the work on these projects started again, but instead of the amelioration of the Niksic valley, the concept of building a reservoir within the range of the hydropower system at "Perucica" was the project that gained approval [3, 9].

Within the scope of exploring water stream use for producing power, delivering the stream of the River Piva into the Zeta was also examined. Single-stage solutions and two-stage solutions were examined with profiles for dams at Duzi and Konzila. Later, this idea was abandoned because of the concerns raised by Serbia and Bosnia and Herzegovina, so instead of delivering Piva water into the Zeta, TPP Piva was created. During the 1960s, the base project for entire Drina basin was executed, in order to solve the problem of delivering water streams from one basin into another. There were several attempts to find a mutually acceptable solution, but the problem remained unsolved [3, 9].

4 The "Perucica" Hydropower System in the Upper Zeta Basin

The "Perucica" hydropower system on the Zeta is very complex and was created because of the circumstances of multiple natural features within the area of the Upper Zeta basin [7, 10].

The Niksic karst field, at an altitude of 600–630 m above sea level, is an area that is rich hydrographically, because it is intersected by the Zeta and its major tributaries the Gracanica, the Mostanica, and the Opacica. These river streams drain a significantly larger area of the basin (about 1,170 km^2) than merely the topographic basin, which is about 850 km^2 [7]. Because of that, as well as because of the fact that this basin has annual rainfall of more than 2,000 mm, at its highest flow, a specific concentration is realized in the relatively narrow area of the Niksic valley, which is on average over 50 m^3/s with a runoff, both from surface and underground sources, of over $1{,}500 \times 10^6\ m^3$ annually [12]. The superficial velocity of the existing water streams is 32 m^3/s approximately or only around 2/3 of the total flow (average), while 1/3 runs off by underground routes, not appearing in the Niksic field at all [4, 7].

Over the rather short distance of only 4.5 km of aerial distance, the Zeta falls about 550 m, because it goes underground at an elevation of 600 m above sea level approximately, only to reappear again at the spring of Glava Zete at an elevation of about 50 m above sea level. That is an extraordinary feature, unique in this area [3, 7]. Thanks to the high concentration of both fall and flow over such a short distance, all charts of linear hydropower potential prove that the zone of Zeta's shallow hole has the highest concentration of disposable hydropower potential, not only within the former Yugoslavia (without exception) but perhaps even in Europe [1, 2, 7].

Bearing in mind these circumstances, the project of the system of "Gornja Zeta" was conceived as a multifunctional system, with its primary purposes being the creation of hydropower, arranging the water regime and the amelioration of the water system and supplying industry with water. The "Perucica" hydropower plant was a very complex construction project, having several natural distinctive features to regulate, which resulted in its complex configuration and nonstandard solution for HPP. These include the karstification of the system zone wider area, which cannot be realized by a unique facility (reservoir) for regulating flow; the shallow hole at Slivlje which creates the fall concentration; the unusually uneven water regimes in terms of temporarily, with flood flows and long dry periods, which require the temporal rearranging of water by reservoirs; and the complexity of arranging the water regimes due to their issue, so that the Niksic valley area could be used for a variety of purposes without disturbing the urban planning of the town [3, 7].

The nonstandard solution of system called "Gornja Zeta" had its origins in such geophysical circumstances. The shallow hole at Slivlje was barred by a cylindrical dam, so as to create minor retention and make intake construction possible in this zone. Due to the unequal water regime, the problem of karstification with different water sustainability within the terrain in the wider field zone, as well as the spatial position and elevation of Niksic as a city, the problem of regulating the flow was solved by reservoirs organized on the cassettes principles [3, 7] (Fig. 4).

Water is used from those reservoirs which lose water massively despite sealing works initially so that it can be kept longer in reservoirs where water sustainability is better. A system of channels with concrete layers is in place to prevent water loss during its transport from facilities upstream to the intake construction. Near the

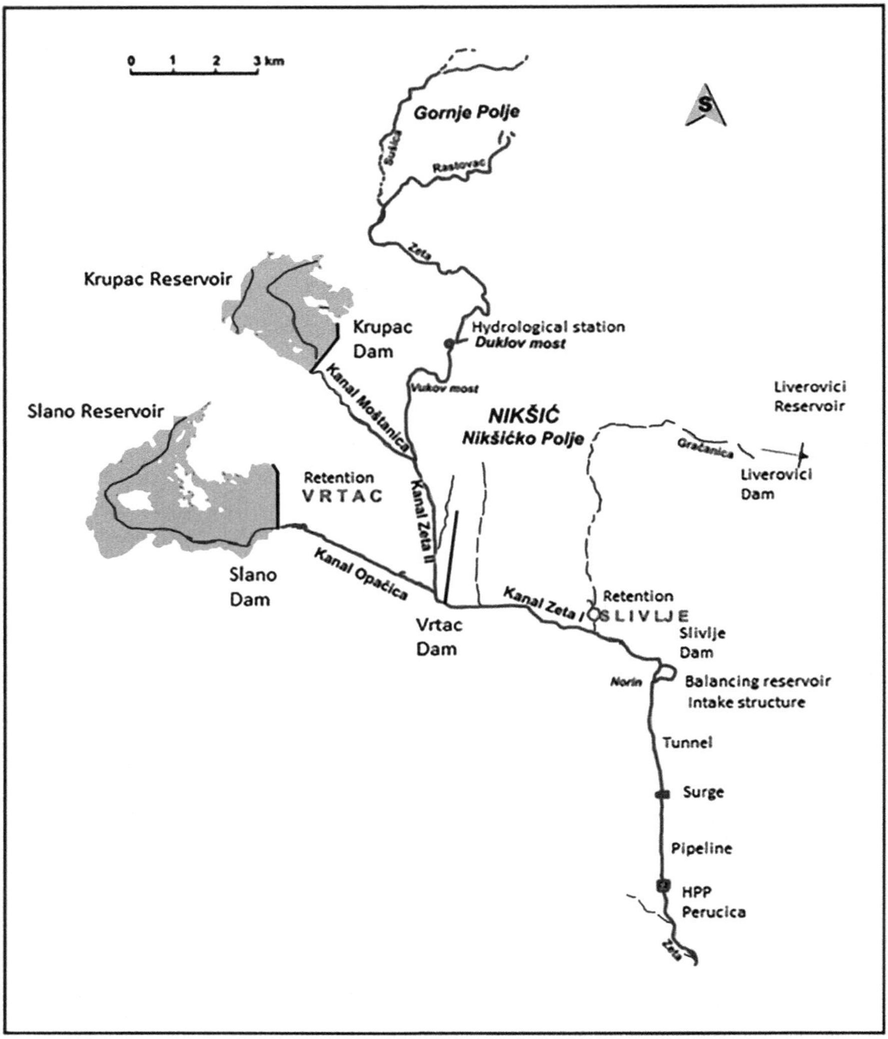

Fig. 4 Scheme of the water reservoir in system of HPP Perucica. Modified after [3]

cylindrical dam at Slivlje, there is an intake construction; then the water is transported by tunnel and pipeline derivation to the machine building, which was built on a river terrace by the Perucica spring close to a wider spring zone called Glava Zete, at an elevation of 65 m above sea level. In this way, the fall concentration of about 544 m is guaranteed by a short derivation which consists of under pressure tunnel which is 3.3 km long and a pipeline 1.8 km in length [3, 9]. HPP Perucica is thus located in a place with one of the naturally most favorable dispositions for derivational hydropower plants in the world, with a fall index concentration of $\kappa = (544\ \text{m}/5.1\ \text{km}) \approx 107\ \text{m/km}$ [3, 9].

The dam at Slivlje is unique in the world. Since the shallow hole is 20 m in diameter, it is isolated by a concrete cylindrical dam that is 50 m in diameter and 17 m high, with a *flood* elevation of 612 m above sea level, and cylinder walls that only 60 cm thick, relying on a basis threshold at an elevation of 610 m above sea level. The Zeta basin on its inflow into the shallow hole contains a bottom outlet. For the evacuation of flood flows from the cylinder, free overflow is realized with a capacity of 120 m^3/s, which ensures the sustainability of the shallow hole [3, 4].

The dam at Vrtac, as a downstream reservoir cassette, is formed by a rock-fill dam, 2,386 m long and 16 m high, with a crown elevation of 616.5 m above sea level. The frontal overflow with three flooding fields, with valve floodgates, forms a retention area up to 614 m above sea level which is 72×10^6 m^3 in volume. The overflow is dimensioned for the evacuation of 300 m^3/s, and the bottom outlet has a capacity of 68 m^3/s [3, 4].

The Slano Reservoir is created by a dam, 21 m high and 1,630 m long. The crown elevation is 623 m above sea level, and the overflow capacity of 150 m^3/s has an elevation of 621 m above sea level. The reservoir volume is 107×10^6 m^3 [3, 4]. The bottom outlet with a capacity of 51 m^3/s is in accordance with the capacity of the Opacica – Mostanica channel – which brings the water to the channel of the Zeta and further towards the intake construction. Some works were executed in sealing the shallow hole by deep closing, as well as gallery injection curtains, especially south of the reservoir.

Krupac Reservoir is formed by dam on the River Mostanica, 1,480 m long and 20 m high. The crown dam elevation is 622 m above sea level, and the overflow is at an elevation of 620 m above sea level with a capacity of 180 m^3/s. The reservoir volume is 35×10^6 m^3 [3, 4]. Its bottom outlet has a capacity of 12 m^3/s, which is also the sustainability capacity of the channel which delivers water to the Zeta channel. Works on reservoir also include complex works intended to deep seal the Krupac shallow hole, which proved successful, because the Krupac Reservoir has operated with full backwater ever since its completion in 1958.

In order to achieve no water losses when specific reservoirs release their water on its way to intake construction, a system of channels was built, having a length of about 40 km in total [3, 7]. In front of the intake construction, there is a precipitation for the purpose of purifying the water of silt particles larger than 0.2 mm. The water in the intake construction is brought to the surge shaft (with a bottom elevation of 570 m above sea level) by an under pressure tunnel 3.3 km long which is 4.8 m in diameter, with a concrete layer having a sustainability rate of 80.75 m^3/s [3, 7]. After bifurcation, where the pipeline has three branching pipes and a surged shaft seal, the pipeline continues in the form of three pipes having a length of 1.8 km, which brings the water into a special gallery created as an annex to the machine building, which has further branches towards the turbines (Fig. 5).

The machine building is located on a plateau near the Zeta headwaters. It is above ground, with seven generators with Pelton wheels (a combination of two turbines which move each generator), with a total installed flow of 80 m^3/s and a power

Fig. 5 View of the inflow pipeline and machine building of HPP Perucica. From www.epcg.com/o-nama/hidroelektrana-perucica

output of 307 MW[2] [1, 12]. The project predicted the installation of an eighth generator (rated at 65 MW), so that the installed capacity of the hydropower plant increases to 365.5 MW [1, 12]. Average production is 900 GWh/annually, which can be increased by an extra generator by another 13–15 GWh/annually [3, 7]. Since the plant has been working from 1960, now, after almost 60 years of exploitation, its renovation and the potential installation of further generators have become a significant strategic issue.

5 The "Piva" Hydropower Plant

The highest concentration of linear potential in Montenegro, even across a wider area, can be found in the wider zone of the Rivers Piva and Tara, which is frequently called the "Scepan Polje node" [1]. That is the reason why projects so far have

[2]The plant does not use this power for its work, due to the insufficient capacities of the inflow and outflow sections. During the period 1998–2001, HPP Perucica worked with a power of 260 MW, and it increased later on to 285–290 MW. After examinations conducted in 2001, the official agreement that the plant could work safely at the power of 285 MW was approved. During the works executed to renovate the plant facility, work to remove the operational limitations should be carried out, especially considering that this is a priority task for the Electric Power Industry of Montenegro.

planned three large reservoir hydropower plants in this zone: HPP Mratinje on the Piva (which was later called simply HPP "Piva") with a normal backwater elevation of 675 m above sea level, HPP Bijeli Brijeg on the Tara with a backwater elevation of 660 m above sea level, and HPP Buk Bijela on the Drina, with a backwater elevation of 500 m above sea level. Since the construction of HPP Bijeli Brijeg on the Tara was abandoned for entirely justified ecological reasons (because it required the sinking of a larger part of canyon), this node has functioned using HPP Piva and HPP Buk Bijela on the Drina, with backwater elevations harmoniously arranged as part of a single hydropower unit [2, 4].

As the most appropriate form, a profile in an asymmetrical "V" shape was chosen, in a canyon about 900 m deep, at a stationary location 10 + 600 from the Piva and Tara composition, about 2.85 km downstream from the Mratinje stream's confluence with the Piva. The adopted solution was a concrete dam with an arch and an underground dam power plant next to the dam in the left corner. In this way, the solution was based on the creation of one of the highest arch dams in the world, with a constructive height of approximately 220 m and a backwater elevation of 675 m above sea level. The average flow in that particular profile of the Piva was estimated to be 74.4 m^3/s during the planning process [4, 7] (Fig. 6).

The topographic basin is smaller than the hydrological basin, which is estimated to be around 1,784 km^2. That hydrogeology fact, which is a result of the

Fig. 6 View of the Piva arch dam on the River Piva; intake facilities are positioned in the left corner. From [3]

karstification of the basin, was obviously indicated by spring at Sinjac ("Pivsko oko"), with a topographic area of only 46 km^2, while the average flow was about 20 m^3/s, with a very wide flow range of slightly less than 1 m^3/s during the period of low waters (even up to around 200 m^3/s) [2, 7]. Large flows in periods of flood flows show the highly developed underground hydrography which makes the fast concentration and sustainability of flood flows possible.

The dam project was one of the most complex global engineering accomplishments [4, 11]. Along the upstream contour of the dam foundation, a stepped injection gallery was constructed, from which the consolidation injection was performed, as well as part of the curtain in the dam domain. In height, at each 40 m, starting from 482 m above sea level, horizontal galleries for the revision and positioning of instruments of auscultation system were installed. They are connected by a vertical maintenance hole in the central console. The dam safety is monitored by over 1,200 mechanical and electrical auscultation instruments [11].

The hydraulic reliability of the dam is secured by three evacuators, positioned vertically on the central part of the dam. In the middle part of the dam crown, with a cutoff wall at 670 m above sea level, overflow with three discharge fields of 13 $\times$ 15 m is positioned, with segment backwaters with automatic regulations. At the mid-height of the dam (input elevation 592 m above sea level, output elevation 580.45 m above sea level), three middle discharges with sealing point were created, and in the deepest part of the reservoir, on the central console at an elevation of 503 m above sea level, two bottom outlets were set. The overflow enables the evacuation of 1,670 m^3/s, the three middle discharges at a backwater elevation of 675 m above sea level have a capacity of 373 m^3/s, and both bottom outlets at a backwater elevation can evacuate 240 m^3/s. In this way, the full hydraulic reliability of dam is established [3, 11].

The narrow canyon and large dimensions of the machine building were prerequisites for the positioning of the power plant, in the left corner of the canyon. For each of three generators, a special inlet branch was created, consisting of an intake at 586.5 m above sea level, an inlet sealing point located in a steep adit, an inlet tunnel of Ø 500 cm with a concrete layer, and a steep tunnel of Ø 400 cm with a steel layer, up to seals in front of the turbines [3, 11]. The machine hall was built into a solid rock mass. The generators contain Francis turbines, with vertical shafts. The command space and gallery with transformers are on the same axis as the generator hall, so the total length of the whole space is 145.4 m. The discharge is solved as the tunnel in the shape of horseshoe section, where $B = H = 10.2$ m.

The backwater elevation of 675 m above sea level provides a total reservoir volume of 880 $\times$ 10^6 m^3. The elevation of the minimum working level is 595 m above sea level, and the usable volume of the reservoir is 790 $\times$ 10^6 m^3. The degree of regulating the flow is $\beta \approx 0.33$, which means that this is a reservoir with annual regulation. The installed flow $Q_{in} = 3 \times 80$ m^3/s $= 240$ m^3/s. The level of installation is $Q_{in}/Q_{av} \approx 3.2$ [3, 11]. According to the contemporary criteria for the installation for reservoirs with annual regulation, that represents insufficient installation (now such parameters for reservoirs with mixed electroenergy system adopt values that are twice as high, which can go up to $Q_{in}/Q_{av} \approx 8$–10) [6]. That is why one of the major tasks is to question the requirements and possibility of increasing the installed capacities of HPP Piva, which would make HPP Piva one of the most important regulation HPPs in this part of Europe. The highest fall height is 186 m, and the lowest is 104 m [11]. The installed capacity is $3 \times 120 = 360$ MW, and the average annual production is 750 GWh/annually, so 93% of it can be placed in the upper parts of the load diagram. The number of plant hours at an annual level where the maximum capacity is reached is 2,100 h. The power value of the reservoir on its own fall is 260 GWh, on the fall of the Drina power plants downstream, it is 300 GWh, and for the whole downstream built system, the fall produces about 472 GWh [3, 4].

After the construction of HPP Piva was completed, cleaning the basin in the zone of low waters of the material which appeared there, during the process of construction, was not in fact undertaken to a sufficient degree. That is why the elevation of the low water was raised, which decreased the power performances of the plant comparing to its maximum possible value. The project of lowering the level of the low waters of HPP Piva [1, 3] shows that by removing 7,000 m^3 and channelling the Piva basin, additional capacity of about 3.2 MW and the production of about 19.7 GWh/annually can be achieved. At a price of only 0.72 €c/kWh and a ratio $B/C = 8.65$, it is the one of most profitable hydropower sources in Europe, and doubts have been raised about why this has not so far been done. These works do not require any additional use of space, and they are certainly justified from an ecological standpoint [1].

6 Montenegrin Hydropower Potential at Facilities Outside Its Borders

One part of the hydropower potential that belongs to Montenegro – because it is part of its linear potential – cannot be realized within Montenegrin territory. That refers to the downstream parts of the Rivers Tara and Piva near the border as well as part of the River Lim on the border with Serbia [2, 7]. The furthest downstream parts of the Tara and the Piva could be very successfully realized should the facility at HPP Buk Bijela be built, within a project which was thoroughly examined and demonstrated to be one of the best proposed hydropower projects in Europe. In the case of the River Lim, this potential could be used by HPP Gostun, which could be successfully constructed in Serbia, while the larger part of the backwater would be provided by the downstream part of the Lim in Montenegro [1].

In the case of the realization of the potential of the Tara and the Piva, following the principles regularly applied in such cases globally, Montenegro could count on having 380 GWh/annually out of 1,140 GWh/annually from average power production [4]. Bearing in mind the peak-load purpose of the power plant, about 90% of the produced energy could be categorized as peak-load energy.

In the case of the realization of the potential of the River Lim downstream, and the proper matching of the facility of HPP Gostun within the surrounding area, a facility offering quite acceptable performance is certainly possible, with an installed capacity of over 20 MW and with an average production of over 90 GWh/annually [1, 4]. Certainly, this is highly usable potential, which could be easily divided on the principle of the linear potential of the river running through the two countries, Montenegro and Serbia.

7 Opportunities and Principles for the Construction of Small Hydropower Plants

In practice worldwide, the construction of small hydropower plants is allowed on all river streams but under the following conditions [1]:

1. Building occurs only in locations at which these plants do not disturb the realization of integral system water using planned facilities.
2. If appropriate, examinations of the effects on the environment prove that the proposed small hydropower plants do not disturb the ecological systems of these smaller river streams.
3. Derivational dispositions must provide for the releasing of a guaranteed flow in a reliable way (discharges which are constantly open and dimensioned to guaranteed ecological flows).
4. In the disposition schemes, there are guarantees that the proposed plants do not devastate the ambient values of protected areas.
5. There is a clause that the license for exploitation will be automatically withdrawn in cases where the user neglects the prescribed conditions for securing the guaranteed ecological flow of the river or stream.

It is estimated that the maximum opportunity limit in using the hydro-potential of small hydro plants is not larger than 450 GWh/annually, so that value can be used in the future in any summative analysis of the hydropower potential of Montenegro [1, 2, 12].

In all the considered dispositions of small hydro plants in Montenegro, the key energy performances are achieved by high fall concentration over quite long derivations. For the fall concentration, the morphological features of mountain streams are used, with lengthwise basin falls. Most commonly, these are tributaries in the corner valleys of major rivers, which have grand lengthwise falls, so that the power concentration is achieved by using tunnel and pipeline derivations. That is a very serious problem, because measures to secure the guaranteed ecological flows must be applied in a strictly disciplined manner, in order to avoid the ecological devastation of these rare hydrographic systems.

On most of the considered water streams and basins, modest-installed power output can be achieved even when applying a very long derivation, even below

1 MW in a large number of cases. Larger power performances can be achieved only within a small number of basins. In particular, the following streams stand out: Vrbnica with a possible production of all planned small hydro plants of 66.5 GWh/annually, Komaraca at 54.9 GWh/annually, Zlorecica at 48.8 GWh/annually, Bistrica at 45.4 GWh/annually, and Bijela totalling 30 GWh/annually [1, 4]. The other smaller river streams have modest power performances. Thus, the River Murinska with two long derivations (1,250 m + 2,560 m) can achieve only 6.9 GWh/annually, the River Vinicka (derivations 960 m + 1,500 m) only 3.2 GWh/annually, and the River Grabovica only 2.7 GWh/annually [1, 4, 13]. Such small hydropower plants have such low energy incomes that it raises the question of whether they could return primarily means and the power used for their construction during their expected working life span. For some of the considered small hydro plants, some approximate power analysis has been prepared, based on the enclosed bill of quantities, which showed that more of the primary means are spent in building such power plants (because of their use of high demand materials – steel, cement, plastic) than these plants can produced in the 30 years after their construction. This confirms that such facilities cannot be treated as producing either renewable or ecologically pure energy [1, 13].

The total average production of the 50 considered small hydro plants is quite modest – it is only about 417 GWh/annually, which means slightly more than 8 GWh/annually per plant on average [5]. It is obvious that such facilities have a small concentration of power in production or, to be more precise, they have a large dispersion of that kind of hydropower in space. Since the average production capacities of large- and medium-scale hydropower plants in Montenegro would be about 5,650 GWh/annually and the small hydro plants considered here would produce only 417 GWh/annually (about 14 times less) [1] – it is an absolute nonsense to claim that Montenegro can solve its electro-power issues by building only small hydropower plants. In reality, only 450 GWh/annually can be realized from small hydro plants altogether across Montenegro, and given that these types of plants are accompanied by certain ecological issues, they can represent only a small additional source of hydropower and absolutely cannot be the main source of this kind of power within Montenegro [1, 2].

8 Review of Hydropower Solutions in Montenegro Under New Conditions

Montenegro has been subjected to some significant changes in its environment, comparing to the time when, in the early 1970s, the conceptual solutions for using its hydropower potential were formulated [1, 12]. Some quite poorly populated river valley, free in infrastructure, has had a real expansion in population growth, with the area of the upstream section of the River Lim as the most obvious example [1]. The attitude about the necessity of protecting the special ecological value of the river ecosystem has changed significantly, which has brought new limitations in terms of possible concepts in using the water potential of some rivers (especially the River Tara). Global standards regarding the protection of water ecosystems have started to be consistently applied in Montenegro as well, which inevitably affects the approaches that can be taken in relation to some of the rivers, above all the Tara. The strategic interest of Montenegro should be in activating its hydropower potential as quickly and as comprehensively as possible [2, 4].

Montenegro has such a lot of hydropower potential at its disposal that it could be (economically speaking) among the top hydropower producers in the world, while retaining control over the ecological and social environments [11, 14]. From about 9.8 TWh/annually of linear potential on larger water streams, about 5.5–6.1 TWh/annually is already categorized (or soon will be) as economically usable potential [2, 7, 11]. Since the energy requirements for electrical power in 2012 were estimated at about 6.9–7 TWh/annually, Montenegro is considered to be one of the small groups of countries which have the opportunity to satisfy the majority of their electrical power requirements from their own hydropower resources, which gives this country an extraordinary advantage on the European power market [1].

The gross potential on the nine major rivers is 9,846 GWh/annually, while the gross hydropower potential on smaller water streams can be estimated at about 800–1,000 GWh/annually [2, 12, 13]. The largest hydropower potential is offered by the River Tara (2,225 GWh/annually), followed by the Zeta (2,007 GWh/annually), the Moraca (1,469 GWh/annually), the Lim (1,438 GWh/annually), and the Piva (1,361 GWh/annually), while the potential of the other rivers is less significant [2, 4, 11].

9 Up-to-Date Recommendations

General Concept The strategy for the use of the hydropower potential of Montenegro [4] cannot yet be considered as having been precisely determined, regardless of the already examined strategic variants within the approach to waterpower in Montenegro [11]. Indeed, only in circumstances where there is a very difficult power situation in decades to come, and decisions on the development and use of other power sources have been taken, will the prerequisites for final decisions on the overall strategy regarding the integral use of waters be fulfilled. Until that time, there are two possible global options: using the water potential under the conditions of the natural regimes (i.e., using the natural stream course) or adopting a strategy of delivering waters for the purposes of creating the most favorable power concentration of potential [1, 2]. In any analysis of these potential pathways, both options should be treated equally.

Since the strategies for the use of the water found in the rivers of the Drina basin could not be agreed, there remain two groups of variants [4, 11] in the conception of the use of the hydropower potential of Montenegro: (a) using the potential in the natural direction of the streams of all the rivers, (b) a variant involving delivering part of the water from the Tara into the Moraca [15], and the construction of a large HPP (RHPP) at Kostanica, which would take on the role of one of the largest regulation hydropower plants in this part of the European EES [8, 9, 16]. The total usable potential technically planned under the variant using natural streaming (the natural stream of the River Tara, without delivering part of its water into the Moraca) is (a) on the nine major rivers, using a realistic option which estimates leaving the whole of the Tara canyon downstream of Ljutica exclusively for rafting and tourism, the total potential is 3,570 GWh/annually, and (b) using the power plant at Buk Bijela, including the part that belongs to Montenegro, which would provide 380 GWh/annually [4, 7, 9]. Thus, the total usable hydropower potential of Montenegro, including those facilities that are already built (which provide 1,670 GWh/annually), would provide ca 5,600 GWh/annually (5.6 TWh/annually), without including the potential of small hydro plants, which is estimated to be an additional maximum of 450 GWh/annually [7] (Fig. 7).

The River Piva The possible further development of the Piva system is to a large degree determined by the position and performances of HPP Piva [3, 7]. The proposed facility – HPP Komarnica – logically continues the disposition, because the upstream water (UW) of HPP Piva represents the downstream water (DW) of HPP Komarnica. The backwater of HPP Komarnica should be at 811 m above sea level [1, 7], so that it will not disturb the Nevidio canyon. The disposition should be created so that the works in the Komarnica canyon cannot affect the morphological and ecological features of the Komarnica canyon that make this area unique. The desired disposition involves using a maintenance hole for lowering the equipment from the upper plateau and passing the heaviest equipment by using water of Lake Piva. Increasing the installed capacity of HPP Piva by building additional generators would also have a significant developmental effect, because in such circumstances, it would become the most important regulative hydropower plant in this part of Europe [1].

The River Tara The Tara is the most important Montenegrin river in terms of energy (its linear potential is about 2,260 GWh/annually) [7, 15]. The starting point for the determination of acceptable options is the following:

1. The option including HPP Trepca is rejected, because that part in the middle of the river canyon downstream from the Ljutica profile must not be disturbed in any way.
2. Both usage options (i.e., both delivering part of the water from the Tara into the Moraca and not doing so) should still be considered and seen as possible [2, 3].

The option of including a system on part of the stream from Ljutica to Mojkovac would need to be completely in accordance with the requirements for preserving the ecological and tourist functions of the downstream part of the river. That includes:

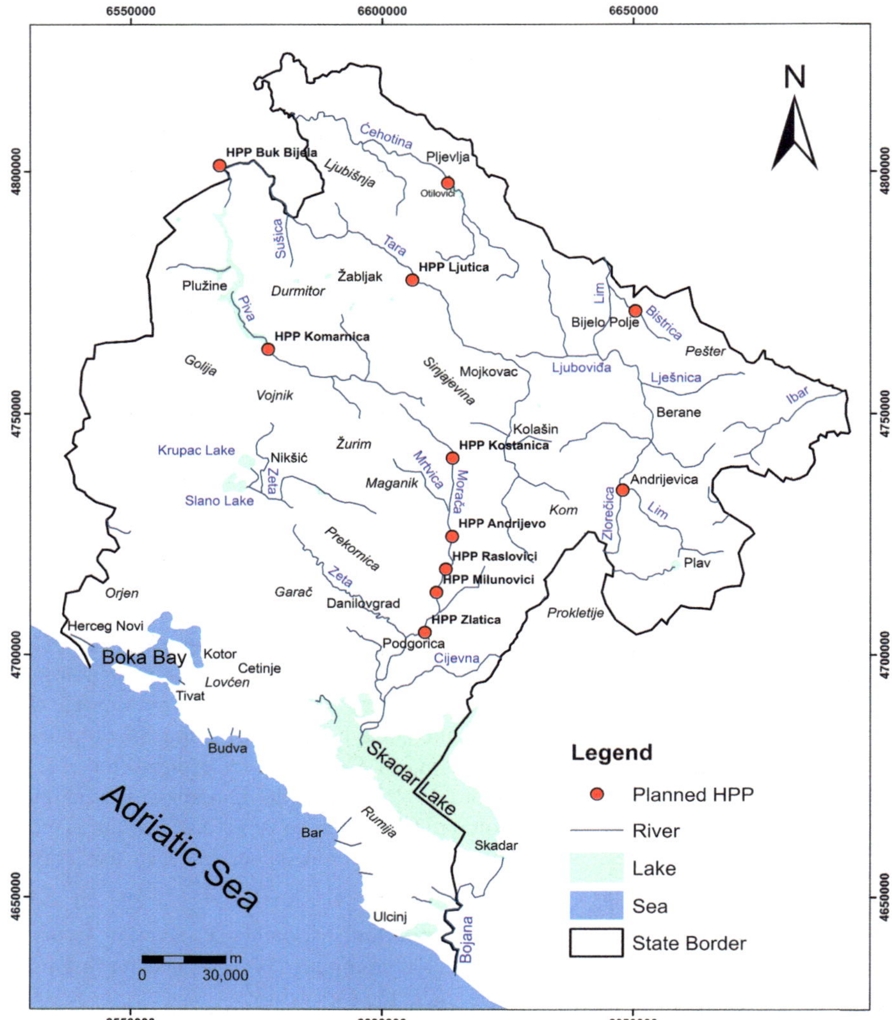

Fig. 7 Planned HPP on major rivers in Montenegro

1. Increasing/improving the flow of low waters, especially during the warmer periods of the year, when the lower part of the stream is used for tourist purposes.
2. Managing the temperature and oxygen regimes of the stream and maintaining them (i.e., keeping the most favorable ecological condition), as well as the condition which requires that the stream be undisturbed in terms of its tourism function [2].

3. The protection of the monastery at Dobrilovina and the protected area at Crne Pode; the protection of Donja Polja downstream of Mojkovac, as a ground resource on which Mojkovac relies upon. The option of a system upstream of Mojkovac should be flexible, so it can provide for the use of water potential entirely, even in the conditions of delivering part of water from the Tara into the Moraca, in case such an agreement is reached [7].

For options involving the delivery of Tara water into the Moraca [7, 10], the following requirements need to be fulfilled:

1. The solution needs to provide the conditions for the harmonious development of Kolasin as a town, as well as its urban connection with the new lake aquatories which are supposed to be placed nearby [7, 15].
2. Protection measures should be envisioned to ensure that the lakes on that part of stream, as well as the other ones on the Tara, permanently retain their oligotrophic and mesotrophic conditions, which are key to the sport, recreational, and tourist use of these aquatories. Only the first option should be considered, with a central management facility for the managing water regimes of the reservoir at HPP Ljutica [3, 7, 10] (Fig. 8).

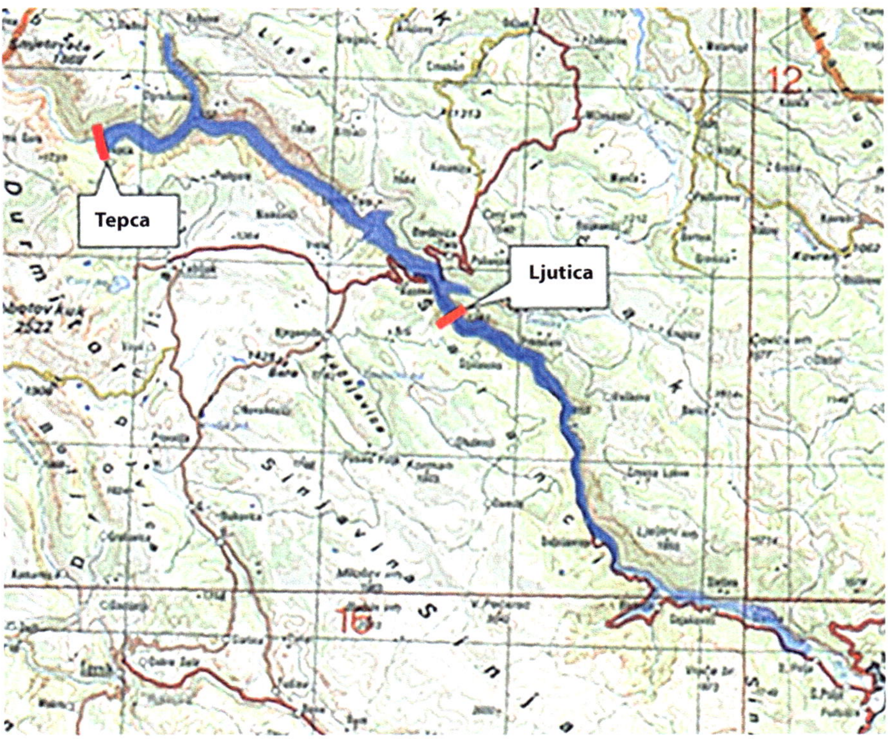

Fig. 8 Planned HPP on the River Tara. Modified after [7]

The key facility for the development of the other concept of using the hydropower of Montenegro, which considers delivering of part of the water of the Tara from the node around Kolasin into the River Moraca, is the derivation of HPP Kostanica [7]. That concept is more favorable energetically because it enables higher fall concentration. In this case, the total energy potential of Montenegro, along with HPP Buk Bijela and the existing facilities, goes beyond 6,100 GWh/annually, not counting small hydropower plants [1, 3, 7]. The project has demonstrated the technical possibility of its plan, as well as the economic justification for it, but the certainty of its realization from the point of view of its international dimensions and agreement in Montenegro is debatable. If that facility were considered a reversible hydropower plant (RHPP) with seasonal regulation, it would be much more acceptable to the whole downstream area [3, 7]. Even so, such a solution involving RHPP Kostanica would enable the preservation of water balances on the Tara and the Drina, and there would be an opportunity to achieve significant improvements in the water regimes during the year – lowering flood flows and increasing low water levels (Fig. 9).

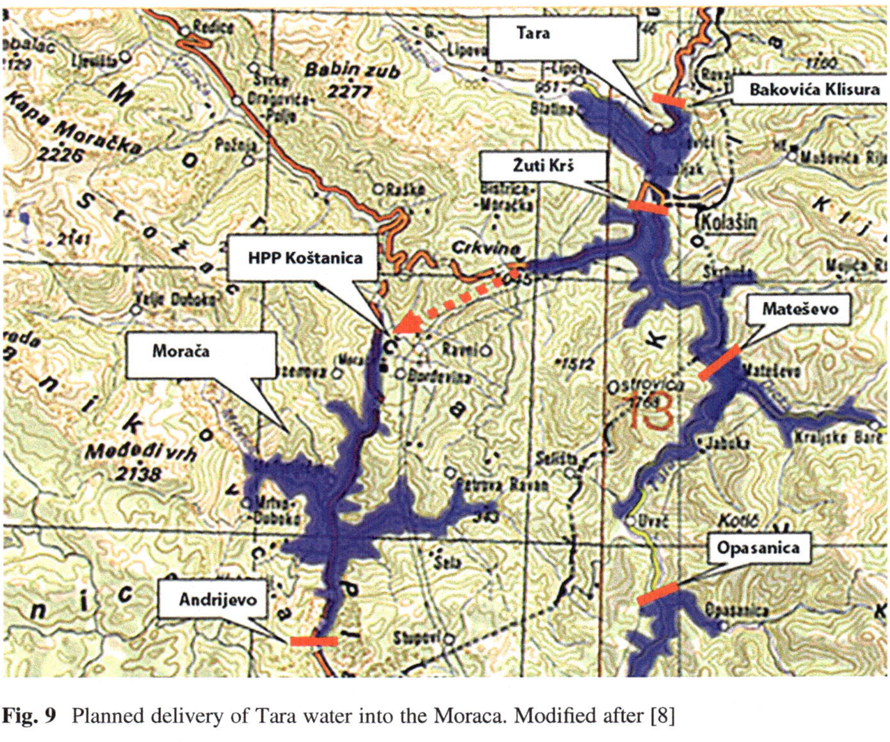

Fig. 9 Planned delivery of Tara water into the Moraca. Modified after [8]

The River Lim Bearing in mind the significant special limitation (urban, transport, and others) in the valley of the Lim, the suggested variant [1] fits best into urban, transport, social, ecological, and other environments. A low step, 10–15 m in the solid trough of the Lim, could ensure that backwater lines were not transferred to the riverbank environment, even during the evacuation of flood flows. In the zone which could be affected by backwaters, only those components which are currently positioned in the zone of the backwater flood flows backwater are found so that they should be removed, reconstructed (with the correction of small parts of the route), or protected from flooding, regardless of whether the hydropower system is constructed or not [1]. According to this option, the low steps by the dams should be constructed within the base trough for flood flows, as typical facilities both in terms of their construction facilities and equipment [5]. Such serial construction, switching equipment for building and mechanics from one facility to another, would be very rational [1]. That would require the central maintenance of the whole system from one facility. In order to realize that, the trough for flood flows should be protected against the unwanted entry of any contents which could ruin this solution by special plan along the whole length of the River Lim. That means normalizing a new approach to the management of the whole river so that even new bridges and roads would be higher than the elevation of backwater during flood flows $Q_{0.1\%}$, with reservoirs of not less than 2 m [1].

The River Cehotina Water from the River Cehotina can be used for two reservoir hydropower plants: "Gradac" and "Milovci." They fit best into the complex solution for the whole upper basin of the Drina, because the regulation of flow in their large usable reservoir volumes improves the conditions of work for all other downstream hydropower plants on the Drina. HPP Milovci is especially significant because of its usable volume, so that it is one of the key facilities in the Drina basin, and in new conditions, it would be one of the facilities which would be of interest to downstream countries in terms of financial investment [7, 10].

The River Ibar On the Ibar, the best solution for using its hydropower potential can be achieved by building a reservoir at HPP Bac [7, 10]. The interest of Serbia in investing in this property and project would rapidly increase [7, 10].

The River Moraca So far, three system options have been considered for Moraca River [16]. Giving up Option 1 [7, 8] is advised, which considered the elevation of the normal backwater in the reservoir at HPP Andrijevo to be 285 m above sea level. From the point of all system goals on the Moraca, this is the best variant, but the "psychological effect" makes it difficult because of the high elevation of the normal backwater (being only a few meters lower than the plateau on which the Moraca Monastery is located). Even though some efficient measures for the protection of this river terrace have been considered (which is largely endangered now, especially in the case of seismic activity), it was evaluated that for this variant, public consensus for the project would not be forthcoming, so this idea was abandoned. Variant 2, with a normal backwater elevation of 250 m above sea level, is appropriate in environmental terms. It practically does not create backwater in the monastery area, and its special value is in the frontal reservoir at Dubravica [2, 7], which provides the conditions for the annual regulation of streams and improving the water regimes. Recently, a potential concession partner elaborated their idea, Variant 3, which is a technically possible solution, with the disadvantage that is does not consider the frontal reservoir which would enable the management of the water regimes [8] (Fig. 10).

10 Conclusion

From the elaboration provided above, it can be concluded that Montenegro has extraordinary hydropower potential at its disposal, which could be a key resource and driving power for accelerating the development of the country. Systems for using this potential can be made to harmoniously fit into the social, ecological, and other environments. Even though Montenegro possesses such extraordinary natural conditions regarding hydropower potential, the reality is that the use of hydropower facilities lags well behind the potential [2]. Only two important hydropower plants have been built – HPP Perucica and HPP Piva on the Piva River, as well as an insignificant number of small hydropower plants. With only 28% of its hydropower potential in use [1, 4], Montenegro has a strategically disturbing backlog in using its most precious natural resource – its water potential [2, 3].

The global tendency regarding the use of hydropower is widely known and is rapidly developing into a sustainable category of economically usable potential (in terms of being a technically usable potential). Therefore, it is rather likely to be expected that the major part of the technically usable potential of Montenegro could become economically usable potential in the near future. Indeed, some of the planned facilities, including HPP Buk Bijela, HPP Kostanica, HPP Komarnica, the cascade HPP on the Moraca, and several others, fall within the category of the most attractive facilities in all of Europe regarding hydro-energy production as a driver of economic development [3, 7].

Since Montenegro has a really solid basis for the realization of large regulation hydropower plants with seasonal regulating [4], it has the opportunity to place top energy – which is more and more in demand and more and more expensive on the global market – and in exchange with other countries from the area, this country can achieve larger quantities of constant power from thermo-power plants arranged in advance [4, 7].

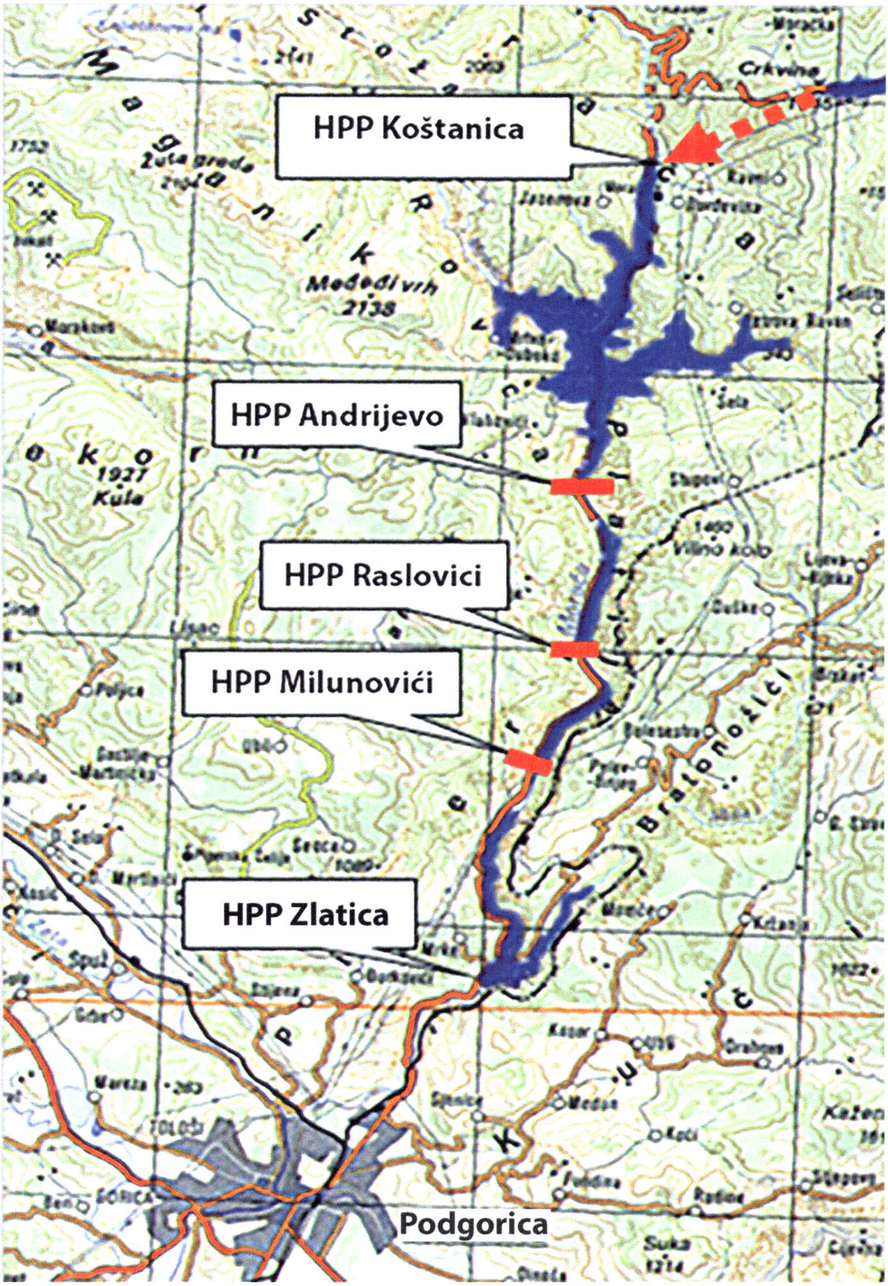

Fig. 10 Planned HPP system on the River Moraca. Modified after [8]

References

1. Đorđević B, Sekulić G (2015) Utilization of Montenegro's water resources in accordance with the renewable energy policy. Montenegrin Academy of Science and Arts, special issues (Monographs and Studies), Book 116, Podgorica
2. Đorđević B, Sekulić G, Radulović M, Šaranović M (2010) Water potentials of Montenegro. Montenegrin Academy of Sciences and Arts, special issues (Monographs and Studies), Book 74, Podgorica (in Serbian)
3. Đorđević B, Šaranović M (2007) Hydropower potential of Montenegro (in Serbian). Montenegrin Academy of Science and Arts, special issues (Monographs and Studies), Book 55, Podgorica
4. Energy Strategy of the Republic of Montenegro until 2025 (2006) Book D: development plan of the power system of the Republic of Montenegro – master plan. Institute for Research in Energy, Ecology and Technology, Ljubljana and Energy Institute "Hrvoje Pozar", Zagreb
5. EURELECTRIC WG Hydro (2011) Hydro in Europe: powering renewables. https://www3.eurelectric.org/media/26690/hydro_report_final-2011-160-0011-01-e.pdf. Accessed 30 Aug 2019
6. Sekulić G, Radulović M (2019) The hydrology and hydrogeology of Montenegro. In: Pešić V, Paunović M, Kostianoy A (eds) The rivers of Montenegro. Springer, Cham
7. IJC (Institute Jaroslav Cerni) (2001) Water master plan of Montenegro. Government of Montenegro, Podgorica. (in Serbian)
8. MSPE (2010) Detailed spatial plan for the area of multi-purpose reservoirs on the River Moraca. Ministry for Spatial Planning and Environment, Podgorica. (in Montenegrin)
9. Šaranović M (2003) Energy potential of Montenegro. In: Time and progress. Montenegrin Academy of Sciences and Arts, Podgorica, pp 99–107
10. MARD (2017) Strategy for water management in Montenegro. Ministry for Spatial Planning and Environment, Podgorica. (in Montenegrin)
11. Group of Authors (1974) Announcements from the symposium on the construction of HPP Mratinje, Niksic
12. Đurović M (2004) Renewables sources and Montenegro – new energy sources in Montenegro. Montenegrin Academy of Sciences and Arts, special issues (Monographs and Studies), Book 9, Podgorica
13. IHSM (Institute of Hydrometeorology and Seismology of Montenegro) (2016) Hydrology. http://www.meteo.co.me/sektori.php?sektor=2. Accessed 30 Aug 2019
14. MARD (2018) Report on the implementation of the strategy for water management. Ministry of Agriculture and Rural Development, Podgorica. (in Montenegrin)
15. Milić S, Orlić D (2003) Energy economic analysis of feasibility of shifting works of water from Tara to Moraca. Energoprojekt-Entel, Belgrade
16. Živaljević R (2006) Possibilities and limitations in providing conditions for utilization of the remaining technically usable potential of the main flows of Montenegro. Water Manag 1–3:97–12

Drainage Basins of Montenegro Under Climate Change

Andrey G. Kostianoy, Evgeniia A. Kostianaia, and Vladimir Pešić

Contents

Abstract Regional climate change has an important impact on the climate and weather in Montenegro and on the intermittent rivers and their drainage basins, in particular. Since the early 1990s, it has resulted in an increase of the frequency, duration, and intensity of extreme hydrological events in Montenegro. For example, in the last two decades, Montenegro has experienced severe droughts in 2000, 2003, 2007, and 2012 and has suffered losses from damaging floods, most notably in 2010. Higher air temperatures, stronger evaporation, the increased frequency, and duration of a drought period increase the share of intermittent rivers in the regional hydrological network by reducing the amount of water and shallowing the rivers. All this affects the river ecosystems which change accordingly. On the other hand, a higher frequency and abundance of floods on the rivers accelerate bank erosion,

A. G. Kostianoy (✉)
P.P. Shirshov Institute of Oceanology, Russian Academy of Sciences, Moscow, Russia

S.Yu. Witte Moscow University, Moscow, Russia
e-mail: kostianoy@gmail.com

E. A. Kostianaia
P.P. Shirshov Institute of Oceanology, Russian Academy of Sciences, Moscow, Russia
e-mail: evgeniia.kostianaia@gmail.com

V. Pešić
Faculty of Sciences, University of Montenegro, Podgorica, Montenegro
e-mail: vladopesic@gmail.com

Vladimir Pešić, Momir Paunović, and Andrey G. Kostianoy (eds.), *The Rivers of Montenegro*, Hdb Env Chem (2020) 93: 69–82, DOI 10.1007/698_2020_481,

increasing suspended sediment load into the rivers, which leads to a reduction of a diversity among aquatic biota and their communities. Finally, the changes in local hydrology (reduced river runoff, higher intermittency) as a result of regional climate change may also reduce the hydropower capacity of exposed rivers. In the present chapter, we discuss the ongoing regional climate change over drainage basins in Montenegro based on the analysis of interannual variability of key meteorological parameters for intermittent rivers.

Keywords Air temperature, Atmospheric precipitation, Drainage basin, Global warming, Intermittent rivers, Montenegro, Regional climate change, Rivers, Snow cover

1 Introduction

Climate change in the Mediterranean resulted in a significant change of regional climate and weather in Montenegro [1, 2]. Recent analysis of the observed regional climate change in the region surrounding the Boka Kotorska Bay basing on a set of key meteorological parameters: near-surface air temperature, sea-level pressure, atmospheric precipitation, snow cover, wind speed, and cloud cover over the past 60 years (1950–2010) showed significant changes in the seasonal and interannual variability, as well as in the anomalies calculated both for warm and cold seasons [1].

During 60 years between 1950 and 2010, the average air temperature in the region grew by more than 1.3°C, which, apparently, is a consequence of global climate warming. Sea-level pressure has increased from about 1,016 to 1,017 hPa [1]. Atmospheric precipitation over the investigated region in terms of “total column rain water” showed a little decreasing trend of about 0.00006 kg/m^2. Snow cover in the region clearly shows a decreasing trend as well. Snowy winters became much rare and snow depth during these events became about two times less than in 1950–1970 [1]. Interannual variability of zonal and meridional wind components showed that westward and southward wind components prevailed during the observed time period. It means that the general air mass transfer was directed from north-east to south-west. Cloud cover in the region demonstrated a strong seasonal variability, when in winter it is as large as 75%, and in summer it is often less than 25%. A notable decreasing trend was detected, which means that cloud cover, in average, had decreased by about 4% during 60 years [1].

Regional climate change, which has markedly increased since the early 1990s, has had a particularly significant impact also on the frequency, duration, and intensity of extreme weather and hydrological events in Montenegro [2, 3]. The first attempt at attributing an extreme weather event to climate change was published in 2004, analyzing the 2003 European summer heat wave that killed tens of thousands of people [4]. Today, the Intergovernmental Panel on Climate Change (IPCC) Reports state that the global and regional climate change is accompanied by the frequency, duration, and intensity of extreme weather events, which is proved by numerous weather anomalies observations in the World [5].

For example, in the last two decades, Montenegro has experienced severe droughts in 2000, 2003, 2007, and 2012 and has suffered losses from damaging floods, most notably in 2010 [3, 6]. The increased frequency and duration of drought periods affects river ecosystems and their biota by reducing the flow in the lotic habitats [3]. Moreover, the rise of ambient air temperature, especially in the lentic parts, leads to changes in water temperature that consequently leads to changes in biological processes in the rivers [6]. On the other hand, the higher frequency and abundance of floods due to the accelerated bank erosion leads to an increase of suspended sediment load into the rivers resulting in a decrease in diversity of aquatic biota and their communities [6]. Recently, Vukašinović-Pešić et al. [7] proved that the rivers in the Black Sea Basin exhibit an increase in the frequency of anomalous and extreme events in water temperature over a 10-year period from 2009 to 2018.

As it has been emphasized by numerous studies, further climate change will be expected to increase extent of intermittent rivers in the global river network [8–11]. At the global level, the intermittent rivers and ephemeral streams represent over half of the river network [12, 13], but in arid and semi-arid regions, such as the southern part of Montenegro, their share may be significantly higher [6], which requires the use of accurate and effective tools for their mapping. Kostianoy et al. [14] showed that Google Earth can be a useful tool for mapping the spatial extent of intermittent and perennial river sections, but its accuracy depends on the presence of forested area and river width [14].

There are important economic consequences of regional climate change related to rivers in Montenegro. The problem is that in the last decade, the building of small hydropower plants has significantly increased, causing a number of conflicts with local communities and their interests [3]. For instance, on the largest number of small rivers and streams in northern part of Montenegro where small hydropower plants were built, the water courses do not have enough water in the summer months to meet the prescribed minimum to function. As a result, at the rivers, where small hydropower plants have been installed, often remain long stretches of river with disturbed natural flow (or even a dry), which prevents the local community from exploiting the water but also devastating habitat for the aquatic biota [3, 6]. Apart of the present conflicts related to the planning and construction of small hydropower plants, there is a big concern in their efficiency in the coming decades due to a regional climate change which will result in a lack of water. Thus, it is important to continue the research on regional climate change and its consequences in Montenegro to make the investments more effective.

In this chapter we investigate seasonal and interannual variability of air temperature, atmospheric precipitation, and snow cover as key parameters for drainage basin of the rivers in Montenegro. This chapter may be regarded as a continuation of the research on climate change we performed for 1950–2010 [1, 2], in which data sets are prolonged for the next decade.

2 Data and Methods

The analyses and visualizations of data on air temperature, atmospheric precipitation, and snow cover used in this study were produced with the Giovanni online data system v.4.32, developed and maintained by the NASA Goddard Earth Sciences Data and Information Services Center (GES DISC) [15]. As for air temperature, we used area-averaged of 2-m air temperature (monthly mean) with spatial resolution $0.50 \times 0.625°$ for a region 42°15′–43°15′N; 18°30′–20°00′E, which covers most of the drainage basins in Montenegro (Fig. 1), and time period from January 1980 to November 2019 (MERRA-2 Model M2IMNXASM_5.12.4) [16]. For atmospheric precipitation (mm/month), we used an area-averaged "merged satellite-gauge precipitation estimate final run" (GPM_3IMERGM v06) (monthly mean) model with spatial resolution $0.1 \times 0.1°$ for the same region and time period from June 2000 to September 2019 [17]. For investigation of snow cover, we used "snow depth" parameter which refers to the vertical measurement of accumulated snow on the land surface. For this purpose we used area-averaged (monthly mean) data from the MERRA-2 Model M2TMNXLND_V5.12.4 for the time period from January 1980 to November 2019 with spatial resolution $0.50 \times 0.625°$ for the same region [18]. For these three key parameters for drainage basins in Montenegro, we investigated seasonal and interannual variability, as well as their linear trends.

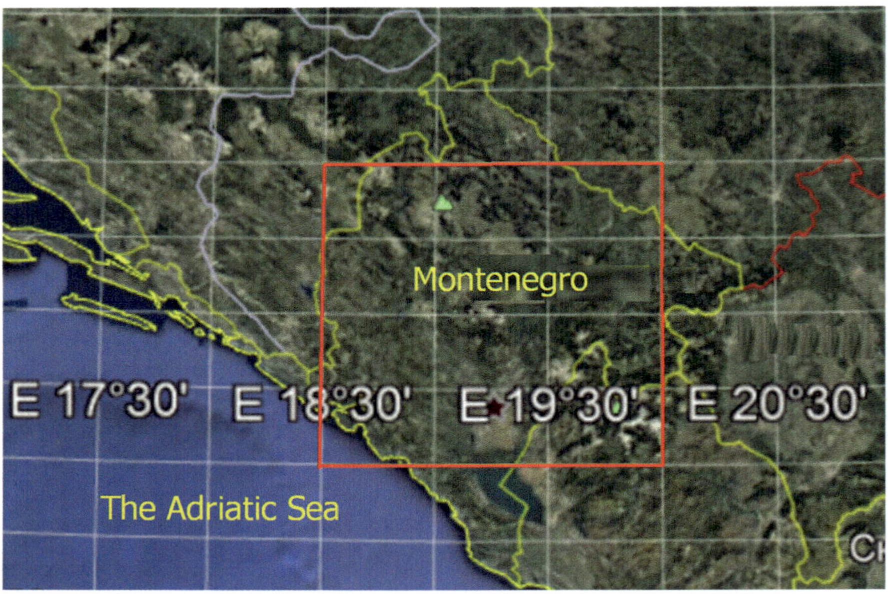

Fig. 1 Map of Montenegro. The chosen research region (42°15′–43°15′N; 18°30′–20°00′E) is shown by the red rectangle

The Modern-Era Retrospective analysis for Research and Applications version 2 (MERRA-2) is a NASA atmospheric reanalysis for the satellite era using the Goddard Earth Observing System Model, Version 5 (GEOS-5) with its Atmospheric Data Assimilation System (ADAS), version 5.12.4. The MERRA project focuses on historical climate analyses for a broad range of weather and climate time scales and places the NASA EOS suite of observations in a climate context [15, 16, 18].

3 Results

3.1 Air Temperature

Kostianoy et al. [1] have analyzed seasonal and interannual variability of the mean monthly air temperature near the surface (height of 2 m) for the period 1950–2010 and the corresponding linear trend for the large area surrounded the Boka Kotorska Bay. It was found that during these 60 years, the average air temperature in the region grew by more than 1.3°C, which, apparently, is a consequence of global climate warming [1, 5]. The coldest winters were observed in the fifties and sixties of the past century, then they became rare with exceptional events in 1981 and 2003. Warm summers with monthly average air temperature over 22°C were observed about 10 times in 1950–1975 and more than 20 times from 1985 to 2010, i.e., almost yearly. The summer of 2003 was the warmest in the 1950–2010 period with the monthly mean air temperature exceeding 26°C [1].

In the present analysis, we focused on the 40-year long period from 1980 to 2019 and on the area of the whole Montenegro. This is why absolute values of air temperature may differ from the previous analysis as it was focused mainly on the coastal zone where weather and climate is different from the northern mountain regions of Montenegro. Figure 2 shows seasonal and interannual variability of air temperature in the region under investigation (see Fig. 1 for location) from January 1980 to November 2019 according to MERRA-2 Model [15, 16].

Linear regression shows a steady rise of air temperature with a rate of 0.05°C/year, thus a yearly average air temperature has increased by 2°C during the past 40 years (Fig. 2). An absolute minimum of −7.4°C was observed in winter 2003, and an absolute maximum of 22.2°C was registered in 2012. It is interesting that in the past 10 years in four cases the highest air temperatures have exceeded the pick of the summer 2003 which was the warmest in the 1950–2010 period. Thus, summers become warmer and warmer, as well as winters, but periodically very cold winters continue to be registered, for example, in 2012, 2017, and 2019 when monthly mean air temperature was below −4°C (Fig. 2).

It is clear that a rate of warming is different in different seasons, which is especially important for intermittent rivers which usually dry up in summertime. Figure 3 shows interannual variability of an average air temperature for June–July–August from 1980 to 2019 for the same region. We observe that for a summertime

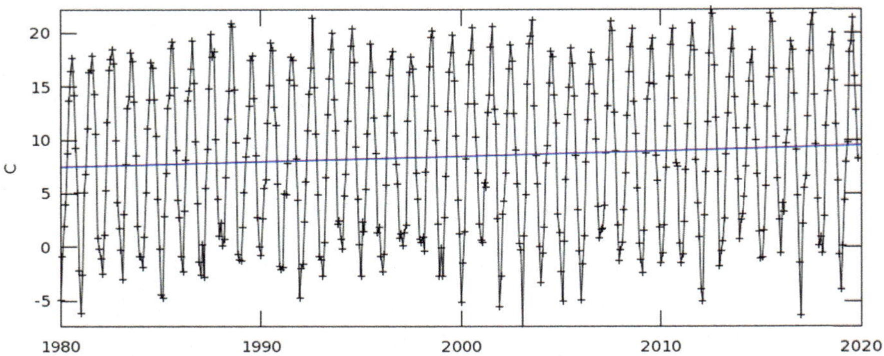

Fig. 2 Seasonal and interannual variability of air temperature (°C) in the region 42°15′–43°15′N; 18°30′–20°00′E from January 1980 to November 2019 according to MERRA-2 Model [15, 16]. Blue line is a linear trend

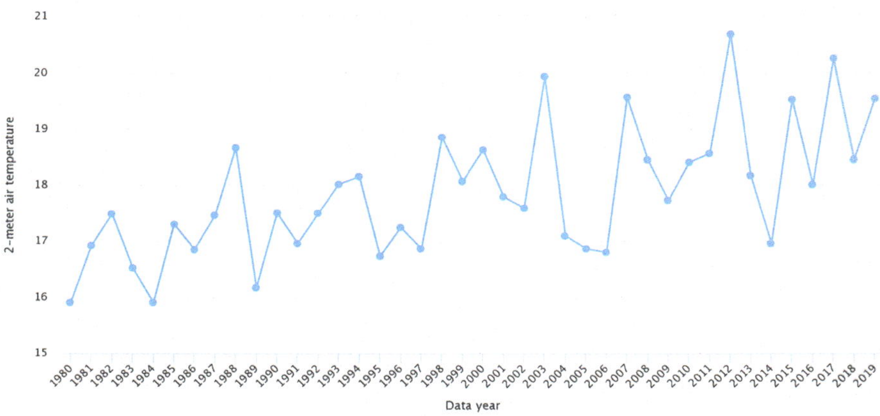

Fig. 3 Interannual variability of air temperature (°C) for June–August in the region 42°15′–43°15′N; 18°30′–20°00′E from 1980 to 2019 according to MERRA-2 Model [15, 16]

the rate of warming is a bit higher than for the whole year because a linear regression gives a value of 0.06°C/year (see Fig. 2 for a comparison). An average summer air temperature in 1980 was about 16.6°C and by 2019 it has increased to 19.0°C. Since 2003 we have observed the following hot summers in 2003, 19.9°C; 2007, 19.6°C; 2012, 20.7°C (absolute maximum); 2015, 19.5°C; 2017, 20.3°C; and 2019, 19.5°C, and the following cold summers in 1980, 15.9°C and 1984, 15.9°C (absolute minimum); 2004, 17.1°C; 2005, 16.9°C; 2006, 16.8°C; and 2014, 17.0°C (Fig. 2).

We repeated the same procedure for the winter months – December, January, and February and got the following results (Fig. 4). We observe that for a wintertime the rate of warming is two times less than for the whole year because a linear regression gives a value of 0.026°C/year (see Fig. 2 for a comparison). An average winter air temperature in 1980 was about −1.2°C, and by 2019 it has increased

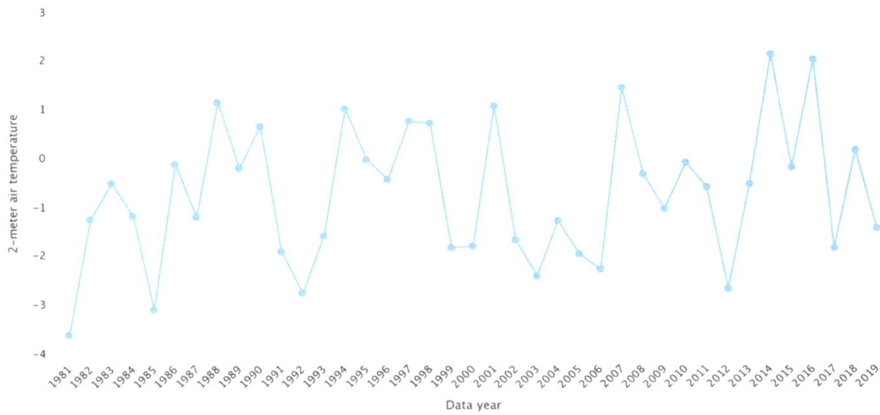

Fig. 4 Interannual variability of air temperature (°C) for December–February in the region 42°15′–43°15′N; 18°30′–20°00′E from 1980 to 2019 according to MERRA-2 Model [15, 16]

to −0.2°C, i.e., by one degree only. Warm winters were observed in 2007, +1.4°C; 2014, +2.1°C; and 2016, +2.0°C, while cold winters were recorded in 1981, 3.6°C; 2014, 2.7°C; and 2017, 1.9°C. In 1980–2005 only seven winters had average air temperatures over 0°C, but from 2006 to 2019, we already have observed four warm winters.

3.2 *Atmospheric Precipitation*

Kostianoy et al. [1] have analyzed interannual and seasonal variability of atmospheric precipitation over the region surrounding the Boka-Kotorska Bay in terms of monthly mean "total column rain water" measured in kg/m^2. It was found a little decreasing trend of about 0.00006 kg/m^2 during the 60-year period (1950–2010), which is a decrease of about by 2% during these 60 years from about an average value of 0.003 kg/m^2. Seasonal and interannual variability of precipitation shows peaks which are irregular in time and amplitude. These peaks can be four times larger than the average value of precipitation and exceed 0.012 kg/m^2. From year to year variability of anomalies showed that they are much larger in cold seasons (both positive and negative anomalies) than in warm seasons [1].

In the present analysis, we focused on the past 20 years long period from 2000 to 2019 and on the area of the whole Montenegro. This is why some peaks of atmospheric precipitation may differ from the previous analysis as it was focused mainly on the coastal zone where weather and climate is different from the northern mountain regions of Montenegro. Figure 5 shows seasonal and interannual variability of atmospheric precipitation (mm/month) in the region under investigation (see Fig. 1 for location) from June 2000 to September 2019 according to GPM_3IMERGM v06 Model [15, 17].

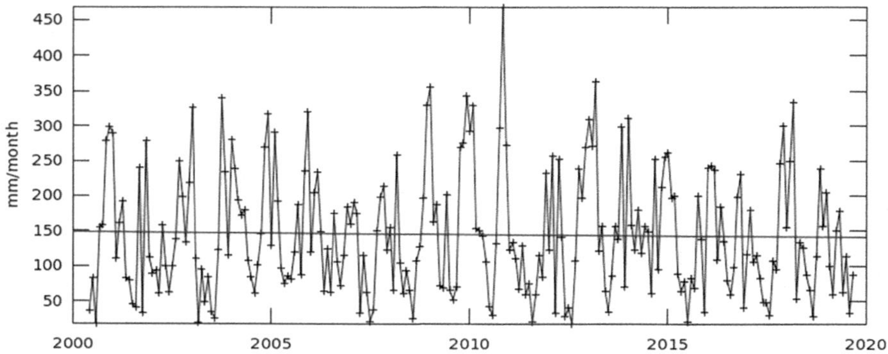

Fig. 5 Seasonal and interannual variability of atmospheric precipitation (mm/month) in the region 42°15′–43°15′N; 18°30′–20°00′E from June 2000 to September 2019 according to GPM_3IMERGM v06 Model [15, 17]. Blue line is a linear trend

Linear regression shows a slow decrease of atmospheric precipitation with a rate of 10 mm/month during 20 years, from an average value of 150 mm/month in 2000 to 140 mm/month in 2019 (Fig. 5). An absolute minimum of 17 mm/month was observed in August 2000 (summertime is a driest period), while an absolute maximum of 468 mm/month was registered in November 2010 (wintertime is a wet period). We have to note that in 2000–2010 precipitation has reached 300 mm/month and more 10 times (10 months), while during the past decade only 6 times. It means that the most heavy rains become a rare event in Montenegro (Fig. 5).

Satellite imagery shows that many rivers in Montenegro are dry in the summertime when weather is hot and dry [6, 14]. This is why there is a special interest for the analysis of interannual variability of atmospheric precipitation namely during the summer months. We repeated the same analysis for June–July–August for 2000–2019 and found a positive linear regression of 0.18 mm/month per year or a little total increase in precipitation of 3.6 mm/month (or +5%) during the past 20 years (Fig. 6). We have to note that minima in precipitation observed in summer 2003, 2007, 2012, and 2017 correspond exactly to the maxima in air temperature during summer of the same years (see Fig. 3).

3.3 Snow Cover

Snow cover is one more important parameter for water capacity over drainage basin. Kostianoy et al. [1] have analyzed snow cover in the large region surrounding the Boka Kotorska Bay and found a clear decrease in snow cover during 60 years (1950–2010). Snowy winters became much rare and snow depth during these events became about two times less than in 1950–1970, i.e., less water is accumulated during wintertime. Frequency analysis did not demonstrate any notable periods of snow cover variability, which means that this process is chaotic like accumulated precipitation in the same region, and it is not related to any global indices like NAO, ENSO, etc. [1].

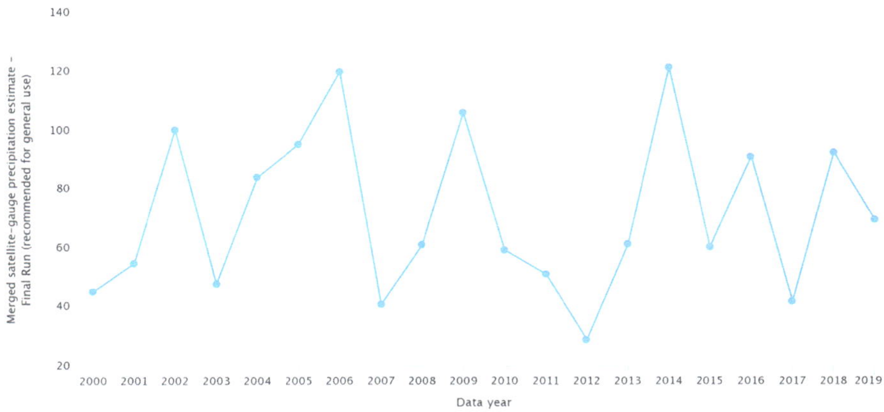

Fig. 6 Interannual variability of atmospheric precipitation (mm/month) for June–August in the region 42°15′–43°15′N; 18°30′–20°00′E from 2000 to 2019 according to GPM_3IMERGM v06 Model [15, 17]

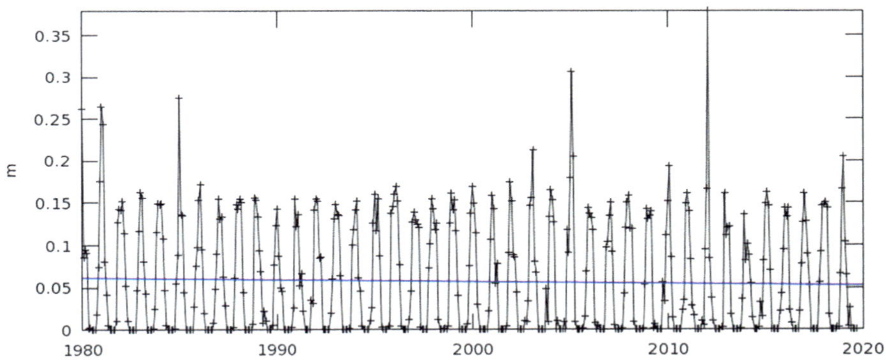

Fig. 7 Seasonal and interannual variability of snow depth (m) in the region 42°15′–43°15′N; 18°30′–20°00′E from January 1980 to November 2019 according to MERRA-2 Model [15, 18]. Blue line is a linear trend

In the present analysis we focused on the past 40 years long period from 1980 to 2019 and on the area of the whole Montenegro (see Fig. 1 for location). This is why some peaks of snow depth in the intersected period (1980–2010) may differ from the previous analysis as it was focused mainly on the coastal zone where weather and climate is different from the northern mountain regions of Montenegro. Figure 7 shows seasonal and interannual variability of snow depth (m) in the region under investigation from January 1980 to September 2019 according to the MERRA-2 Model M2TMNXLND_V5.12.4 [15, 18].

A decreasing trend of 0.025 cm/year or during the past 40 years a yearly average snow depth has decreased from 6 to 5 cm, i.e., by 20% which is a significant value. An absolute maximum of snow depth (38 cm) was registered in February 2012,

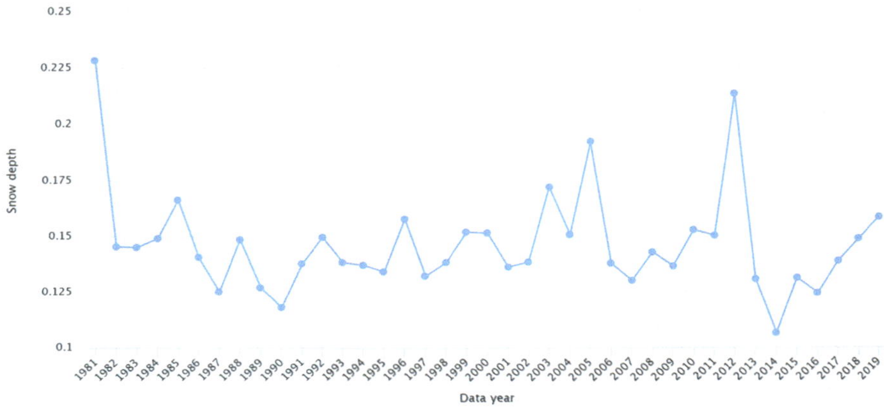

Fig. 8 Interannual variability of snow depth (m) for December–February in the region 42°15′–43°15′N; 18°30′–20°00′E from 1980 to 2019 according to MERRA-2 Model [15, 18]

while usually in winter it reaches about 15 cm (Fig. 7). During the whole time period, it was only six cases when snow depth exceeded 20 cm in 1981, 1985, 2003, 2005, 2012, and 2019. If we take only winter months, December–January–February, an average snow depth in 1981 (23 cm) was even higher than in 2012 (21 cm) (Fig. 8). In the past decade, two snow depth minimums of 2014 (11 cm) and 2016 (12 cm) are explained by relative anomalously warm winters (see Fig. 4).

4 Discussion

Recent studies stressed that due to the regional climate change, likely more and more rivers in Montenegro, their functions and services, as well as their aquatic biota will be affected. Consequently, this will require the development of adequate management and policy strategies for their long-term protection [3, 6]. The results of the present study confirm that the impact of regional climate change on the rivers of Montenegro especially when it comes to sustainable planning and preventing potential conflicts over the water resources becomes more and more dramatic. This statement is supported by the fact that we observe an accelerated warming of the area of Montenegro, because air temperature has increased from 1950 to 2010 by 1.3°C, but from 1980 to 2019 already by 2.0°C. During the past 40 years, air temperature is rising with a rate of about 0.05°C/year. For a summertime the rate of warming is even a bit higher than for the whole year because a linear regression gives a value of 0.06°C/year. Thus, summer months become warmer quickly than the other months of the year. If we add an increasing trend in wind velocity and the decreasing trend in the cloud cover [1], we will understand that evaporation from land is rising (Fig. 9), which reduce amount of water available for rivers.

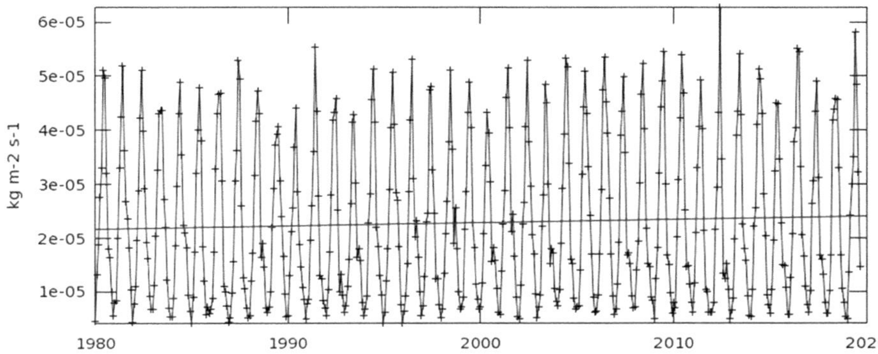

Fig. 9 Seasonal and interannual variability of evaporation from land (kg m^{-2} s^{-1}) in the region 42°15′–43°15′N; 18°30′–20°00′E from January 1980 to November 2019 according to MERRA-2 Model M2TMNXLND v5.12.4 [15, 19]. Blue line is a linear trend

Second negative factor is atmospheric precipitation (a direct source of water for rivers) which is decreasing with a rate of 10 mm/month during 20 years, from an average value of 150 mm/month in 2000 to 140 mm/month in 2019. In wintertime snow depth, as a potential source of water for rivers, is also decreasing with a rate of 0.025 cm/year, or during the past 40 years, a yearly average snow depth has decreased from 6 cm to 5 cm, i.e., by 20% which is a significant value.

Third negative factor are extreme weather events, in particular, in air temperature, which during the past 60 years (1950–2010) have increased seriously [2]. According to the results obtained by Kostianoy et al. [2] the amplitude of positive anomalous events has increased over the past 60 years by approximately 0.1°C, while the amplitude of negative anomalous events remained unchanged, indicating a general increase in temperature anomalies in the region. The amplitude of extreme positive events also demonstrates growth from about 5 to 7°C, while the amplitude of negative extreme phenomena remains unchanged for the period under consideration at the level of 8°C [2]. As a consequence of global warming, the annual number of negative anomalous events has greatly decreased from an average of 27 at the beginning of the 1950s to 14–16 at the beginning of the 2010s, while the number of positive anomalous events has increased from 15–16 in the 1950s to 28–30 in the 2000s. Also, the number of positive extreme events increased from 2 to 4.5 per year, while the negative ones remained unchanged [2]. The duration of anomalous temperature events shows an increase from 2.1 to 2.5 days for positive events and from 2.25 to 2.4 for negative ones [2]. Thus, extreme (anomalously warm) air temperature events became stronger in amplitude and much more frequent and last longer.

All these factors have also a negative effect on rivers in Montenegro and their services. Moreover, it is to be expected that changes in air temperature and atmospheric precipitation will have an impact on the aquatic biota and its communities, both in the perennial and intermittent rivers, as many taxa may not have enough time to adapt to changing conditions especially in the summer months which now become warmer more quickly than before.

Forecasts for regional climate changed in the region of Montenegro should be thoroughly analyzed and compared with the observed tendencies because as it was shown above it has a significant impact on the drainage basins in Montenegro, water availability for rivers, and the state of the related ecosystems. As concerns economic aspects, investigations on regional climate change, and water capacity in rivers should be taken into account when taking a decision in planning and construction of small hydropower plants in Montenegro.

Acknowledgments Analyses and visualizations used in this study were produced with the Giovanni online data system v.4.32 developed and maintained by the NASA Goddard Earth Sciences Data and Information Services Center (GES DISC). A.G. Kostianoy was partially supported in the framework of the P.P. Shirshov Institute of Oceanology RAS budgetary financing (Project N 149-2019-0004).

References

1. Kostianoy AG, Serykh IV, Kostianaia EA, Mitrović L, Ivanov M (2017) Regional climate change in the Boka Kotorska Bay. In: Joksimovich A, Djurovic M, Semenov AV, Zonn IS, Kostianoy AG (eds) The Boka Kotorska Bay Environment. The Handbook of Environmental Chemistry. Springer Nature Switzerland AG, Cham, pp 473–493
2. Kostianoy AG, Serykh IV, Kostianaia EA (2018) Climate change in the Lake Skadar/Shkodra region. In: Pešić V, Karaman G, Kostianoy A (eds) The Skadar/Shkodra Lake Environment. The Handbook of Environmental Chemistry. Springer Nature Switzerland AG, Cham, pp 63–88
3. Pešić V, Paunović M, Kostianoy AG (2020) The rivers of Montenegro: introductory remarks. In: Pešić V, Paunović M, Kostianoy AG (eds) The Rivers of Montenegro. The Handbook of Environmental Chemistry. Springer Nature Switzerland AG, Cham. https://doi.org/10.1007/698_2019_416
4. National Academies of Sciences, Engineering, and Medicine (2016) Attribution of extreme weather events in the context of climate change. The National Academies Press, Washington. https://doi.org/10.17226/21852
5. IPCC Climate Change (2013) The physical science basis. Contribution of WG1 to the V assessment report of the IPCC. Cambridge, UK and New York, NY, USA, 1535 pp
6. Pešić V, Pavićević A, Savić A, Hadžiablahović S (2020) The intermittent rivers of South Montenegro: ecology and biomonitoring. In: Pešić V, Paunović M, Kostianoy AG (eds) The Rivers of Montenegro. The Handbook of Environmental Chemistry. Springer Nature Switzerland AG, Cham
7. Vukašinović Pešić V, Blagojević N, Savić A, Tomić N, Pešić V (2020) The change in the water chemistry of the rivers of Montenegro over a 10-year period. In: Pešić V, Paunović M, Kostianoy AG (eds) The Rivers of Montenegro. The Handbook of Environmental Chemistry. Springer Nature Switzerland AG, Cham
8. Datry T, Larned ST, Tockner K (2014) Intermittent rivers: a challenge for freshwater ecology. Bioscience 64(3):229–235. https://doi.org/10.1093/biosci/bit027
9. Acuna V, Datry T, Marshall J, Barcelo D, Dahm CN, Ginebreda A, McGregor G, Sabater S, Tockner K, Palmer MA (2014) Why should we care about temporary waterways? Science 343(6175):1080–1082. https://doi.org/10.1126/science.1246666
10. Larned ST, Datry T, Arscott DB, Tockner K (2010) Emerging concepts in temporary-river ecology. Freshw Biol 55:717–738. https://doi.org/10.1111/j.1365-2427.2009.02322.x

11. Shumilova O, Zak D, Datry T et al (2019) Simulating rewetting events in intermittent rivers and ephemeral streams: a global analysis of leached nutrients and organic matter. Glob Change Biol 25:1591–1611. https://doi.org/10.1111/gcb.14537
12. Datry T, Singer G, Sauquet E et al (2017b) Science and management of intermittent rivers and ephemeral streams (SMIRES). Res Ideas Outcomes 3:e21774. https://doi.org/10.3897/rio.3.e21774
13. Datry T, Foulquier A, Corti R, von Schiller D, Tockner K, Mendoza-Lera C et al (2018) A global analysis of terrestrial plant litter dynamics in non-perennial waterways. Nat Geosci 11:497–503. https://doi.org/10.1038/s41561-018-0134-4
14. Kostianoy AG, Soloviev DM, Pešić V (2020) Application of Google Earth in mapping intermittent rivers of Montenegro. In: Pešić V, Paunović M, Kostianoy AG (eds) The Rivers of Montenegro. The Handbook of Environmental Chemistry. Springer Nature Switzerland AG, Cham
15. Acker JG, Leptoukh G (2007) Online analysis enhances use of NASA earth science data. Eos Trans AGU 88(2):14–17
16. Global Modeling and Assimilation Office (GMAO) (2015) MERRA-2 instM_2d_asm_Nx: 2d, monthly mean, single-level, assimilation, single-level diagnostics V5.12.4. Goddard Earth Sciences Data and Information Services Center (GES DISC), Greenbelt. Accessed 10 Jan 2020. https://doi.org/10.5067/5ESKGQTZG7FO
17. Huffman GJ, Stocker EF, Bolvin DT, Nelkin EJ, Tan J (2019) GPM IMERG final precipitation L3 1 month 0.1 degree x 0.1 degree V06. Goddard Earth Sciences Data and Information Services Center (GES DISC), Greenbelt. Accessed 11 Jan 2020. https://doi.org/10.5067/GPM/IMERG/3B-MONTH/06
18. Global Modeling and Assimilation Office (GMAO) (2015) MERRA-2 tavgM_2d_lnd_Nx: 2d, Monthly mean, time-averaged, single-level, assimilation, land surface diagnostics V5.12.4. Goddard Earth Sciences Data and Information Services Center (GES DISC), Greenbelt. Accessed 12 Jan 2020. https://doi.org/10.5067/8S35XF81C28F
19. Global Modeling and Assimilation Office (GMAO) (2015) MERRA-2 tavgM_2d_lnd_Nx: 2d, monthly mean, time-averaged, single-level, assimilation, land surface diagnostics V5.12.4. Goddard Earth Sciences Data and Information Services Center (GES DISC), Greenbelt. Accessed 14 Jan 2020. https://doi.org/10.5067/8S35XF81C28F

The Change in the Water Chemistry of the Rivers of Montenegro over a 10-Year Period

Vesna Vukašinović-Pešić, Nada Blagojević, Ana Savić, Nevenka Tomić, and Vladimir Pešić

Contents

V. Vukašinović-Pešić (✉) and N. Blagojević
Faculty of Metallurgy and Technology, University of Montenegro, Podgorica, Montenegro
e-mail: vesnav@ucg.ac.me; nadab@ucg.ac.me

A. Savić
Department of Biology, University of Niš, Niš, Serbia
e-mail: anka@pmf.ni.ac.rs

N. Tomić
Institute for Hydrometeorology and Seismology in Montenegro, Podgorica, Montenegro
e-mail: nena.tomic@meteo.co.me

V. Pešić
Department of Biology, University of Montenegro, Podgorica, Montenegro
e-mail: vladopesic@gmail.com

Vladimir Pešić, Momir Paunović, and Andrey G. Kostianoy (eds.), *The Rivers of Montenegro*, Hdb Env Chem (2020) 93: 83–110, DOI 10.1007/698_2019_417,

Abstract In this chapter, we provide an overview of changes in the chemical parameters most commonly used in water quality assessment within the framework of the National Surface Water Monitoring Programme of Montenegro. We chose the parameters (water temperature, pH, electroconductivity, dissolved oxygen, biological oxygen demand, chemical oxygen demand, alkalinity, water hardness, and quantities of ammonium, chlorides, sulphates, phosphates and nitrates) which were monitored over a 10-year period from 2009 to 2018. We briefly review the main results of the monitoring of the water quality of the watercourses in Montenegro in the investigated period. The quality index results revealed an increase in water quality since 2012. The results of our study show significant differences between the Adriatic and Black Sea basins in terms of linear trends and interannual variability, notably in the frequency of occurrence of anomalous and extreme values of the investigated chemical parameters over a 10-year period.

Keywords Montenegro, Pollution, Rivers, Water chemistry, Water quality monitoring

1 Introduction

Montenegro is rich in water resources, which represent one of the fundamental potential development opportunities for the country. In terms of its mean annual specific runoff, which is around 43 L s^{-1} km^{-2} [1], it is one of the richest countries in the world. The spatial distribution of the surface and groundwater, as well as their relationship, is determined primarily by the morphological and hydrogeological conditions of the karst terrains which dominates the majority of the Montenegrin landscape.

All water flows within the territory of Montenegro go in one of two directions: either towards the Adriatic Sea or towards the River Danube and the Black Sea. The total area of the Danube/Black Sea basin is about 7,545 km^2 or 54.6% of the total Montenegrin territory [2]. The waters of the Ibar empty into the Zapadna Morava, while the waters of the Tara, Piva, Lim and Ćehotina flow into the River Drina. The total area of the Adriatic basin of Montenegro is about 6,268 km^2 or 45.4%. The waters of the River Morača with its most important tributary the Zeta and Rijeka Crnojevića all empty into Lake Skadar which is connected to the Adriatic Sea by the River Bojana/Buna, with an outflow on average of 304 m^3 s^{-1} [3].

Water resources in Montenegro are subject to significant pressures which cause the deterioration of the water quality. The global climate warming may accelerate decreasing water quality, for example, by increasing eutrophication and by raising nutrient input from fertilized fields [4]. These changes in the external nutrient loads as a result of various anthropogenic factors can affect the chemical and biological processes occurring in the rivers that, in turn, may lead to a deterioration in water quality. The impact of these changes is determined by a range of factors, such as changes in the water flow, and the hydromorphological characteristics of rivers and theirs basins, but also the ecological status of the river which is characterized by the relative homoeostasis of the river continuum where these changes occur [5].

Monitoring physical and chemical indicators is usually defined as the first step in water quality assessment, aimed at providing information on the factors degrading the quality of surface waters. Council Directive 75/440/EEC of 16 June 1975 [6] lists 46 biological and chemical parameters for defining surface water quality. Regardless of the continuous monitoring conducted in Montenegro, it should be noted that data regarding concentrations of most biological and chemical parameters are either scarce or not available. As is the case with other countries in the region, Montenegro has declared that it follows the Water Framework Directive (WFD) in order to achieve the ultimate aim of this standard, which is the "good ecological status" of its water [7]. Water quality assessment requires a definition of "reference conditions" and the identifying of reference sites, based on a number of attributes, often related to the absence of any form of anthropogenic pressure (for a discussion see [8]). In practice, the identification of "reference conditions", although crucial to water quality, is difficult and requires the availability of long-term data and appropriate meta-analysis studies.

In this chapter, we provide an overview of the changes in the chemical parameters most commonly used in water quality assessment within the framework of National Surface Water Monitoring Programmes. For the purpose of this study, only basic parameters monitored over a minimum of 10 years were chosen. The results of this study aim to describe and predict the observed effects of the change in water chemistry in river systems in order to improve the ongoing programmes for the monitoring of surface water in Montenegro.

2 Monitoring Surface Water in Montenegro

2.1 The Legislative Framework

Surface Water Monitoring in Montenegro is regulated by the following documents – The Law on Environment [9], The Water Law [10], The Decree on Classification and Categorization of surface and Groundwater of Montenegro [11] and The Decree on the National List of Environmental Indicators [12]. The Water Law, which is in line with EU requirements, as the primary water quality legislation in Montenegro within the water management sector, defines the basic objectives for water protection

Table 1 Classification definitions of water quality in Montenegro (based on [11])

A, A1, A2, A3	Water that can be used for drinking and food production
S, Š and C	Water for fish and shell production
K1 and K2	Bathing waters

and management, as well as the conditions for surface, groundwater and coastal water quality classes and water body categories in Montenegro.

The Decree on the Classification and Categorization of Surface and Groundwater [11] categorizes water into classes according to the permissible limit values of certain groups of parameters. The regulation defines the manner and dynamics of sampling, the analytical methodology and the requirements of the assessment of water quality. The definition of classification is given in Table 1. Further, the regulation gives a precise categorization of water, by which the waters are classified into three categories: I (Class A1, S and K1), II (Class A2, C and K2) and III (Class A3).

Water quality monitoring based on physico-chemical parameters is performed by the Institute for Hydrometeorology and Seismology of Montenegro (Montenegrin: Hidrometeorološki zavod Crne Gore – HMZ). Monitoring is performed four times a year from April to December at each site (Fig. 1), meaning that all four seasons are covered, as well as the low water period when water pollution is expected to be higher [13].

The Annual Report on Water Quality (as part of the regular Environmental Status Report) has been published every year since 2009 by the Environmental Protection Agency of Montenegro (Montenegrin: Agencija za zaštitu prirode i životne sredine – EPA) and is available at www.epa.org.me. These reports provide an analysis of all the individual watercourses, with an emphasis on the affiliation of the individual river (and site) to a particular class of water according to the Decree on the Classification and Categorization of Surface and Groundwater. In particular, those parameters that deviate from a particular class as well as watercourses and measuring points that are most polluted are emphasized.

Moreover, the EPA has published two 4-year reports on water quality, for 2009–2012 [14] and for 2013–2016 [15], respectively, based on the indicators defined by the Decree on the National List of Environmental Indicators [12]. The definitions of the indicators used are given in Table 2.

2.2 The Water Quality of the Rivers of Montenegro

The largest sources of surface water pollution in Montenegro are municipal sewage, which is most often discharged into the recipient directly without treatment, followed by pollution of agricultural origin as well as the chemical pollution generated by industry [14, 15]. Based on the Annual Report on Surface Water

Fig. 1 Map of Montenegro with marked river quality monitoring stations. Adriatic Sea basin. The River Morača (1, Pernica; 2, Zlatica; 3, Momišići city beach; 4, below the effluent outlet from the sewage treatment plant in Podgorica; 5, Grbavci; 6, Vukovci). The River Zeta (7, Vidrovan; 8, Duklov Most; 9, Danilovgrad; 10, Vranjske Njive). the River Cijevna/Cemi (11, Trgaja; 12, the mouth into the River Morača). Crnojevića Rijeka (13, Brodska Njiva). The River Bojana/Buna (14, Fraskanjel). Black Sea basin. The River Lim (15, Plav; 16, Andrijevica; 17, Skakavac; 18, Zaton; 19, Bijelo Polje; 20, Dobrakovo). The River Tara (21, Crna Poljana; 22, below the Kolašin town; 23, Trebaljevo; below the Mojkovac town; 24, Đurđevića Tara; 25, Šćepan Polje). The River Piva (27, Šćepan Polje). The Ibar River (28, upstream from Rožaje town; 29, Bać). The River Ćehotina (30, Rabitlje; 31, below the confluence with Vezišnica River; 32, below the Pljevlja town; 33, Gradac). The Vezišnica River (34, upstream from the mouth into the River Ćehotina). 35: The River Grnčar. Kutska Rijeka (36, upstream from the Andrijevica town)

Table 2 Description of indicators according to the Decree on the National List of Environmental Indicators of Montenegro [12]

Indicator	Description	Water quality categories
Surface nutrients	Concentrations of orthophosphate and nitrate	
Biological oxygen consumption	Biological oxygen demand (BOD5) and concentration of total ammonium nitrogen	
Water quality index (WQI)	Dissolved oxygen saturation, biochemical oxygen demand, ammonium ion, pH, total nitrogen oxides, content of orthophosphates, total suspended solids, water temperature, electrical conductivity and coliform bacteria	Excellent: WQI value 90–100
		Very good: WQI value 84–89
		Good: WQI value 72–83
		Poor: WQI value 39–71
		Very poor: WQI value 0–38

Monitoring for the period 2009–2018, it can be concluded that the impact of transport development on surface water quality has increased [13–15] especially in recent years as a result of the construction of certain large infrastructure projects, such as the development of the Bar-Boljare highway.

Based on the water quality index (WQI), which has been shown to be an effective monitoring tool for assessment, poorer water quality was observed for the River Ćehotina and the River Vezišnica, as well as the River Morača below the effluent outlet from the sewage treatment plant in Podgorica and the River Ibar below Rožaje (and the site of Bać) [13–15]. The WQI result revealed that most watercourses with degraded water quality belong to the Black Sea basin. There has evidently been an increase in water quality since 2012 (Fig. 2). In 2012, about 45% of watercourses were rated as good, 30% as very good and 25% as poor in terms of water quality [14], while in 2016 as many as 83% of watercourses had water quality rated as excellent [15].

2.2.1 The Black Sea Basin

The Black Sea basin in Montenegro covers 7,545 km^2 [16]. The rivers of the Black Sea basin are characterized by more water than those of the Adriatic Sea basin, which affects the dynamics of their chemical parameters as well as pollutants and consequently their performance in water quality assessment. All five main river basins are included in the annual monitoring of surface water. In total eight rivers (the Lim, the Grnčar, the Kutska (Zlorečica), the Ibar, the Tara, the Piva, the Ćehotina and the Vezišnica) and 22 sites are monitored [13–15]. Based on the available data regarding water quality monitoring, the River Piva had the best water quality among the rivers in the Black Sea basin, while the most polluted

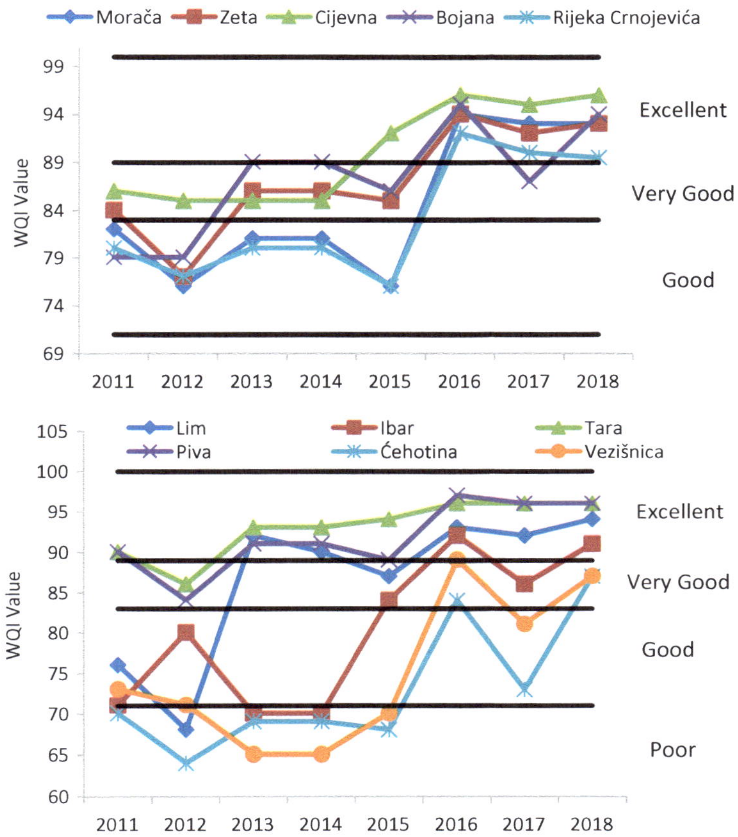

Fig. 2 The water quality index (WQI) of the rivers of the Adriatic and Black Sea basins in Montenegro over the period 2011–2018

watercourses were the Vezišnica, the Ćehotina (downstream of Pljevlja) and the Ibar (below Rožaje) [13–15].

The Rivers Ćehotina and Vezišnica The water of the River Ćehotina is sampled at four sites, while its tributary the River Vezišnica is sampled at one site. The Ćehotina upstream of Pljevlja (Rabitlja) should belong to the class A1, S, K1, while downstream of the mouth of the River Vezišnica and the town of Pljevlja should belong to class A2, C, K2 [13]. The River Vezišnica should belong to the class A1, S, K1 [13].

The watercourse of the River Ćehotina and its tributary the River Vezišnica are often cited as the most polluted rivers in Montenegro [14, 15, 17]. For example, the percentage of the determined classes that were outside the required class in 2016 was 46.1% (of which 27.3% was out of class), while in 2017, as much as 54.7% of the determined classes were outside the required class (of which 34.4% was out of class). In the River Vezišnica, the percentage of the determined classes that were outside the required class was even higher than for the River Ćehotina, being 75.0% in 2017 (of which 21.8% was out of class). The water quality is under the influence of wastewater from the Pljevlja Thermal Power Plant, a coalmine, and the "Šuplja Stijena" lead and zinc mine [18, 19]. The worst situation is observed at the sites (below the town and the mouth of the Vezišnica, respectively, and at Gradac) downstream from Pljevlja itself [13–15]. However, the WQI results reveal improving water quality in the River Ćehotina (and the River Vezišnica, see Fig. 1) in terms of the latest available data with a value of 62.5% of the water classes that were determined being appropriate in 2018 [13].

The River Lim The water of the River Lim is sampled at six sites (Fig. 1). The water of the river, upstream from the town of Berane (Plav and Andrijevica), should belong to the class A1, S, K1 [13–15]. On the other hand, sites (Skakavac, Zaton, Bijelo Polje and Dobrakovo), which are downstream of Berane, should be assigned to the class A2, C, K2. The main sources of pollution affecting the water quality of the River Lim include the agro-industrial activities of Zora (in Berane) and Mesopromet (in Bijelo Polje) plants, a cellulose and paper production plant (Celuloza) and the presence of deposited sites affected by the communal waste dump, landfill sawdust and offcuts of wood [18, 19]. The middle part of the watercourse near urbanized centres is more loaded with nutrients, resulting in an increase of the percentage of the determined classes that were outside the required class. For example, at the site of Skakavac (near Berane), 28.1% of the determined classes were found to be outside the required class [20]. In general, however, the WQI results revealed that the water quality of the River Lim has improved since 2012 (Fig. 2).

The River Ibar The water of the River Ibar is sampled at two sites (Fig. 1). The water upstream from the town of Rožaje should belong to the class A1, S, K1, while at the site of Bać, located downstream from the town, it should belong to the class A2, C, K2 [13–15]. The most important sources of pollution influencing the water quality of the River Ibar are wastewater outfalls and sewage from Rožaje [18]. Poor condition was particularly visible at the site of Bać where 43.7% (of which 28.1% was out of class altogether) in 2016 [20] and 56.3% (of which 31.3% was out of class altogether) in 2017 [21] of the determined classes were outside the required class, indicating that this site is one of the most polluted among all the studied sites of the Black Sea basin [13–15]. Figure 2 shows that the water quality of the River Ibar has improved since 2014.

The River Tara The water of the River Tara is sampled at six sites and should belong to the class A1, S, K1. The main part of the River Tara watercourse is located within a National Park [22], which should help in achieving better water quality, but as has recently been shown, this is difficult to maintain. For example, the percentage of determined classes that were out of the required class in 2016 was 29.7% [20], while in 2017 this percentage increased to 38.0% [21]. A particularly significant decrease in water quality was recorded in the upper part of the river, which was influenced by turbidity and construction activities related to a priority section of the highway that crosses the River Tara near Mateševo. Poor condition was especially visible through the content of phosphate, which was in class A3 at all sites in 2016, while in Kolašin it was even outside the required class altogether [20]. The most important source of pollution influencing the water quality of the River Tara are the wastewater outfalls and sewage from of Kolašin and Mojkovac [23] as well as the municipal waste deposit sites in these municipalities, as well as wastewater from the Trudbenik plant that produces wood sawdust [19].

The River Piva The River Piva is sampled from one location (Šćepan Polje), and its waters should belong to class A2, C, K2. Based on the results of the monitoring survey over the last 10 years, it can be concluded that this river has the best water quality in Montenegro [13–15]. The range of values of the class that were determined as appropriate varies from 81.3% in 2017 [21] to 96.6% in 2013 [24]. The River Piva WQI results from 2011 to 2018 revealed excellent water quality for all years, except in 2012 when water quality was classified as very good [14].

2.2.2 The Adriatic Sea Basin

The Adriatic Sea basin in Montenegro covers 6,268 km^2 [16] and includes three main hydrological basins: the Lake Skadar/Shkodra basin, the Trebišnjica river basin and the Montenegrin Coastal Area. Five rivers that belong to the Adriatic Sea basin (the Morača, the Zeta, the Cijevna, the Bojana and the Rijeka Crnojevića) examined through 14 sampling sites are included in the regular monitoring of the water quality, and the main results are discussed below.

The River Bojana The water of the Bojana is sampled at one place (Fraskanjel) and should belong to the class A2, C, K2 [13–15]. The WQI results over the studied period reveal that the water quality has improved since 2012 (Fig. 2). The percentage of the determined classes that were within the required class over the studied period varied from 78.1% in 2015 [25] and 2017 [21] to 90% in 2014 [26].

The River Morača The water of the Morača is sampled at six sites. The sites upstream of Podgorica (Zlatica and Pernica) should belong to the class A1, S, K1, while the sites in Podgorica (i.e. Momišići city beach and site below the effluent outlet from the sewage treatment plant in Podgorica) as well as those downstream of Podgorica (Grbavci and Vukovci) should belong to the class A2, C, K2 [13–15]. Poor conditions were particularly visible at the site below the effluent outlet from the sewage treatment plant in Podgorica where, for example, 40.6% of the determined classes in 2017 [21] were outside the required class. This is the result of wastewater outfalls and sewage from Podgorica that causes excessive contamination of the water at this site.

Downstream from Podgorica the water quality improves, due to the hydromorphological characteristics of the lower part of the River Morača. As a consequence, at the sites of Grbavci and Vukovci, the percentage of the tested water that were outside the required class declined (e.g. 31.3% at Grbavci and 18.8% at Vukovci in 2017 [21]) indicating an improvement in water quality.

The River Zeta The water of the Zeta is sampled from four locations. The upper part of the river above Brezovik (at the site of Vidrovan) should belong to the class A1, S, K1, while the sites downstream of Brezovik towards the confluence with the River Morača should belong to the class A2, C, K2 [13–15]. In the upper part of the Zeta, the highest percentage of the determined classes that were outside the required class, indicating a deterioration in water quality, was observed in 2017 at the site of "Duklov Most" where 37.6% of the determined classes were outside the required class [21]. In the middle and lower parts of the watercourse, the water of the Zeta, after sinking and receiving water from the hydroelectric power plant, remains at a good level. The range of values of the determined classes that were in line with the required class at the sites of Danilovgrad and Vranjske Njive was 87.6–84.4% in 2016 [20] and lower in 2017, being between 71.9 and 81.3% [21]. The WQI results reveal that the water quality of the River Zeta has improved since 2012 (Fig. 2).

The River Cijevna The water of the Cijevna is sampled from two sites and should belong to the class A1, S, K1 [13–15]. The available data reveals that the percentage of the determined classes that were outside the required class at Trgaja, located on the upper part of the Cijevna, varied from 30% (in 2018) [13] to 34.4% (in 2017) [21]. This was higher than at the second site point (the mouth into the River Morača) located downstream, where the percentage of the determined classes that were within the required class varies from 56.7% (in 2013) [24] to 83.2% (in 2018) [13]. It is worth emphasizing that the lower part of the Cijevna towards its mouth into the River Morača dries up in the summer months, resulting in a lower frequency of sampling (in this case, only one or two times per year).

Rijeka Crnojevića The water of Rijeka Crnojevića is sampled from one site (Brodska Njiva) and should belong to the class A1, S, K1 [13–15]. The water quality is affected by the wastewater outfalls and sewage from Cetinje that causes a deterioration in water quality [13]. The available data reveals that the percentage of determined classes that were within the required class varied from 56.7% (in 2013) [24] to 86.7% (in 2014) [26]. Poor condition was especially visible in terms of the content of phosphate, which was out of all classification [13].

3 Data Collection and Methodological Comments

The data for this paper are a part of the data collection of the Montenegrin Programme of Monitoring Surface Waters carried out by the Hydrometeorological Institute of Montenegro and contain results from a 10-year period (2009–2018). Those parameters monitored for a minimum of 10 years were chosen, while other parameters were excluded from analysis. Data were available for 13 parameters, from the 13 rivers and 36 sites, which mean values over the monitored period are presented in Tables 3 and 4.

Table 3 The mean 10-year (2009–2018) values ± standard deviation of individual parameters (temperature, pH, electroconductivity, dissolved O_2, biological oxygen demand (BOD5), chemical oxygen demand (COD), alkalinity) at the at the river quality monitoring stations in the Adriatic and the Black Sea basins

Basin	River	Monitoring stations	Temperature, °C	pH	Electroconductivity, μS cm^{-1}	Dissolved O_2, mg L^{-1}	BOD-O_2, mg L^{-1}	COD_{KMnO4}, mg L^{-1}	Alkalinity, mg L^{-1}
The Adriatic Sea basin	Morača River	Pernica	14.39 ± 3.53	8.13 ± 0.21	223.83 ± 26.46	10.60 ± 0.87	1.45 ± 1.14	1.43 ± 0.53	136.63 ± 11.80
		Zlatica	14.51 ± 3.39	8.13 ± 0.14	226.12 ± 19.59	11.18 ± 1.35	1.64 ± 1.06	1.51 ± 0.61	138.95 ± 11.63
		Momišići beach	15.26 ± 2.79	8.03 ± 0.16	258.07 ± 20.00	10.27 ± 1.12	1.56 ± 1.00	1.67 ± 0.66	155.61 ± 14.92
		Below the effluent outlet from the sewage treatment plant	15.59 ± 2.99	7.82 ± 0.29	280.36 ± 30.15	9.15 ± 2.00	6.03 ± 2.41	3.17 ± 1.73	168.98 ± 18.21
		Grbavci	16.77 ± 3.41	8.03 ± 0.18	259.20 ± 17.76	11.18 ± 2.37	2.63 ± 1.48	2.05 ± 0.91	158.41 ± 14.22
		Vukovci	17.17 ± 3.55	8.04 ± 0.19	258.69 ± 16.59	10.55 ± 1.32	1.95 ± 0.96	2.00 ± 0.67	157.45 ± 12.75
	Zeta River	Vidrovan	8.29 ± 1.04	8.05 ± 0.18	209.24 ± 14.81	11.80 ± 0.47	1.88 ± 0.81	1.60 ± 0.67	132.24 ± 10.74
		Duklov Most	15.04 ± 4.50	7.96 ± 0.21	289.10 ± 29.08	9.20 ± 1.69	2.30 ± 1.69	2.27 ± 1.04	174.17 ± 16.96
		Danilovgrad	15.00 ± 2.59	7.95 ± 0.18	275.27 ± 38.81	10.11 ± 1.20	1.66 ± 1.04	1.85 ± 0.81	167.95 ± 22.28
		Vranjske Njive	16.32 ± 3.22	7.99 ± 0.18	264.48 ± 29.79	10.04 ± 0.96	2.17 ± 1.67	2.09 ± 1.12	162.33 ± 16.96
	Cijevna/ Cemi River	Trgaja	15.52 ± 3.32	8.20 ± 0.14	209.05 ± 15.88	11.21 ± 0.99	1.57 ± 0.91	1.55 ± 0.62	134.34 ± 15.52
		The mouth into the River Morača	15.75 ± 4.08	8.24 ± 0.13	209.05 ± 36.51	10.96 ± 1.01	2.53 ± 3.55	1.62 ± 0.72	127.37 ± 11.96
	Bojana/ Buna River	Fraskanjel	19.48 ± 3.46	8.09 ± 0.18	247.17 ± 32.91	9.72 ± 1.17	2.26 ± 2.06	1.86 ± 0.70	130.12 ± 9.53
	Crnojevića River	Brodska Njiva	12.36 ± 0.92	8.20 ± 0.15	289.60 ± 26.66	12.04 ± 1.15	1.90 ± 1.14	1.75 ± 0.74	170.71 ± 16.19
	Lim River	Plav	14.00 ± 3.77	8.12 ± 0.16	199.20 ± 20.69	10.77 ± 1.24	1.67 ± 0.75	1.64 ± 0.54	127.56 ± 11.78
		Andrijevica	13.45 ± 3.24	8.21 ± 0.13	209.24 ± 19.62	10.76 ± 0.94	1.65 ± 0.99	1.57 ± 0.59	132.90 ± 13.19
		Skakavac	13.17 ± 3.05	8.15 ± 0.14	224.52 ± 22.28	10.75 ± 0.94	1.99 ± 1.17	1.63 ± 0.72	139.90 ± 14.28
		Zaton	13.70 ± 3.14	8.17 ± 0.15	228.71 ± 29.39	10.79 ± 1.19	1.75 ± 0.80	1.65 ± 0.61	142.02 ± 14.55
		Bijelo Polje	13.86 ± 3.30	8.14 ± 0.14	240.37 ± 21.70	10.38 ± 1.02	2.06 ± 0.78	1.89 ± 0.68	146.02 ± 11.99
		Dobrakovo	13.88 ± 3.17	8.12 ± 0.16	240.99 ± 44.27	10.12 ± 1.29	2.70 ± 1.05	1.88 ± 1.03	149.64 ± 12.17
The Black Sea basin	Grnčar River	Gusinje	12.33 ± 1.93	7.95 ± 0.16	278.88 ± 36.01	9.35 ± 0.83	1.24 ± 1.05	1.41 ± 0.56	170.93 ± 17.79
	Kutska River	Upstream from Andrijevica	10.42 ± 2.04	8.18 ± 0.14	213.29 ± 18.33	11.18 ± 0.90	1.45 ± 0.82	1.52 ± 0.60	134.93 ± 12.91
	Ibar River	Upstream from Rožaje	11.52 ± 4.00	8.23 ± 0.15	217.68 ± 11.24	10.53 ± 1.07	1.44 ± 0.85	2.00 ± 0.75	139.71 ± 12.11
		Bać	13.81 ± 4.40	8.21 ± 0.20	274.17 ± 37.56	10.63 ± 1.55	2.60 ± 1.55	2.79 ± 1.46	164.48 ± 20.63

(continued)

Table 3 (continued)

Basin	River	Monitoring stations	Temperature, °C	pH	Electroconductivity, μS cm^{-1}	Dissolved O_2, mg L^{-1}	BOD-O_2, mg L^{-1}	COD_{KMnO4}, mg L^{-1}	Alkalinity, mg L^{-1}
	Tara River	Crna Poljana	15.41 ± 3.71	8.21 ± 0.12	216.25 ± 18.97	9.51 ± 0.93	1.02 ± 0.80	1.51 ± 0.58	133.80 ± 13.12
		Below the Kolašin	13.85 ± 3.13	8.17 ± 0.12	218.23 ± 21.63	9.84 ± 1.56	1.31 ± 0.89	1.56 ± 0.71	133.28 ± 11.74
		Trebaljevo	13.67 ± 3.03	8.17 ± 0.14	219.00 ± 18.77	10.27 ± 0.75	1.08 ± 0.60	1.45 ± 0.56	134.00 ± 14.71
		Below the Mojkovac	15.48 ± 4.12	8.19 ± 0.11	222.10 ± 20.98	10.19 ± 1.07	1.24 ± 1.00	1.60 ± 0.60	133.32 ± 10.73
		Đurđevića Tara	12.63 ± 2.44	8.10 ± 0.14	234.38 ± 14.40	10.86 ± 0.77	1.45 ± 0.79	1.56 ± 0.57	141.43 ± 12.39
		Šćepan Polje	11.52 ± 2.06	8.06 ± 0.12	240.41 ± 15.22	11.25 ± 0.68	1.48 ± 0.74	1.45 ± 0.64	146.90 ± 13.59
	Piva River	Šćepan Polje	8.72 ± 1.46	n.a.	n.a.	12.24 ± 0.83	1.87 ± 0.89	1.57 ± 0.66	140.59 ± 12.65
	Ćehotina River	Rabitlja	12.59 ± 2.19	8.07 ± 0.14	304.53 ± 36.17	10.50 ± 0.76	1.60 ± 1.41	1.97 ± 0.64	179.50 ± 14.75
		Below the Pljevlja	13.34 ± 2.09	7.85 ± 0.20	375.15 ± 52.43	8.76 ± 1.34	5.16 ± 1.63	3.05 ± 2.08	212.66 ± 17.04
		Below the confluence with Vezišnica River	13.77 ± 2.30	7.96 ± 0.19	379.59 ± 42.82	8.75 ± 1.16	3.81 ± 1.39	3.35 ± 3.83	218.68 ± 18.71
		Gradac	13.96 ± 2.70	8.14 ± 0.19	361.03 ± 30.06	10.26 ± 1.92	2.76 ± 1.50	2.40 ± 0.97	210.85 ± 14.99
	Vezišnica River	Upstream from the mouth into the River Ćehotina	14.32 ± 2.99	8.40 ± 0.20	390.12 ± 36.63	9.16 ± 0.92	2.25 ± 1.39	2.66 ± 0.93	212.24 ± 13.96

Table 4 The mean 10-year (2009–2018) values ± standard deviation of individual parameters (water hardness, ammonium, chlorides, sulphates, phosphates and nitrates) at the river quality monitoring stations in the Adriatic and the Black Sea basins

Basin	River	Monitoring stations	Water hardness, dH°	NH_4^+, mg L^{-1}	Cl^-, mg L^{-1}	SO_4^{2-}, mg L^{-1}	PO_4^{3-}, mg L^{-1}	NO_3^-, mg L^{-1}
The Adriatic Sea basin	Morača River	Pernica	7.01 ± 0.67	0.05 ± 0.06	3.14 ± 1.75	7.20 ± 3.75	0.03 ± 0.03	0.56 ± 0.36
		Zlatica	7.08 ± 0.69	0.05 ± 0.07	3.27 ± 1.41	5.52 ± 2.24	0.06 ± 0.09	0.86 ± 0.43
		Momišići beach	8.06 ± 0.77	0.05 ± 0.06	3.84 ± 1.71	5.73 ± 3.12	0.04 ± 0.04	1.49 ± 0.58
		Below the effluent outlet from the sewage treatment plant	8.50 ± 0.81	0.94 ± 1.50	5.5 ± 3.02	7.74 ± 3.52	0.38 ± 0.38	2.55 ± 2.56
		Grbavci	8.08 ± 0.77	0.13 ± 0.16	3.92 ± 1.58	5.96 ± 2.61	0.11 ± 0.07	1.97 ± 0.67
		Vukovci	8.09 ± 0.81	0.08 ± 0.08	4.13 ± 2.11	5.83 ± 2.42	0.07 ± 0.04	2.03 ± 0.72
	Zeta River	Vidrovan	6.78 ± 0.66	0.04 ± 0.06	3.77 ± 2.85	3.67 ± 3.79	0.04 ± 0.03	0.99 ± 0.57
		Duklov Most	8.93 ± 0.82	0.08 ± 0.10	4.99 ± 3.27	6.30 ± 4.48	0.07 ± 0.13	1.99 ± 1.02
		Danilovgrad	8.51 ± 1.17	0.03 ± 0.04	4.03 ± 1.89	5.22 ± 2.19	0.07 ± 0.09	3.20 ± 3.34
		Vranjske Njive	8.27 ± 0.88	0.06 ± 0.05	3.78 ± 1.87	4.95 ± 2.25	0.05 ± 0.04	1.78 ± 0.70
	Cijevna	Trgaja	6.70 ± 0.79	0.03 ± 0.04	3.27 ± 1.81	4.02 ± 2.69	0.03 ± 0.04	1.13 ± 0.56
		The mouth into the River Morača	6.53 ± 0.83	0.04 ± 0.06	2.77 ± 1.61	3.30 ± 1.54	0.04 ± 0.03	0.92 ± 0.52
	Bojana River	Fraskanjel	7.35 ± 0.63	0.09 ± 0.07	7.19 ± 8.61	17.99 ± 8.64	0.04 ± 0.03	1.22 ± 0.74
	Crnojevića River	Brodska Njiva	8.39 ± 0.92	0.04 ± 0.06	6.72 ± 2.03	6.37 ± 2.97	0.37 ± 0.36	5.49 ± 3.35
	Lim River	Plav	6.42 ± 0.85	0.06 ± 0.10	3.31 ± 1.91	5.12 ± 2.68	0.04 ± 0.04	0.53 ± 0.42
		Andrijevica	6.68 ± 0.77	0.05 ± 0.09	3.16 ± 1.99	7.74 ± 2.22	0.05 ± 0.08	0.67 ± 0.41
		Skakavac	7.10 ± 0.88	0.07 ± 0.10	3.36 ± 1.79	8.10 ± 2.93	0.06 ± 0.07	1.35 ± 1.64
		Zaton	7.38 ± 0.90	0.07 ± 0.09	3.50 ± 1.84	9.53 ± 4.86	0.05 ± 0.03	1.58 ± 3.15
		Bijelo Polje	7.52 ± 0.79	0.10 ± 0.10	3.40 ± 1.72	9.62 ± 3.09	0.05 ± 0.05	1.13 ± 0.64
		Dobrakovo	7.67 ± 0.80	0.11 ± 0.11	3.78 ± 2.29	10.04 ± 5.42	0.11 ± 0.21	1.27 ± 0.66

(continued)

Table 4 (continued)

Basin	River	Monitoring stations	Water hardness, dH°	NH_4^+, mg L^{-1}	Cl^-, mg L^{-1}	SO_4^{2-}, mg L^{-1}	PO_4^{3-}, mg L^{-1}	NO_3^-, mg L^{-1}
The Black Sea basin	Grnčar River	Gusinje	8.51 ± 1.13	0.03 ± 0.05	3.12 ± 1.88	10.64 ± 11.78	0.06 ± 0.06	0.81 ± 0.50
	Kutska River	Upstream from Andrijevica	6.92 ± 0.81	0.06 ± 0.17	2.91 ± 1.49	8.50 ± 4.19	0.04 ± 0.04	0.48 ± 0.39
	Ibar River	Upstream from Rožaje	7.24 ± 0.92	0.03 ± 0.05	3.10 ± 2.10	6.55 ± 2.33	0.04 ± 0.03	0.78 ± 0.62
		Bać	8.55 ± 1.14	0.27 ± 0.47	5.64 ± 2.10	9.60 ± 4.45	0.23 ± 0.28	2.48 ± 1.50
	Tara River	Crna Poljana	6.98 ± 0.80	0.03 ± 0.05	3.47 ± 2.23	8.38 ± 3.13	0.04 ± 0.03	0.67 ± 0.58
		Below the Kolašin	7.04 ± 0.86	0.03 ± 0.05	3.42 ± 1.76	7.48 ± 2.38	0.06 ± 0.05	0.78 ± 0.56
		Trebaljevo	7.09 ± 0.82	0.03 ± 0.06	3.35 ± 2.17	7.65 ± 3.97	0.04 ± 0.03	0.91 ± 0.53
		Below the Mojkovac	7.19 ± 0.85	0.02 ± 0.03	3.40 ± 2.22	10.02 ± 7.99	0.05 ± 0.05	0.85 ± 0.57
		Đurđevića Tara	7.60 ± 0.84	0.05 ± 0.13	3.26 ± 2.12	8.82 ± 3.97	0.04 ± 0.05	1.03 ± 0.59
		Šćepan Polje	7.73 ± 0.78	0.02 ± 0.04	3.13 ± 1.73	7.16 ± 5.59	0.04 ± 0.03	1.23 ± 0.44
	Piva River	Šćepan Polje	7.47 ± 0.86	0.02 ± 0.03	3.22 ± 1.82	5.63 ± 4.89	0.05 ± 0.04	1.02 ± 0.44
	Ćehotina River	Rabitlja	9.74 ± 0.99	0.05 ± 0.09	3.36 ± 1.93	11.98 ± 13.63	0.04 ± 0.04	1.54 ± 0.73
		Below the Pljevlja	11.70 ± 1.18	0.67 ± 0.43	6.49 ± 9.42	19.33 ± 6.81	0.27 ± 0.17	2.90 ± 1.08
		Below the confluence with Vezišnica River	11.69 ± 0.98	0.74 ± 0.97	5.09 ± 1.99	24.44 ± 13.26	0.47 ± 1.23	4.21 ± 5.78
		Gradac	11.10 ± 1.03	0.27 ± 0.45	4.75 ± 2.27	19.27 ± 6.36	0.21 ± 0.27	3.89 ± 1.71
	Vezišnica River	Upstream from the mouth into the River Ćehotina	11.07 ± 0.94	0.23 ± 0.14	7.80 ± 7.21	27.53 ± 10.14	0.15 ± 0.13	2.25 ± 1.05

The largest number of sampling sites is located on the Rivers Morača, Tara and Lim (6), Zeta and Ćehotina (4), followed by the Cijevna and the Ibar with two each, while the Bojana, Rijeka Crnojevića, the Grnčar, Kutska Rijeka, the Piva and the Vezišnica each have only one sampling site (see Fig. 1 for the site positions). Sampling frequency varied; most of the sites for characterizing water quality were sampled four times per year.

In this research, we focused on the determination of interannual changes over the studied period in the frequency of the occurrence of anomalous and extreme values of the studying parameters, which is a novel innovation in this study. All the anomalies and extremes were calculated relative to the average trend for the 10-year investigated period (2009–2018). We considered events "anomalous" when the parameter value exceeded one standard deviation and "extreme" when anomalies exceeded two standard deviations. The similar methodological approach was used in analysing the variability of extreme climate events (see, e.g. [27]). Regression analysis was conducted to analyse the linear trend variability of both the anomalous and extreme events experienced by the selected chemical parameters. The analyses were conducted using SPSS 17.0 (SPSS Inc., Chicago, IL).

4 Results

4.1 Water Temperature

The mean 10-year water temperature values of the monitored sites in the Adriatic and the Black Sea basins over the studied period are given in Table 3. The lowest mean temperature value in the Adriatic Sea basin was found at Vidrovan on the upper part of the River Zeta (8.29 $\pm$ 1.04°C), while the highest mean temperature was observed at the locality of Fraskanjel on the River Bojana (19.48 $\pm$ 3.46°C). In the Black Sea basin, the lowest mean temperature value was found at the confluence of the River Piva at Šćepan Polje (8.72 $\pm$ 1.46°C), while the highest value was observed on the River Tara near Mojkovac (15.48 $\pm$ 4.12°C).

The mean annual number of anomalous events per site for the Adriatic basin was 1.16 $\pm$ 1.029, which was slightly lower than in the Black Sea basin (1.33 $\pm$ 0.951). On the other hand, the average annual number of extreme events per site for the Adriatic basin (0.06 $\pm$ 0.274) was significantly lower ($p = 0.003$) than for the Black Sea basin (0.16 $\pm$ 0.371).

The results of the regression analysis revealed that the mean annual number of anomalous events decreases by 0.5 events per year for the studied localities in the Adriatic Sea basin, while in the Black Sea basin, it increases by 1.15 events per year. The mean annual number of extreme events in the Adriatic basin decreased by 0.03 events per year, while in the Black Sea basin, it increased by 0.74. We found that in the investigated period, there were years in both basins in which more anomalies occurred on an annual basis. The largest number (with the mean annual number of anomalies per site in parentheses) of anomalous events in the Adriatic Sea basin

(1.79) was observed in 2011, with the largest number of extreme cases (0.21) occurred in 2016. In the Black Sea basin, the largest number of anomalous (2.68) and extreme events (0.73) was registered in 2015 and 2016, respectively, indicating the strong interannual variability of water temperature events.

4.2 *pH*

The mean annual pH values of the monitored sites in the Adriatic and Black Sea basins indicate neutral to slightly alkaline water (see Table 3). The average yearly number of anomalous events per site for localities from the Adriatic basin was slightly higher (1.07 $\pm$ 1.075) than for those sites belonging to the Black Sea basin (1.03 $\pm$ 0.893). The average annual number of extreme events per site was the same for both basins (0.17 $\pm$ 0.417 in the Adriatic Sea basin and 0.17 $\pm$ 0.379 in the Black Sea basin).

The results of the regression analysis revealed that the average number of anomalous events decreased by 0.88 events per year for the Adriatic basin, while in the Black Sea basin, it increased by 1.18 events per year. On the other hand, the average number of extreme events in the Adriatic basin decreased by 0.48, while in the Black Sea basin, it increased by 0.06 events per year. The largest number (with the mean annual number of anomalies per site in parentheses) of anomalous events (1.86) in the Adriatic basin was observed in 2011, with the highest number of extreme events (0.5) registered in 2010. In the Black Sea basin, the largest number of anomalous events (1.71) was registered in 2018, while the largest number of extremes (0.57) occurred in 2011.

4.3 *Electroconductivity*

The mean 10-year values of the water electroconductivity of the studied sites from the Adriatic and Black Sea basins are given in Table 3. The highest values, exceeding >300 μS cm^{-1}, were observed in the River Ćehotina and its tributary the Vezišnica, which are the both strongly influenced by the wastewater from the Pljevlja TPP. The high conductivity at these sites can be attributed mainly to long-term industrial sewage discharge.

The mean annual number of anomalous events per site for the Adriatic basin (1.08 $\pm$ 1.070) was slightly higher in comparison with the Black Sea basin (1.03 $\pm$ 1.035). Moreover, the average annual number of extreme events per site for the Adriatic basin was higher (0.23 $\pm$ 0.530) than for the Black Sea basin (0.18 $\pm$ 0.442). The results of the regression analysis revealed that average number of anomalous events in both the Adriatic and Black Sea basins decreased by 1.51 and 0.43 events per year, respectively. Furthermore, the average number of extreme events in both the studied basins decreases by 0.65 and 0.52 events per year, respectively. The largest number (with the mean annual number of anomalies per site in parentheses) of anomalous (2.57 and 2.71, respectively) and extreme events (0.86 and 0.81, respectively) in the Adriatic Sea and the Black Sea basins was found in 2009.

4.4 Dissolved Oxygen

The mean 10-year values of dissolved oxygen at the monitored sites from the Adriatic and Black Sea basins are given in Table 3. The values of dissolved oxygen were quite uniform in both basins, varying in the range 9.15–12.04 mg L^{-1} for the Adriatic Sea basin and 8.75–12.24 mg L^{-1} for the Black Sea basin.

The mean annual yearly number of anomalous events per site for the Adriatic basin was slightly higher (1.21 $\pm$ 0.993) in comparison with the Black Sea basin (1.16 $\pm$ 0.915). The mean annual number of extreme events per site was 0.18 $\pm$ 0.420 for the Adriatic basin, being slightly lower in comparison with the Black Sea basin (0.19 $\pm$ 0.390).

The results of the regression analysis revealed that the average number of anomalous events in the Adriatic basin decreased by 0.43 per year, while in the Black Sea basin, it increased by 0.52. On the other hand, the average number of extreme events in the Adriatic basin decreased by 0.36 events per year, while in the Black Sea basin, it increased by 0.20. The largest number (with the mean annual number of anomalies per site in parentheses) of anomalous (1.71) and extreme (0.79) events in the Adriatic basin was registered in 2011. In the Black Sea Basin, the largest number of anomalous events (1.77) was registered in 2017, while the largest number of extremes (0.54) was registered in 2015.

4.5 Biological Oxygen Demand (BOD5)

The mean 10-year BOD5 values of the water from the monitored sites in the Adriatic and Black Sea basins are given in Table 3. The range of BOD5 values does not differ significantly between the two basins, except at one site in each basin, where the values were above the expected range. In the Adriatic Sea basin, the range of BOD5 values at all sites varies between 1.45 and 2.70 mg L^{-1}, except for one site on the River Morača below the sewage effluent outlet from the treatment plant in Podgorica, where the mean 10-year value was 6.03 $\pm$ 2.41 mg L^{-1}. In the Black Sea basin, the BOD5 values were ranged between 1.02 and 2.76 mg L^{-1}, except at two sites on the River Ćehotina, below Pljevlja (5.16 $\pm$ 1.63 mg L^{-1}) and below the confluence with the Vezišnica (3.81 $\pm$ 1.39 mg L^{-1}), where the mean 10-year values were much higher.

The mean annual number of anomalous events per site for the Adriatic basin was slightly lower (0.94 ± 0.907) than in the Black Sea basin (1.12 ± 0.958). On the other hand, the average annual number of extreme events per sites was the same in both the studied basins (0.18 ± 0.437 for the Adriatic and 0.18 ± 0.395 for the Black Sea basin).

The results of the regression analysis revealed that the mean number of anomalous events in the Adriatic basin decreased by 1.39 events per year, while in the Black Sea basin, it increased by 0.73 events per year. The mean number of extreme events in both the Adriatic and the Black Sea basin decreased, by 0.48 and 0.01 events per year, respectively. The largest number (with the mean annual number of anomalies per site in parentheses) of anomalous (1.71) and extreme (0.43) cases in the Adriatic basin were registered in 2009. In the Black Sea basin, the largest number of anomalous events (1.54) was registered in 2017, while the largest number of extremes (0.41) occurred in 2009.

4.6 Chemical Oxygen Demand (COD)

The mean 10-year values of chemical oxygen demand at the monitored sites in the Adriatic and Black Sea basins are given in Table 3. As can be seen from the Table, the range of mean values does not significantly differ between the two basins, except at the most polluted sites in both basins, where the mean values were higher. In the Adriatic Sea basin, the mean COD values were in the range 1.43–2.27 mg L^{-1}, except on the River Morača below the discharge from the sewage treatment plant in Podgorica, where the mean value was higher (3.17 ± 1.73 mg L^{-1}). In the Black Sea basin, the mean COD value was in the range 1.41–2.00 mg L^{-1}, except at the following sites: Bać (2.79 ± 1.46 mg L^{-1}), the sites along the River Ćehotina, downstream from Pljevlja (2.40–3.35 mg L^{-1}) and the River Vezišnica (2.66 ± 0.93 mg L^{-1}) where the mean values were >2.0 mg L^{-1}.

The mean annual number of anomalous events per site for the Adriatic basin was slightly lower (1.14 ± 1.088) than in the Black Sea basin (1.17 ± 1.046). On the other hand, the average annual number of extreme events for the Adriatic basin was slightly higher (0.19 ± 0.442) than in the Black Sea basin (0.18 ± 0.383).

The results of the regression analysis revealed that the average number of anomalous events in both the Adriatic and the Black Sea basin decreased, by 2.43 and 2.74 events per year, respectively. On the other hand, the average number of extreme events in the Adriatic basin decreases by 0.33 events per year, while in the Black Sea basin, it increased by 0.04 events per year. The largest number (with the mean annual number of anomalies per site in parentheses) of anomalous (2.10) events in the Adriatic basin was in 2009, while the largest number of extreme cases (0.86) was registered in 2012. In the Black Sea basin, the largest number of anomalous events (2.18) was registered in 2010, while the largest number of extremes (0.27) occurred in 2011 and 2012.

4.7 Alkalinity

The mean 10-year values of the water alkalinity at the monitored sites in the Adriatic and Black Sea basins are given in Table 3. The mean values of alkalinity were similar in both basins and were in the range 127.56–174.17 mg L^{-1} for the Adriatic basin and 133.28–179.50 mg L^{-1} for the Black Sea basin. The only exceptions were the River Ćehotina downstream from Pljevlja and the River Vezišnica, both belonging to the Black Sea basin, where the average values were higher, exceeding 200 mg L^{-1}.

The mean annual number of anomalous events per site for the Adriatic basin was lower (1.19 $\pm$ 0.944) than in the Black Sea (1.29 $\pm$ 1.129). On the other hand, the mean annual number of extreme events per sites was the same in both the studied basins (for the Adriatic Sea basin 0.17 $\pm$ 0.432 and the Black Sea basin 0.17 $\pm$ 0.387).

The results of the regression analysis revealed that the average number of anomalous events in the Adriatic basin decreased every year by 1.04, while in the Black Sea basin, it increased by 0.93. The average number of extreme events in both the Adriatic and the Black Sea basin decreased, by 0.30 and 0.18 events per year, respectively. The largest number (with the mean annual number of anomalies per site in parentheses) of anomalous events (1.86) in the Black Sea basin was registered in 2018, while the largest number of extremes (0.41) occurred in 2010. In the Adriatic basin, the largest number of anomalous (2.36) and extreme cases (0.57) was both detected in 2010.

4.8 Water Hardness

The mean 10-year values of water hardness at the monitored sites from the Adriatic and Black Sea basins are given in Table 4. As with alkalinity, the mean values of water hardness were similar in both the basins and ranged from 6.42 to 8.93 dH° for the Adriatic Sea basin and 6.92–8.55 dH° for the Black Sea basin. The only exceptions were the Ćehotina and the Vežišnica, where the 10-year mean range was higher (9.74–11.70 dH°).

The mean annual number of anomalous events per site for the Adriatic basin was higher (1.39 $\pm$ 1.090) than in the Black Sea basin (1.29 $\pm$ 0.968). On the other hand, the average annual number of extreme events for the Adriatic basin was slightly lower (0.16 $\pm$ 0.426) than for the Black Sea basin (0.17 $\pm$ 0.421).

The results of the regression analysis revealed that the average number of anomalous events in both the Adriatic and the Black Sea basin increased each year, by 0.58 and 2.08, respectively. Moreover, the average number of extreme events in both the Adriatic and the Black Sea basin also increased, by 0.33 and 1.00 events per year, respectively. The largest number (with the mean annual number of anomalies per site in parentheses) of anomalous events (2.21) in the Adriatic basin was observed in 2013, with the largest number of extreme cases (0.86) appeared in 2016. In the Black Sea basin, the largest number of anomalous events (2.18) was registered in 2017, while the largest number of extremes (0.59) occurred in 2018.

4.9 Ammonium (NH_4^+)

The mean 10-year values of ammonium nitrogen in the water at the monitored sites in the Adriatic and Black Sea basins are given in Table 4. The mean 10-year ammonium nitrogen values for most sites in the Adriatic Sea basin varied in the range 0.03–0.13 mg L^{-1}, except at the site below the effluent outlet from the sewage treatment plant in Podgorica where the mean NH_4^+ concentration was 0.94 ± 1.50 mg L^{-1}. In the Black Sea basin, the mean 10-year NH_4^+ values at most sites were in the range 0.03–0.06 mg L^{-1}, except at Bać below Rožaje (0.27 ± 0.47 mg L^{-1}); the River Ćehotina, downstream of Pljevlja (0.27–0.74 mg L^{-1}); and the River Vezišnica (0.23 ± 0.14 mg L^{-1}).

The mean annual number of anomalous events per site for the Adriatic Sea basin was higher (0.39 ± 0.607) than in Black Sea basin (0.34 ± 0.673). Similarly, the mean annual number of extreme events per site for the Adriatic basin was slightly higher (0.21 ± 0.461) in comparison with the Black Sea basin (0.20 ± 0.474). The available data from the monitoring for period 2009–2016 revealed that the content of ammonium nitrogen in water of monitored sites decreased from 2009.

The results of the regression analysis revealed that the mean number of anomalous events in the Adriatic and the Black Sea basin increases by 0.15 and 0.39 events per year, respectively. On the other hand, the mean number of extreme events increases in the Adriatic basin by 0.53 and in the Black Sea basin by 0.80 events per year. In the Black Sea Basin, the largest number (with the mean annual number of anomalies per site in parentheses) of anomalous events (0.5) was registered in 2015 and 2017, while the same number of extremes (1.04) was registered in 2017. In the Adriatic basin, the largest number of anomalous and extremes, 0.71 and 1.07 events per site, respectively, occurred in 2017.

4.10 Chlorides

The mean 10-year chloride values in the water at the monitored sites in the Adriatic and Black Sea basins are given in Table 4. The range of mean values of chloride was similar in both basins and was 2.77–7.19 mg L^{-1} for the Adriatic and 2.91–7.80 mg L^{-1} for the Black Sea basin. The higher values of chloride in the water are a result of the dissolution of chloride-containing minerals (Fraskanjel on the River Bojana) and/or the influence of wastewater (at Vezišnica and on the River Ćehotina below Pljevlja; see Table 4).

The mean annual number of anomalous events per site for the Adriatic Sea basin was higher (0.91 ± 1.135) than in the Black Sea basin (0.75 ± 1.014). On the other hand, the mean annual number of extreme events per site for the Adriatic Sea basin was slightly lower (0.19 ± 0.507) than in the Black Sea basin (0.20 ± 0.451).

The results of the regression analysis revealed that the average number of anomalous events in both the Adriatic and the Black Sea basin decreased by 2.12 and 4.03 events per year, respectively. Moreover, the average number of extreme events in the Adriatic and the Black Sea basin decreases by 0.67 and 1.47 events per year, respectively. The largest number (with the mean annual number of anomalies per site in parentheses) of anomalous (2.14 and 2.09, respectively) and extreme (0.93 and 0.73, respectively) events in the Adriatic and the Black Sea basins was observed in 2010.

4.11 Sulphates

The mean 10-year values of sulphates at the monitored sites in the Adriatic and the Black Sea basins are given in Table 4. Based on the available data, it can be seen that the mean sulphate value over the studied 10-year period was lower at most of the studied sites in the Adriatic Sea basin in comparison to those in the Black Sea basin. In the Adriatic Sea basin, the mean sulphate values were in the range 3.30–10.04 mg L^{-1}, with the exception of the site at Fraskanjel (on the River Bojana), where it was notably higher, being 17.99 ± 8.64 mg L^{-1}. In the Black Sea basin, the mean sulphate value varied in the range 6.55–11.98 mg L^{-1}, except at three sites along the River Ćehotina, downstream of Pljevlja (19.33–24.44 mg L^{-1}), and on the River Vezišnica (27.53 ± 10.14 mg L^{-1}) where mean value range for the 10-year period was higher.

The mean annual number of anomalous events per site in the Adriatic basin (1.11 ± 1.104) was significantly higher ($p = 0.018$) than in the Black Sea basin (0.82 ± 0.961). On the other hand, the mean annual number of extreme events per site (in parentheses) for the Adriatic basin was slightly higher (0.22 ± 0.450) than in the Black Sea basin (0.20 ± 0.434), but no statistical significance was found.

The results of the regression analysis revealed that the mean number of anomalous events in both the Adriatic and Black Sea basins decreased, by 1.16 and 0.68 events per year, respectively. Moreover, the mean number of extreme events decreased by 1.02 and 0.38 events per year in the Adriatic Sea and Black Sea basins, respectively. The largest number (with the mean annual number of anomalies per site in parentheses) of anomalous (2.29 and 1.64, respectively) and extreme events (0.93 and 0.59, respectively) in both the Adriatic and Black Sea basins was in 2009.

4.12 Phosphates

The mean 10-year values of phosphates at the monitored sites in the Adriatic and Black Sea basins are given in Table 4. The range of mean value for the studied period mainly overlaps for both basins, showing values of in the Adriatic Sea basin of 0.03–0.11 mg L^{-1} and in the Black Sea basin of 0.04–0.11 mg L^{-1}. The only exception in the Adriatic basin where the mean value exceeded the range was at the site on the River Morača below the effluent outlet from the sewage treatment plant in Podgorica (0.38 $\pm$ 0.38 mg L^{-1}) and Rijeka Crnojevića (0.37 $\pm$ 0.36 mg L^{-1}). In the Black Sea basin, higher mean values were noted at the site of Bać on the River Ibar (0.23 $\pm$ 0.28 mg L^{-1}), at the sites along the River Ćehotina, downstream from Pljevlja (0.21–0.47 mg L^{-1}), and on the River Vezišnica (0.15 $\pm$ 0.13 mg L^{-1}). The role of municipal and industrial sewage inflow can be recognized as the main causes of the high-phosphate content.

The mean annual number of anomalous events per site in the Adriatic basin was lower (0.81 $\pm$ 0.934) than in the Black Sea basin (0.90 $\pm$ 0.947). Similarly, the average annual number of extreme events per site in the Adriatic basin was lower (0.14 $\pm$ 0.371) than in the Black Sea basin (0.17 $\pm$ 0.391).

The results of the regression analysis revealed that the average number of anomalous events in the Adriatic basin decreased by 0.76, while in the Black Sea basin, it increased by 0.05 events per year. On the other hand, the average number of extreme events in the Adriatic basin increased by 0.04, while in the Black Sea basin, it decreased by 0.08 events per year. The largest number of anomalous events (1.29 per site) in the Adriatic basin was observed in 2009 and again in 2016, with the largest number of extreme cases (0.43 per sites) appeared in 2016. In the Black Sea basin, the largest number of anomalous events (1.23 per site) was registered in 2018, while the largest number of extremes (0.36 per site) occurred in 2014.

4.13 Nitrates

The mean 10-year values of nitrates at the monitored sites in the Adriatic and Black Sea basins are given in Table 4. The mean nitrate values of most sites in the Adriatic Sea basin were in the range 0.56–2.03 mg L^{-1}, except on the River Morača below the discharge from the sewage treatment plant in Podgorica (2.55 $\pm$ 2.56 mg L^{-1}), on the River Zeta at Danilovgrad (3.20 $\pm$ 3.34 mg L^{-1}) and at Rijeka Crnojevića (5.49 $\pm$ 3.35 mg L^{-1}). In the Black Sea basin, the mean nitrate values of most sites were in the range 0.67–1.54, except on the River Ibar at Bać, below Rožaje (2.48 $\pm$ 1.50 mg L^{-1}); the River Ćehotina, downstream of Pljevlja (2.90–4.21 mg L^{-1}); and the River Vezišnica (2.25 $\pm$ 1.05 mg L^{-1}) where the 10-year mean values were higher.

The mean annual number of anomalous events per site in the Adriatic basin was slightly higher (0.91 ± 0.951) than in the Black Sea basin (0.88 ± 0.921). Similarly, the mean annual number of extreme events per site in the Adriatic basin was slightly higher (0.19 ± 0.474) in comparison with the Black Sea basin (0.17 ± 0.445).

The results of the regression analysis revealed that the mean number of anomalous events in both the Adriatic and the Black Sea basins decreased, by 1.50 and 1.74 events per year, respectively. On the other hand, mean number of extreme events in the Adriatic basin decreased by 0.46, while in the Black Sea basin, it increased by 0.33 events per year. The largest number (in parentheses) of anomalous events (2.21 per site) in the Adriatic basin was observed in 2009, with the largest number of extreme cases (0.57 per site) registered in 2009. In the Black Sea basin, the largest number of anomalous events (1.82 per site) was registered in 2009, while the largest number of extremes (0.64 per site) occurred in 2016.

5 Discussion

The chemical parameters of water quality in river systems are generally subjected to significant fluctuations, associated mainly with the impact of pollutants, but other factors such as water flow regimes determined by atmospheric loads also play important roles. For the purpose of this chapter, we define two types of fluctuations of "anomalous" and "extreme" events, viewing them as parameter values exceeding one and two standard deviations, respectively.

The result of our study reveals the difference in the linear trend of anomalies between the Adriatic and Black Sea basins. Figure 3 shows that the frequency of anomalies in the Adriatic Sea basin decreased from an average of 20.8 in 2009 to 10.4 anomalous events in 2018. On the other hand, this trend is not clearly recorded in the Black Sea basin, which exhibits higher fluctuations in anomalies over the studied period, with a weak linear trend of decreasing the frequency of anomalous events and an increasing trend for extreme events over the 10-year period.

Moreover, for most parameters we found a difference in the frequency of anomalous and extreme events between the two studied basins. In the Adriatic basin, there is an evident decreasing trend in anomalous events over the studied period for all parameters except water hardness and ammonium. In the Black Sea basin, the annual frequency of anomalous events of electroconductivity, COD, chlorides, sulphates and nitrates values decreases, while the other parameters show an increasing trend. In the case of the frequency of extreme events, in the Adriatic Sea basin, all the parameters except water hardness, ammonium and phosphates show a decreasing trend. In the Black Sea basin, the annual frequency of extreme events of electroconductivity, BOD5, alkalinity, chlorides, sulphates and phosphates values decreased, while the other parameters show an increasing trend.

Thus, it is evident that the frequency of anomalies in the Black Sea basin over the studied period increased for a greater number of parameters in comparison with the Adriatic Sea basin. The reason for this may be that the Black Sea basin contains

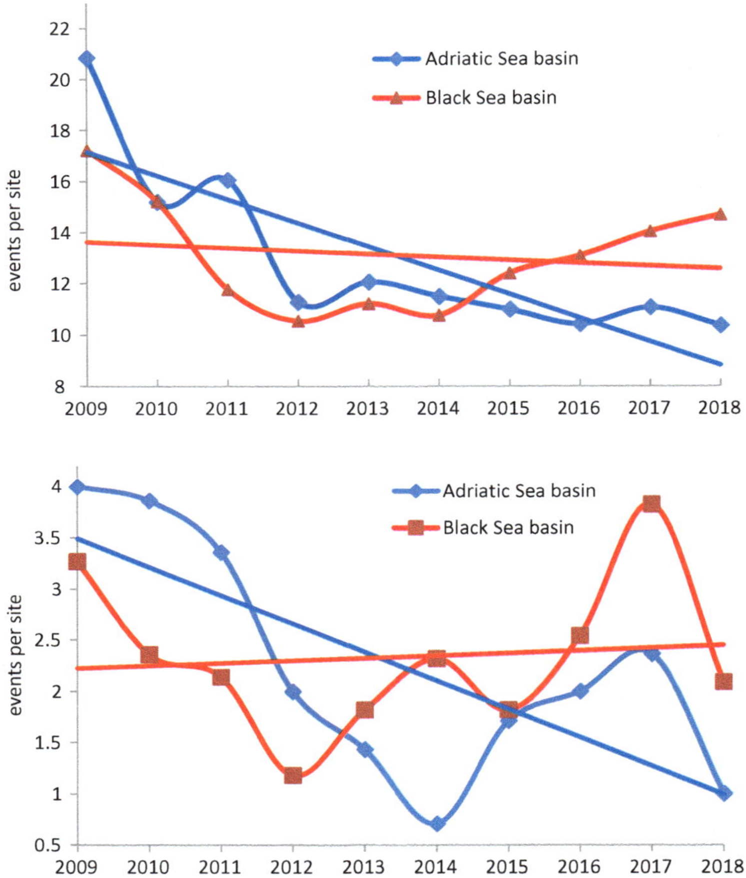

Fig. 3 Annual changes of the mean number of anomalous (upper part) and extreme (lower part) events in the Adriatic and Black Sea basins, respectively, and their liner trends

the most polluted watercourses. However, it is worth noting that the water quality index results over the last 10 years have revealed that the water quality in Montenegrin rivers has been improved since 2012.

The second reason may be because of the considerable interannual variability in the flow between the basins over the studied period [1], which likely affects the occurrence of the amplitude and frequency of both anomalous and extreme values of the monitored chemical parameters. It is know that climatic factors by changing hydrological regime and water temperature can affect the water quality by influencing the sources, migration and transformation of pollutants [28, 29].

The results of our study revealed a strong interannual variability in anomalies within the same basin even the same watercourse. The highest number of anomalous and extreme events in the Adriatic Sea basin was recorded for the period 2009–2011.

In the Black Sea basin, the largest number of anomalous events was registered in 2009 and 2010, while the largest number of extremes occurred in 2009 and 2017 when the frequency exceeded three events per site yearly. According to the Second National Communication on Climate Change of Montenegro [30], the 2001–2010 decade was the warmest since records began in 1949/1951 with the greatest changes observed in the northern mountainous region of +1.40°C and in the coastal region of +1.30°C. In the same period, the changes in precipitation become more extreme [30].

Changes in hydrological regime resulting in the lower flow in the summer months may cause the deterioration of water quality in rivers [31, 32] affecting the chemical parameters such as concentrations of phosphorus and biological oxygen demand (BOD) which are likely to increase, and the dissolved oxygen and ammonium which are likely to decrease, for the latter parameter as the result of an increased nitrification rate [29]. On the other hand, increases in water temperature induced by change of ambient air temperature lead to a reduced concentration of dissolved oxygen and consequently changes in biological processes in the rivers [29, 32] that, in turn, can affect chemical processes and water quality assessment. The analysis of water temperature anomalies shows a strong increase in the frequency of anomalous and extreme events in the Black Sea basin over a 10-year period, indicating that these changes may have an important impact on the physico-chemical equilibriums in rivers of the latter basin.

The insights gained from this study have the potential to improve our understanding of the impact of climate change and to predict its effect on river ecosystems.

References

1. Sekulić G, Radulović M (2020) The hydrology and hydrogeology of Montenegro. In: Pešić V, Paunović M, Kostianoy A (eds) The rivers of Montenegro. The handbook of environmental chemistry. Springer International Publishing AG, Cham
2. Pešić V, Paunović M, Kostianoy V (2020) The rivers of Montenegro – introductory remarks. In: Pešić V, Paunović M, Kostianoy A (eds) The rivers of Montenegro. The handbook of environmental chemistry. Springer International Publishing AG, Cham
3. Pešić V, Karaman GS, Kostianoy AG (2018) Introduction. In: Pešić V, Karaman G, Kostianoy A (eds) The Skadar/Shkodra lake environment. The handbook of environmental chemistry, vol 80. Springer International Publishing AG, Cham, pp 1–10
4. Moss B, Kosten S, Meerhoff M, Battarbee RW, Jeppesen E, Mazzeo N, Havens K, Lacerot G, Liu Z, De Meester L, Paerl H, Scheffer M (2011) Allied attack: climate change and eutrophication. Inland Waters 1:101–105
5. Vannote RL, Minshall GW, Cummins KW, Sedell JR, Cushing CE (1980) The river continuum concept. Can J Fish Aquat Sci 37(1):130–137
6. Council Directive 75/440/EEC (1975) Council Directive 75/440/EEC of 16 June 1975 concerning the quality required of surface water intended for the abstraction of drinking water in the Member States. http://eur-lex.europa.eu/legal-content/EN/TXT/PDF/?uri¼CELEX:31975L0440&from¼en. Accessed 10 Aug 2019
7. European Commission (2000) Directive 2000/60/EC, establishing a framework for Community action in the field of water policy. Off J Eur Comm Brussels L 327:1–71

8. Pešić V, Karaman GS, Kostianoy AG, Vukašinović-Pešić V (2018) Conclusions: recent advances and the future prospects of the lake Skadar/Shkodra environment. In: Pešić V, Karaman G, Kostianoy A (eds) The Skadar/Shkodra lake environment. The handbook of environmental chemistry, vol 80. Springer, Cham, pp 481–500
9. The Law on Environment (2016) Official Gazette of Montenegro No 052/16 (in Montenegrin)
10. Water Law (2007) Official Gazette of Montenegro No. 27/07, 73/10, 32/11, 47/11, 48/15, 52/16, 55/16, 02/17 (in Montenegrin)
11. The Decree on Classification and Categorization of surface and Groundwater of Montenegro (2007) Official Gazette of Montenegro No 2/07 (in Montenegrin)
12. The Decree on the National List of Environmental Indicators (2013) Official Gazette of Montenegro No 019/13 (in Montenegrin)
13. EPA (2018) Information on the state of the environment for 2017. Environmental Protection Agency of Montenegro. https://www.dropbox.com/s/28xb9kpnv5rarqw/Informacija-o-stanju-zivotne-sredine_za_2017_godinu-prosirena-verzija.pdf?dl=0. Accessed 1 Aug 2019
14. EPA (2013) Report on the state of the environment in Montenegro based on indicators. Environmental Protection Agency of Montenegro, Podgorica
15. EPA (2017) Report on the state of the environment in Montenegro based on indicators. Environmental Protection Agency of Montenegro, Podgorica
16. IJC (Institute Jaroslav Cerni) (2001) Water master plan of Montenegro. Government of Montenegro, Podgorica. (in Serbian)
17. Tomović S (2008) Ciljevi vodoprivrednog razvoja Crne Gore. Vodoprivreda 40 (231–233):127–137
18. Radulović VL (1977) The waters of the Socialist Republic of Montenegro, their pollutants and pollution. Ekologija (Acta Biologica Yugoslavica) 12(2):141–171
19. IWRM (2016) Support to water resources management in the Drina River Basin. Montenegro – IWRM Study and Plan – Background Paper. World Bank. http://www.wb-drinaproject.com/pdf/Roof%20Report%20-%20English%20-%20Final.pdf. Accessed 5 Aug 2019
20. EPA (2016) Information on the state of the environment for 2016. Environmental Protection Agency of Montenegro, Podgorica. https://epa.org.me/wp-content/uploads/2018/04/Informacija-o-stanju-zivotne-sredine-za-2016.pdf. Accessed 1 Aug 2019
21. EPA (2017) Information on the state of the environment for 2017. Environmental Protection Agency of Montenegro, Podgorica. https://epa.org.me/wp-content/uploads/2018/11/Skracena-verzija-EPA-IoSZS-_za_2017.pdf. Accessed 1 Aug 2019
22. Nikolić S (2000) Nature and tourism of Montenegro. The Republic Institute for the Protection of Nature, Podgorica
23. Purić M, Mihailović R (1983) Physico-chemical features of waters from the Tara River drainage area with special review on possibility of natural and industrial pollution. Glas Republ Zavoda Zašt Prirode – Prirodnjačkog Muzeja Titograd 16:149–159
24. EPA (2013) Information on the state of the environment for 2013. Environmental Protection Agency of Montenegro, Podgorica. https://epa.org.me/wp-content/uploads/2017/12/Informacija-o-stanju-ziv.sredine-za-2013.pdf. Accessed 1 Aug 2019
25. EPA (2015) Information on the state of the environment for 2013. Environmental Protection Agency of Montenegro, Podgorica. https://epa.org.me/wp-content/uploads/2017/12/izvjestaj-o-stanju-zivotne-sredine-za-2015.pdf. Accessed 1 Aug 2019
26. EPA (2014) Information on the state of the environment for 2014. Environmental Protection Agency of Montenegro, Podgorica. https://epa.org.me/wp-content/uploads/2017/12/Informacija-o-stanju-%C5%BEivotne-sredine-u-Crnoj-Gori-u-2014.pdf. Accessed 1 Aug 2019
27. Kostianoy AG, Serykh IV, Kostianaia EA (2018) Climate change in the Lake Skadar/Shkodra region. In: Pešić V, Karaman G, Kostianoy A (eds) The Skadar/Shkodra lake environment. The handbook of environmental chemistry, vol 80. Springer, Cham, pp 63–88
28. Delpla I, Jung AV, Baures E, Clemen M, Thomas O (2009) Impacts of climate change on surface water quality in relation to drinking water production. Environ Int 35(8):1225–1233

29. Whitehead PG, Wilby RL, Battarbee RW, Kernan M, Wade AJ (2009) A review of the potential impacts of climate change on surface water quality. Hydrol Sci J 54:101–12328
30. The Second National Communication on Climate Change (2015) The Second National Communication on Climate Change of Montenegro to the United Nations Framework Convention on Climate Change (UNFCCC). Ministry of Sustainable Development and Tourism, Podgorica. http://www.mrt.gov.me/ResourceManager/FileDownload.aspx?rid=205823&rType=2. Accessed 1 May 2019
31. Van Vliet MTH, Franssen WHP, Yearsley JR, Ludwig F, Haddeland I, Lettenmaier DP, Kabat P (2013) Global river discharge and water temperature under climate change. Glob Environ Chang 23:450–464
32. Hosseini N, Johnston J, Lindenschmidt KE (2017) Impacts of climate change on the water quality of a regulated prairie river. Water 9(3):199

Impact of Pollution on Rivers in Montenegro: Ecotoxicological Perspective

Margareta Kračun-Kolarević, Stoimir Kolarević, Jovana Jovanović, Jelena Đorđević, Marija Ilić, Karolina Sunjog, Jovana Kostić-Vuković, Aleksandra Divac Rankov, Bojan Ilić, Vladimir Pešić, Branka Vuković-Gačić, and Momir Paunović

Contents

M. Kračun-Kolarević (✉), J. Jovanović, M. Ilić, and M. Paunović
Institute for Biological Research "Siniša Stanković" National Institute of Republic of Serbia, University of Belgrade, Belgrade, Serbia
e-mail: margareta.kracun@ibiss.bg.ac.rs

S. Kolarević and B. Vuković-Gačić
Chair of Microbiology, Center for Genotoxicology and Ecogenotoxicology, Faculty of Biology, University of Belgrade, Belgrade, Serbia
e-mail: stoimirk@bio.bg.ac.rs

J. Đorđević, K. Sunjog, and J. Kostić-Vuković
Institute for Multidisciplinary Research, University of Belgrade, Belgrade, Serbia
e-mail: sunjogkarolina@imsi.rs

A. Divac Rankov
Institute for Molecular Genetics and Genetic Engineering, University of Belgrade, Belgrade, Serbia
e-mail: aleksandradivac@imgge.bg.ac.rs

B. Ilić
Institute of Zoology, Faculty of Biology, University of Belgrade, Belgrade, Serbia

V. Pešić
Faculty of Science and Mathematics, University of Montenegro, Podgorica, Montenegro
e-mail: vladopesic@gmail.com

Vladimir Pešić, Momir Paunović, and Andrey G. Kostianoy (eds.), *The Rivers of Montenegro*, Hdb Env Chem (2020) 93: 111–134, DOI 10.1007/698_2019_425,

Abstract Montenegrin surface water and groundwater are important for the Balkan Peninsula since they are connected by the transboundary Dinaric Karst Aquifer System with the waters of additional five countries. The pollution from the surface water can rapidly infiltrate in aquifer and endanger this sensible ecosystem and the health of humans through drinking water supply. This chapter gives insights in the pressures of pollution on Montenegrin waters and in a limited literature data regarding freshwater ecotoxicological studies in Montenegro. Also, this chapter provides new ecotoxicological data obtained during survey in 2019, with a focus on the sites which are identified as hotspots of fecal pollution. The highest responses of biomarkers which indicate embryotoxic, genotoxic, and phytotoxic effects in zebrafish embryo test and in roots of *Allium cepa* were obtained at Ćehotina – *downstream of Pljevlja*. Similar results were detected at the site *downstream Mojkovac* at Tara, yet this site is affected by different type of pollution. Genotoxic endpoints in zebrafish stressed out sites on Morača and Lim rivers which are under pressures of fecal pollution. The data in this chapter provides an insight into current status obtained by the ex situ bioassays and indicates need for more comprehensive in situ assessment.

Keywords *Allium cepa* root tip assay, Ecotoxicology, FET, Montenegro, Pollution, Rivers

1 Introduction

Montenegro is a Dinaric and Adriatic-Mediterranean country with specific geological and hydrological characteristics. More than 60% of its territory is made of karstified rocks [1, 2] mainly carbonated. Montenegro shares the Dinaric Karst Aquifer System with Italy, Slovenia, Croatia, Bosnia and Herzegovina, and Albania. This Aquifer is one of the largest in Europe and one of the most international ones being transboundary between six countries. One of the highest rainfall rates in Europe is recorded in Montenegro (>2,000 mm annually in the southern part and about 5,000 mm in Boka Kotorska Bay), yet 50–80% of the precipitation infiltrates into the Dinaric Karst Aquifer [3]. High precipitation rate makes this region one of the richest in Europe in water resources which are unequally distributed throughout the year [4].

The quality of surface water and groundwater in Montenegro is threatened by various sources such as insufficiently treated wastewaters, intense mining (lead-zinc, bauxite, barite), quarry, intensive tourism, different water activities on rivers (rafting, kayaking, fly-fishing), presence of active agriculture in river valleys, etc.

Due to geological characteristics of carbonated karst (high porosity), polluted surface water (pollutants from mines, industrial and domestic wastewaters, fertilizers, pharmaceuticals, etc.) can enter rapidly into the Dinaric Karst Aquifer System and deteriorate water quality at a larger geographic magnitude. Since drinking water supply relies on groundwater not only in Podgorica, capital of Montenegro, but also in Tirana (capital of Albania) and Sarajevo (capital of Bosnia and Herzegovina), as well as many other Adriatic coastline cities, the presence of pollutants in groundwater presents danger to humans and to specific aquatic ecosystems in Dinaric Karst region.

The importance of the waters of Montenegro, surface water and groundwater, for the Dinaric region is obvious, yet there are only a few studies dealing with the effects of pollution on the aquatic biota which inhabits these waters.

This chapter represents an overview of ecotoxicological studies related to the rivers of Montenegro. Additionally, an insight in the current condition at the sites which are under the highest pollution pressure is provided by ecotoxicological ex situ experiments with native water samples from five Montenegrin rivers.

2 Rivers of Montenegro

Waters of Montenegro belong to two watersheds, the Black Sea and the Adriatic Sea watershed. The Black Sea drainage is the larger one covering 52.5% (7,260 km^2) of the territory of Montenegro, while 47.8% (6,650 km^2) belongs to the Adriatic watershed.

The southernmost part of *the Black Sea drainage* in Montenegro consists of 48 inland surface water bodies (rivers and lakes). The Montenegrin Black Sea drainage is formed by two watersheds. Rivers Lim, Piva, Tara, and Ćehotina belong to the Drina River Basin (flowing into the Sava River), while the Ibar River flows into the Zapadna Morava. *The Lim River* emerges from the Plavsko Lake, and 98 km of its length belongs to Montenegro with a drainage area of 2,280 km^2. It is a transboundary river shared by Serbia, Montenegro, and Bosnia and Herzegovina. This river has many tributaries, Murinska, Zlorečica, Đurička, Rženička, Velička, Komarača, Krašica, Trebić, Ševarinska, Bistrica, etc. *The Tara River* springs on the Maglić. The right bank of the Tara River is better developed than the left one. There are many tributaries of this river: Opasanica, Pčinja, Plašnica, Ravnjak, Ljutica, Drcka, Skrbuša, Jezerštica, etc. The length of the Tara River is 148 km and the catchment area is 2,040 km^2. *The Piva River* is formed in the high mountains of Durmitor. Along its stream it has several names: in the upper part it is called

Bukovica, and then in Šavnik it is called Pridvorica. After the confluence with river Upper Komarnica, again, it changes name to Komarnica until Pivski Monestery. After the monastery it is called Piva. Together with Tara, Piva forms the Drina River. In 1978, the accumulation on Piva was formed (4 km long, 180 m deep) by the construction of the 220-m-high dam 10 km upstream from Šćepan Polje. The Piva River basin has a length of 85 km to Šćepan Polje and an area of 1,784 km^2. *The Ćehotina River* springs below the Strožer Mountain. It is one of the largest tributaries of the Drina River. Significant tributaries of the Ćehotina River are Koričić, Maočnica, Vezišnia, and Voloder. Open coal pit mine Pljevlja is situated in its vicinity. The length of Ćehotina River within the Montenegro is 99 km, and the drainage area is 809.8 km^2. *The Ibar River* springs on the Hajla Mountain. The main tributaries are Ibarac, Županica, Limnička, Grahovska, Bukovačka, Baltička, Crnja, and Bačka. Specific hydrographic characteristics of the Ibar River allow rapid formation of flood waves. The length of the Ibar River within the Montenegro is 35 km, while the drainage area is 413 km^2.

The Adriatic drainage is made of 41 inland surface water bodies (rivers and lakes). The larger representatives in the Adriatic drainage of Montenegro are the Zeta, Morača, and Bojana rivers. Also there are many smaller rivers like Orahovštica, Crmnička Rijeka, Sutorina, Sjevernica, Mrtvica, Nožica, Mala Rijeka, Sušica, Gračanica, Ribnica, Matica, Sitnica, Cijevna, Rijeka Crnojevića, etc. The majority of these rivers are flowing into the Skadar Lake and by the Bojana River, which makes the connection to the sea, drain into the Adriatic Sea. *The Zeta River* is formed by joining two rivers, Sušica and Rastovca. A part of its riverbed is modified, and the water is transported by tunnel to the hydropower plant "Perućica". It is the right tributary of the Morača River and the most significant one. The length of the Zeta River is 85 km. *The Morača River* springs under the Rzača Mountain. It has many tributaries: Koštanica, Sjevernica, Javorski Potok, Slatina, Ibrištica, Ratnja, Mrtvica, etc. Before flowing into the Skadar Lake, the Morača River flows through the Zeta plain downstream of Podgorica. The length of the Morača River is 113 km. *The Bojana River* springs from the Skadar Lake and makes the connection with the sea. It presents a natural border between Montenegro and Albania. The Bojana River is ranked third in the Mediterranean by the amount of water which is released in the Adriatic Sea. The length of the Bojana River is 41 km. In regard to its hydrology and ecology, it is a very unique river. Smaller but important tributaries of the Skadar Lake are rivers Crnojevića, Orahovštica, and Plavnica and river Kiri in Albania. *The Skadar Lake* is the biggest lake on the Balkans. It covers an area about 400 km^2 at low waters and 525 km^2 at high waters. It is the most important water body in Montenegro regarding hydrology, economy, water management, and tourism.

3 Pressures of Pollution in Rivers of Montenegro

Water resources classify Montenegro as a country rich in waters, yet due to the geological structure, surface watercourses transfer part of their water to groundwater reservoirs. Aquatic biotas that inhabit most of the surface watercourses in Montenegro live in challenging environmental conditions. Specific environmental conditions are caused by considerable differences in water level during the year, from abounding waters during high waters in spring to dry riverbeds during low waters in summer. Natural processes that can lead to a deterioration of aquatic environments are soil and water erosions which occur relatively often in Montenegro. Soil erosion, besides the degradation of land, causes degradation by sedimentation in water bodies and can be a source of pollution by introducing pollutants present in land. Besides these natural pressures, quality of freshwater ecosystems in Montenegro is affected by various anthropogenic activities – tourism, traffic, chemical and pharmaceutical industry, mines and quarries, agriculture and intensive livestock, meat and food industry, wood and stone processing, etc.

The northern and central region of Montenegro, the part that belongs to the Black Sea drainage, has 177,837 inhabitants (28.6% of the total population), while the southern region, the part belonging to Adriatic drainage, has 442,193 inhabitants (71.4% of the total population) [5]. Discharge of wastewaters (municipal and industrial) directly into the surface water bodies through sewage systems is the major point source of pollution in Montenegro. Releasing insufficient treated or untreated wastewaters into aquatic environments is the main way of introduction of various xenobiotic (pharmaceuticals, personal care products, metals, pesticides, veterinary products, etc.) [6–8] with adverse effects (mutagenic, carcinogenic, toxic, morphological malformations, histopathological alterations, etc.) [9–13] in the freshwater ecosystems. In the north, only 3 cities have more than 10,000 people (Bijelo Polje, Pljevlja, and Berane) without wastewater treatment plants. Bijelo Polje and Berane belong to the Lim River Basin, while Pljevlja belongs to Ćehotina River Basin. Three municipalities with active wastewater treatment plants are Mojkovac, Žabljak, and Šavnik. In northern and central Montenegro, pollution has an origin in industry, while the major sources are mines, quarries, transport, bakeries, intensive livestock, meat processing, food industry, metallurgy, chemical and pharmaceutical industry, wood-paper processing, construction, and electricity production [14]. Pljevlja, Plužine, Žabljak, Bijelo Polje, Mojkovac, Kolašin, Berane, Rožaje, and Šavnik are cities affected by industrial pollution [14].

The capital of Montenegro, Podgorica, is situated in the Adriatic River Basin of Montenegro. With 156,200 inhabitants (25% of the entire population of Montenegro), Podgorica is the largest city in Montenegro. In Podgorica and Nikšić (the second largest city by number of inhabitants – 57,290), 41.6% of the total population of Montenegro is settled. There are 7 more cities with more than 10,000 inhabitants, Herceg Novi, Bar, Budva, Cetinje, Kotor, Ulcinj, and Tivat. Taking into account the intensive tourism at the seaside in summer, this region of Montenegro has two groups of population, the permanent and the occasional population. The most important economic activity in Montenegro is tourism. According to MONSTAT (Statistical Office of Montenegro), the number of tourists in 2017 was about 2,000,000 people. As a consequence the main pollution pressures are exerted during summer when water consumption demand is elevated and when larger quantities of waste and wastewater are produced. Cities with wastewater treatment plants are Bar (Virpazar), Budva, Jaz, Cetinje, Herceg Novi, Nikšić, Podgorica, and a mutual one for Tivat and Kotor. The main enterprise types that contribute to the pollution of rivers in the Adriatic River Basin are bauxite production, meat processing, paper and cardboard production, aluminum production, bakery products, chemical industry (hygiene products, pharmaceutics), vineyards, fish, fruits and vegetable processing, intensive livestock, brewery, metal construction, and stone and wood processing. In the most cases, recipient rivers that endure most of these pressures are Morača, Zeta, Cetinje, and Rijeka Crnojevića.

4 Overview of the Ecotoxicological Literature Related to Freshwater Ecosystems in Montenegro

Table 1 summarizes available literature data on ecotoxicological studies related to freshwater ecosystems in Montenegro. Most of the studies are related to the Skadar Lake and the most significant tributaries. In general, available literature data can be divided into two groups. The first group is formed of ecotoxicological studies focused mainly on the sediment extracts covering estrogen activity, toxicity, embryotoxicity, cytotoxicity, genotoxicity, and mutagenicity. The study of Rastal et al. [15] indicated presence of estrogenic activity at five out of six investigated sites at the Skadar Lake pointing to contribution of pollution derived through the significant tributaries to the status of the Lake. Mouths of the rivers Morača and Plavnica were included in the study. In the study of Stesevic et al. [16], toxic effects (on *Myriophyllum aquaticum* and *Lemna minor*) were detected for sediment extracts again from the sites at the mouth of Morača and Plavnica. Additionally, genotoxic and mutagenic potential of the sediment extracts from the mouth of the Morača River was demonstrated in the study of Perović et al. [17]. The same group confirmed mutagenic potential of the sediment extracts collected at the mouths of Rijeka Crnojevića in Perović et al. [18]. Cytotoxic and embryotoxic potential was detected for the same samples in the study of Perović et al. [19]. The study of active biomonitoring was performed by Perović et al. [20] at several sites at the Morača River, starting from the outlet of WW in Podgorica further downstream till it reaches the mouth of the river in Skadar. Study was performed on *Unio pictorum*, and the results indicated a gradual increase of genotoxic potential going from the site in Skadar Lake, whereas the highest genotoxicity was detected in the specimens exposed to the site situated downstream of the outlet.

The second group is formed by the studies which are dealing with heavy metal accumulation in biota/sediment. Seasonal changes in metal accumulation and

Table 1 Overview of the ecotoxicological studies related to freshwater ecosystems in Montenegro

Study	Organism	Endpoint	Medium	Location	Type of study
Rastal et al. [15]	RTL-W1, *Saccharomyces cerevisiae*, *Salmonella typhimurium* TA98/Ta100	Estrogenic and mutagenic potential	Water	Skadar Lake and mouth of tributaries Morača, Plavnica, Rijeka Crnojevića, Raduš	In vitro
Stesevic et al. [16]	*Myriophyllum aquaticum*, *Lemna minor*	Toxicity	Sediment	Skadar Lake and mouth of Morača and Plavnica	In vitro
Perović et al. [17]	RTL-W1, *Salmonella typhimurium* TA98/TA100	Genotoxicity, mutagenicity	Sediment	Skadar Lake and mouth of Morača	In vitro
Perović et al. [18]	*Salmonella typhimurium* TA98/TA100	Mutagenicity	Sediment	Skadar Lake and mouth of tributaries Morača, Raduš, Plavnica, Rijeka Crnojevića	In vitro
Perović et al. [19]	*Danio rerio*, RTL-W1	Embryotoxicity, cytotoxicity	Sediment	Skadar Lake and mouth of Morača	In vitro
Petrović et al. [23]	*Trapa natans*	Accumulation of heavy metals	Sediment biota	Skadar Lake and mouth of tributaries Morača River, Rijeka Crnojevića, Plavnica	In situ
Perović et al. [20]	*Unio pictorum*	Genotoxicity	Biota	Morača River and its mouth in Skadar	In situ
Mesi et al. [24]	*Allium cepa*	Cytotoxicity Genotoxicity	Water	Skadar	In vitro
Kastratovic et al. [21]	*Phragmites australis*	Accumulation of heavy metals	Biota	Skadar Lake and mouth of tributaries Morača, Plavnica, Rijeka Crnojevića, Raduš	In situ

(continued)

Table 1 (continued)

Study	Organism	Endpoint	Medium	Location	Type of study
Kastratović et al. [22]	*Phragmites australis*, *Ceratophyllum demersum*, *Lemna minor*	Accumulation of heavy metals and its seasonal variation in tissue	Water and sediment samples	Skadar Lake tributaries Morača, Plavnica, Rijeka Crnojevića, Raduš	In situ
Rakočević et al. [25]	*Scardinius knezevici*, *Alburnus scoranza*, *Cyprinus carpio*, *Rutilus prespensis*, *Anguilla anguilla*, *Perca fluviatilis*	Accumulation of heavy metals	Biota	Skadar Lake, mouth of Morača and Rijeka Crnojevića	In situ
Vukašinović-Pešić et al. [26]	*Viviparus mamillatus*	Accumulation of heavy metals	Biota	Zeta River, Matica River, Skadar Lake	In situ
Vukašinović-Pešić and Blagojević [27]	Macrophytes, molluscs, fish	Accumulation of heavy metals – review	Biota	Skadar Lake	In situ

distribution in the organs of *Phragmites australis* (common reed) from Skadar Lake were observed in the study of Kastratović et al. [21] where general contamination was noticed at the sites situated close to the confluence of the major tributaries. Observations were confirmed on *Phragmites australis*, *Ceratophyllum demersum*, and *Lemna minor* by Kastratović et al. [22]. The study of Petrović et al. [23] performed on *Trapa natans* indicated that Morača and Rijeka Crnojevića are important pathways for import of heavy metals in the Skadar Lake.

5 Status Quo at the Sites Under the Highest Pollution Pressure

5.1 Samples and Sites

Looking at the overview of the literature, it is evident that most of the studies are related to the Skadar Lake and its tributaries which leaves a significant gap in knowledge for the rest of Montenegro. Led by the results of the study of Kolarević et al. [28], we decided to focus our investigation at the sites on the rivers which are under the highest impact of pollution by wastewaters. Within the Montenegro survey in 2019, water samples were collected from 25 sites covering watersheds from both Adriatic and Black Sea basins. Microbiological indicators were used to assess the level of fecal pollution at the sites, and hotspots of pollution were identified combining obtained data and data from the national monitoring program. Additional water samples (200 mL) for ecotoxicological analyses were collected from each site, immediately frozen and stored at −20°C. Samples were preliminarily screened with the fish embryo test, and the most potent ones were used for more comprehensive studying.

The major focus of the study was on five sites. Besides the site downstream of the wastewaters outlet in Podgorica on the Morača River which was an obvious choice, Ćehotina caught our attention as this river is known as one of the hotspots of pollution originating from the town Pljevlja and nearby situated coal mine and thermal power plant [29–31]. The site situated at the Lim River downstream of Bijelo Polje was chosen as this site was marked as frequently polluted by wastewaters [28].

5.2 *Biomarkers and Bioassays*

Biomarkers are defined as detectable biochemical and tissue-level changes that indicate altered physiology of organism resulting from an agent [32–34]. The best approach to evaluate early response to pollution is to identify the best biomarker or sets of biomarkers, rather than investigate the whole animal response [33–35]. Biomarkers have been classified by the extent that they reflect (1) exposure to environmental stressors, or (2) adverse health effects from contaminant exposures, while some biomarkers can also indicate (3) susceptibility to adverse outcomes from environmental contaminants [34, 36, 37].

In the current study, we decided to use endpoints which point toward the deterioration of the ecosystem quality and also provide insight into harmful effects from the aspects of fisheries and irrigation. Moreover this study included assessment of embryotoxicity and genotoxicity in zebrafish (*Danio rerio*) model and assessment of phytotoxicity in *Allium cepa* model.

5.3 *Embryotoxicity and Genotoxicity in Zebrafish (*Danio rerio*)*

5.3.1 Fish Embryo Test (FET) with the Zebrafish

Under the guidance of the Water Framework Directive [38], fish are used in the routine monitoring of the quality of effluents and surface water. Fish embryo test (FET) with the zebrafish (*D. rerio*) has shown high potential in the assessment of toxicity of wastewater, sediment, and various chemicals [39–42]. Test was originally designed as a substitute for acute fish toxicity testing [43] and later was standardized as ISO test 15088 [44]. Zebrafish eggs, embryos, and adults have shown high sensitivity to a large range of environmental pollutants, and since nonfeeding developmental stages are not characterized as protected stages [45], FET can easily be used for the scientific purposes regarding many national and international regulations [42]. Zebrafish is a cyprinid species whose development from zygote to adults is well-known. Therefore, developmental, physiological, and gene expression and behavioral information/responses gained in the environmental studies make them preferable for toxicological researches [46].

5.3.2 Exposure and Methodology

Zebrafish (wild type; Tübingen) was cultivated in conditions described in Babić et al. [47]. Briefly, zebrafish were maintained at 28 ± 1°C, 12 h light/12 h dark photoperiod, in water with the following chemical characteristics: 5.5 mg KCl 294 mg $CaCl_2{\cdot}2H_2O$, 123 mg $MgSO_3{\cdot}7H_2O$, and 62 mg $NaHCO_3$ per L (ISO water). Every 7 days fish were spawned, and the experiments were conducted if the rate of fertilized eggs was >90%. The experiments were performed in 24-well plates. Ten embryos were placed in each well in 2 mL of native water samples from seven sampling sites on five Montenegrin rivers (Lim, Tara, Morača, Zeta, and Ćehotina). Each water sample was tested in triplicate. As a negative control, ISO water was used. Treatments started at 6 h post-fertilization (hpf).

The embryos were observed under the stereomicroscope Stemi 508 (Carl Zeiss Microscopy, Gottingen, Germany) at 32× and 40× magnification and imaged with camera AxioCam Erc 5s (Carl Zeiss Microscopy, Gottingen, Germany).

Biomarkers such as coagulation, non-detached tail, and lack of somites were assessed as lethal endpoints at 24 hpf and 48 h, while at 72 hpf lack of heartbeat was considered lethal [41]. Embryo hatching rates were assessed at 48 and 72 hpf since hatching in zebrafish normally occurs at this period. Developmental malformations (head malformation, eye and body pigmentation, tail malformation, scoliosis, yolk morphology, heart malformation, pericardial edema, tail circulation, growth retardation) were assessed at 72 hpf.

5.3.3 Results

Significantly higher value of mortality in the FET test was detected when comparing control group of embryos and embryos treated with water from sampling sites *ds Bijelo Polje* (Lim), *ds Mojkovac* (Tara), *ds WW outlet Podgorica* (Morača), *ds Pljevlja* (Ćehotina), and *ds Vezišnica* (Ćehotina) (Fig. 1). The water samples *Vidrovan* (Zeta) and *Rabitlje* (Ćehotina) did not induce significant increase in mortality in comparison with control water. At 24 hpf hatching was observed in the control group of embryos (28.6%), while in the case of treated embryos, there

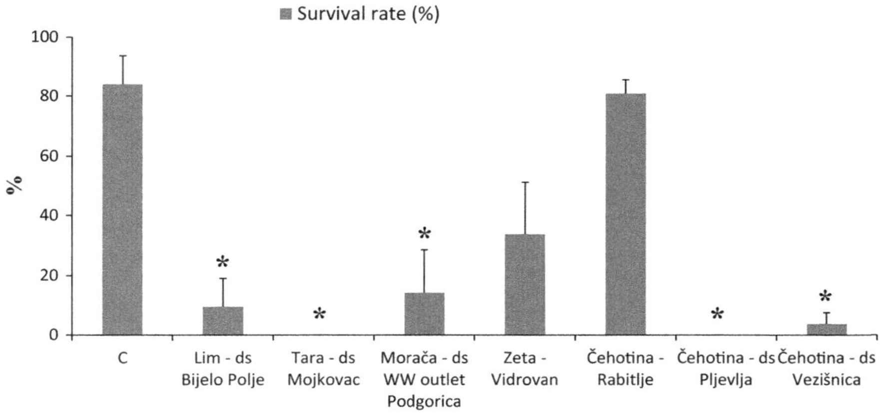

Fig. 1 Survival rates 72 h after exposure to native water samples. Results are presented in percent ± SE; ∗ statistical significance obtained by *t*-test, $p < 0.05$

Table 2 Sublethal endpoints at 72 h hpf for embryos in the control sample and water samples from seven sites on the Lim, Tara, Morača, Zeta, and Ćehotina rivers

Sampling sites	Voluminous yolk sac (%)	Scoliosis (%)	Eye and body hypopigmentation (%)	Body length (mm)
Control	13.9	16.67	13.89	2.93
Lim – *ds Bijelo Polje*	50	0	50	3.00
Tara – *ds Mojkovac*	nh	nh	nh	nh
Morača – *ds WW outlet Podgorica*	100	33.33	100	2.84
Zeta – *Vidrovan*	62.5	12.50	50	2.94
Ćehotina – *Rabitlje*	47.78	16.67	17.78	2.97
Ćehotina – *ds Pljevlja*	nh	nh	nh	nh
Ćehotina – *ds Vezišnica*	100	0	100	2.74

nh – absence of hatched embryos at 72 hpf

was an absence of hatching (0%). At 72 hpf hatching was totally blocked at two sampling sites – *ds Mojkovac* (Tara) and *ds Pljevlja* (Ćehotina).

Developmental malformations observed in treated embryos are summarized in Table 2, while representative images are provided in Figs. 2 and 3. Enlargement of yolk sac and presence of scoliosis were assessed in hatched embryos. Voluminous yolk sacs were detected in all groups exposed to water samples as well as in the control group of embryos. Scoliosis was absent in embryos exposed to water samples from sampling sites: *ds Bijelo Polje* (Lim) and *ds Vezišnica* (Ćehotina). Scoliosis was also present in the control embryos (16.67%). Hypopigmentation of body and eyes was assessed in hatched embryos at 72 hpf. It was observed that embryos with hypopigmentation of body also lack the pigmentation in eyes. The

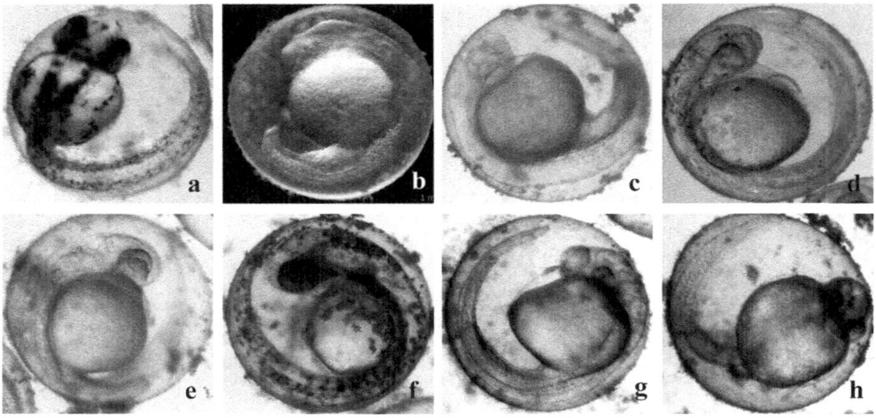

Fig. 2 Representative images of embryos at 24 hpf; embryos were exposed to native water samples from five rivers in Montenegro; (**a**) control group of embryos; (**b**) embryos exposed to water sample from Lim – *ds Bijelo Polje*; (**c**) embryos exposed to water sample from Tara – *ds Mojkovac*; (**d**) embryos exposed to water sample from Morača – *ds WW outlet Podgorica*; (**e**) embryos exposed to water sample from Zeta – *Vidrovan*; (**f**) embryos exposed to water sample from Ćehotina – *Rabitlje*; (**g**) embryos exposed to water sample from Ćehotina – *ds Pljevlja*; (**h**) embryos exposed to water sample from Ćehotina – *ds Vezišnica*

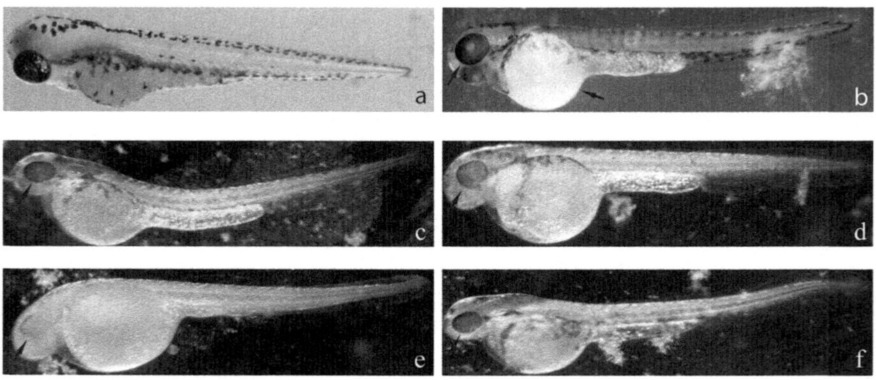

Fig. 3 Representative images of hatched embryos at 72 hpf; embryos were exposed to native water samples from five rivers in Montenegro; (**a**) control group of embryos; (**b**) embryos exposed to water sample from Lim – *ds Bijelo Polje*; (**c**) embryos exposed to water sample from Morača – *ds WW outlet Podgorica*; (**d**) embryos exposed to water sample from Zeta – *Vidrovan*; (**e**) embryos exposed to water sample from Ćehotina – *ds Vezišnica* (only one hatched embryo); (**f**) embryos exposed to water sample from Ćehotina – *Rabitlje*; arrows indicate developmental malformations

highest rates of hypopigmentation (100% – lack of pigmentation in all hatched embryos) were detected in groups exposed to water samples from Morača, *ds WW outlet Podgorica*, and Ćehotina – *ds Vezišnica*. These two sublethal endpoints were also present in control embryos yet in lower rate (13.89%). Body length in treated groups of embryos showed no statistically significant difference when compared to control.

Other developmental malformations (head malformation, heart malformation, pericardial edema, tail circulation) were not observed in treated embryos.

5.4 *Genotoxicity in Zebrafish: Comet Assay*

Comet assay, also known as the single-cell gel electrophoresis (SCGE), is a fast and reliable method which allows the detection of DNA damage in individual cells [48]. Currently, this is one of the most commonly used tests in genotoxicology which is based on quantification of negatively charged DNA fragments which in electrophoresis move through the gel toward a positively charged cathode. Nuclei with fragmented DNA material form comet-like shapes which determined the name of the assay. In our previous research, we have adopted this method and use it in various models starting from mammalian cell lines [49, 50] marine mussels [51, 52], freshwater mussels [54–57], freshwater fish [13, 58–64], caterpillars [65], and freshwater oligochaetes [66, 67].

In the current study, we decided to assess genotoxic potential using zebrafish embryos for several reasons. The zebrafish embryo test has been shown to be a valuable alternative for acute toxicity testing, and in contrast to cellular replacement methods, such as fish cell, the embryo model offers a complex, multicellular system integrating the interaction of various tissues and differentiation processes [68]. Comet assay has been successfully adopted for zebrafish embryos. Jarvis et al. [69] applied comet assay on zebrafish larvae to evaluate the level of DNA damage induced by γ-radiation. Kosmehl et al. [70–72] developed a novel contact assay for testing genotoxicity of sediments employing comet assay on zebrafish embryos. Eleršek et al. [73] developed a protocol for preparation of the comet assay slides by simple squashing of the 24-h-old embryos embedded in agarose.

5.4.1 Exposure and Methodology

In preliminary experiments we have tested two different protocols. Protocol by Eleršek et al. [73] was primary selected as the more rapid method which provides assessment of genotoxicity in a single embryo (Fig. 4a). However, the disadvantage of this protocol was the randomly occurring interaction of the yolk and agarose which made intense background noise on the slides (Fig. 4b). Protocol described by Hollert et al. [74] provided a satisfactory number of cells per slide by using five embryos per group (Fig. 4c). Briefly, after 48 h of treatment, chorion was removed mechanically with tweezers, and embryos were transferred to Eppendorf tube in 50 μL of water. Afterward, 50 μL of 0.1% trypsin was added, and embryos were in parallel with enzymatic treatment squashed mechanically with a spatula. After 5 min

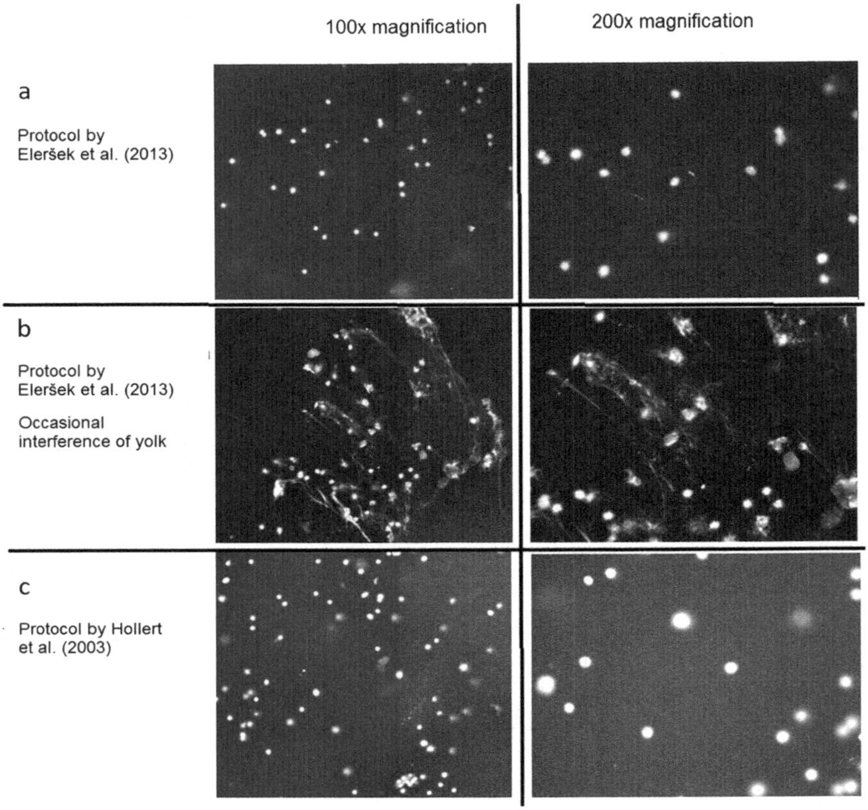

Fig. 4 Comparative display of two methods used for preparation of the slides for comet assay

of trypsinization, 1 mL of 10% fetal bovine serum in 1× PBS was added to the tube. Tubes were centrifuged for 5 min at 1500 rpm. Pellets were eluted in 100 μL of 1× PBS and used for comet assay. Cell viability assessed by differential acridine orange/ ethidium bromide staining [53] was above 80% in all groups.

5.4.2 Results

For the assessment of the genotoxic potential, the same sites used for embryotoxicity were selected. The highest level of DNA damage was recorded in embryos exposed to the water collected from Morača (*ds WW outlet Podgorica*). All samples with the exception of the sample from Zeta induced significant increase of DNA damage. The site *Rabitlje* was chosen to investigate the genotoxic potential of Ćehotina upstream and downstream of Pljevlja. However, significant difference among the sites on Ćehotina was not observed (Fig. 5).

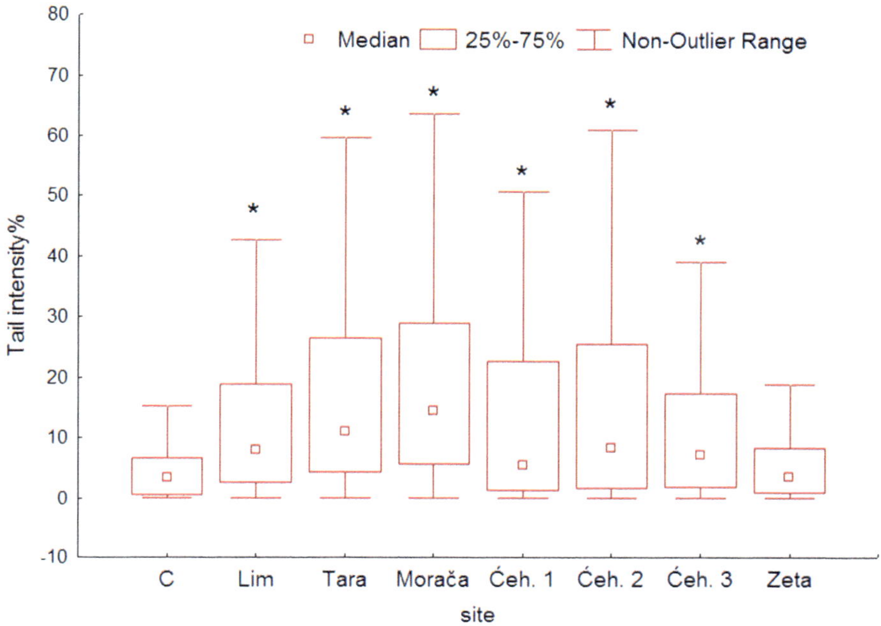

Fig. 5 The level of DNA damage assessed by tail intensity % in cells of zebrafish embryos exposed to water collected from Lim (*ds Bijelo Polje*), Tara (*ds Mojkovac*), Morača (*ds WW outlet Podgorica*), Ćehotina (1-*Rabitlje*, 2-*ds Pljevlja*, 3-*ds Vezišnica*), and Zeta (*Vidrovan*); * significant difference in comparison with control $p < 0.05$ (Mann-Whitney U test)

5.5 Allium cepa *Root Tip Assay*

Allium cepa root tip assay was first introduced by Levan in 1938 [75] as a convenient test to examine the effect of colchicine on cells. The advantages of this test are simplicity, low cost, speed, and sensitivity. It is recommended as a standard in monitoring of the wastewater status and river water ecosystems as part of a test battery whose positive results should be considered as a risk indicator to human health [76, 77]. The test can also be used to prevent and predict the environmental effects of drugs and herbicides [78]. An additional advantage of the test is that it can be used to test wastewater quality without the necessary prior preparation such as purification treatment or sterilization [77]. The sensitivity of the test shows good correlation with the Ames test, the mammalian test system, the human lymphocyte test system, and carcinogenicity tests in rodents [76, 79, 80], which speaks in favor of applying the test without sacrificed or used animals [81]. The assay provides several toxicological endpoints such as phytotoxicity, cytotoxicity/proliferation inhibition, genotoxicity, and mutagenicity [82].

5.5.1 Exposure and Methodology

Exposure was carried out by the protocol described in Vujošević et al. [83] with slight modifications. Briefly *A. cepa* bulbs were placed on test tubes containing clean water with known origin (local bottled water) for at least 2 days until the new roots on the bulbs were formed. Rooted onions (five per group) were transferred on the tubes containing water samples from selected sites, and treatment was performed for 72 h [84]. For negative control, onions were exposed to commercial bottled water and for positive control to solution of cisplatin (33 μM). After treatment, length of roots on each bulb ($n = 10$) was measured to the nearest mm. Root tips were cut with clean scalpel and transferred to tubes with freshly prepared ice-cold Carnoy's solution. After 24 h, fixed root tips were transferred to ice-cold 70% ethanol and stored upon analyses. At the day of analysis, root tips were transferred to 1 M HCl and incubated for 5 min at 60°C. The roots were carefully tweezed into distilled water tubes for 2 min, then transferred to a previously prepared solution of 10 μg/mL acridine orange, and incubated for 2 min. After staining, the roots were transferred to distilled water to remove excess dye and tweezed onto microscopic plates where only the tips of the roots were dissected (2–3 mm from the top) to remove cells that were not in division and to analyze only the meristematic cells and daughter cells of the F1 generation. A drop of distilled water was added to the tops of the roots, and cover glasses were placed, which were then pressed using a block of absorption paper to remove excess water. The preparations thus prepared were observed on a 400× fluorescence microscope (Leica DMLS, Austria). A total of 1,000 cells per microscope plate (5,000 cells per treatment/sample) were analyzed, of which the number of cells in mitosis (Fig. 6) was determined for evaluating cytotoxic potential [85]. The mitotic index was calculated using the following formula [78, 82]: MI = number of cells in mitosis/1,000 cells examined.

5.5.2 Results

For *A. cepa* test, we have selected four sites (Lim, *Dobrakovo*; Morača, *ds WW outlet in Podgorica*; Ćehotina, *ds Pljevlja* and *ds Vezišnica*) which are under evident impact of wastewaters and which were indicated as hotspot of pollution in the study of Kolarević et al. [28]. Additionally, the site *Vidrovan* on Zeta was selected as the site which is under minimal impact of pollution and can be considered as pristine. The site on the Tara River – *ds Mojkovac* – was included as significant effects were observed in the experiments on zebrafish described previously. In comparison with control, significant decrease in root length was detected for the sites Tara, *ds Mojkovac*, and Ćehotina – *ds Pljevlja*. It is interesting that the increase of mitotic index was observed for the sites Lim, *Dobrakovo*; Morača, *ds WW outlet*; and Ćehotina, *ds Vezišnica* (Table 3). Despite the high impact of wastewaters at these three sites, phytotoxic and cytotoxic effect is not present. Moreover, it seems that high nutrient load at these sites even stimulates the growth of the roots [86, 87].

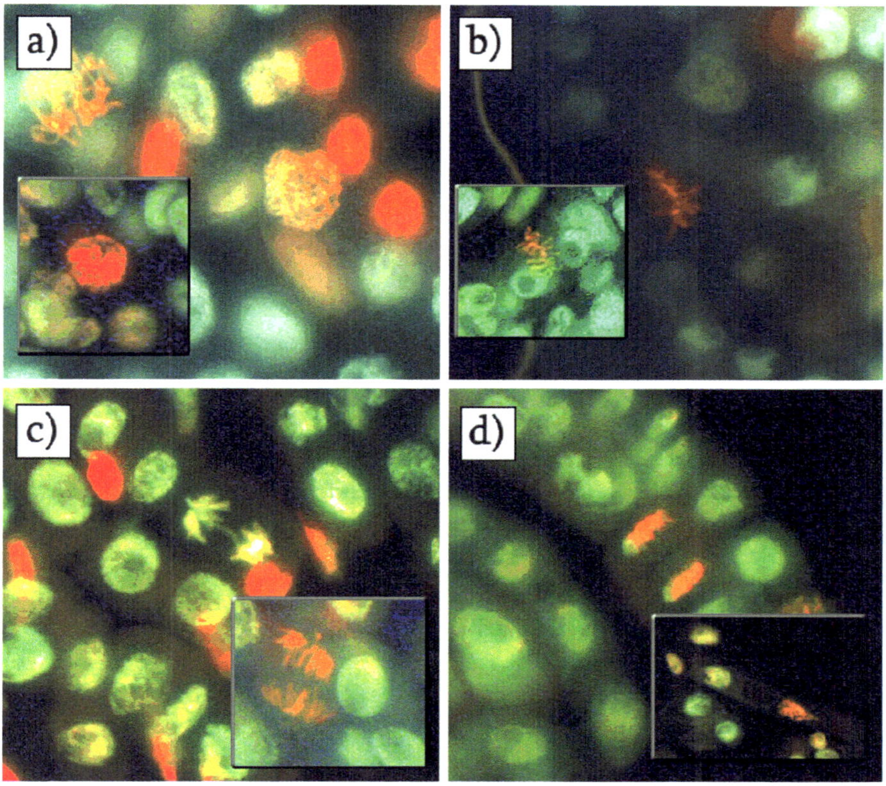

Fig. 6 Meristematic cells in *A. cepa* root tips in mitosis: (**a**) prophase, (**b**) metaphase, (**c**) anaphase, (**d**) telophase

Table 3 Values of mitotic index and root length for *A. cepa* bulbs exposed to the native water samples

		MI ± SD (%)	Root length ± SD (mm)
Negative control		5.5 ± 2.4	22.4 ± 6.7
Positive control		1.5 ± 0.4[a]	7.7 ± 2.5[a]
River	Site		
Zeta	*Vidrovan*	3.5 ± 0.5	21.0 ± 4.5
Tara	*ds Mojkovac*	NA	14.8 ± 4.8[a]
Lim	*Dobrakovo*	9.5 ± 3.0[a]	22.6 ± 5.3
Morača	*ds WW outlet*	15.2 ± 4.8[a]	23.7 ± 5.2
Ćehotina	*ds Pljevlja*	5.2 ± 1.1	14.4 ± 3.3[a]
Ćehotina	*ds Vezišnica*	12.5 ± 4.6[a]	27.8 ± 10.3

[a]Significant difference in comparison with negative control (*t*-test, $p < 0.05$)

6 Conclusions

This chapter provides glance into the ecotoxicological status of the rivers in Montenegro with a focus on the sites which are heavily impacted by wastewaters. Among the sites identified as hotspots of fecal pollution, the site situated *downstream of Pljevlja* at Ćehotina is the most critical one, as the data obtained in used models – embryos of zebrafish and roots of *A. cepa* indicated phytotoxic, embryotoxic, and genotoxic effects. It is interesting that the site situated at Tara *ds Mojkovac* showed similar results having in mind that this site is under impact of a different type of pollution. Impact of wastewaters on Morača and Lim was evident through the genotoxic endpoint but also through the various developmental parameters monitored in embryos. Data presented in this study provides an insight into the status quo at the selected sites, and more comprehensive in situ assessment is needed for proper evaluation of the harmful effects on the ecosystem at these sites.

Acknowledgment The authors are grateful to Luka Gačić who provided improvements to our English.

References

1. DIKTAS (2012) Protection and sustainable use of the Dinaric Karst transboundary aquifer system – Montenegro country report
2. Stevanovic Z, Kukuric N, Treidel H, Pekas Z, Jalovic B, Radojevic D, Pambuku A (2012) Characterization of transboundary aquifers in Dinaric karst – a base study for sustainable water management at regional and local scale. IAH Congress, Niagara Falls Sept
3. Stevanović Z, Pekaš Ž,Jolović B, Pambuku A, Radojević D (2014) Classical Dinaric Karst aquifer – an overview of its past and future. The international conference and field seminar "Karst Withut Boundaries", extended abstract of plenary paper, pp 1–6
4. Bonacci O (1987) Karst hydrology; with special reference to the Dinaric Karst. Springer, Berlin
5. MONSTAT, Census 2011, Statistical Yearbook
6. Richardson SD, Ternes TA (2005) Water analysis: emerging contaminants and current issues. Anal Chem 77(12):3807–3838
7. Kümmerer K (2010) Pharmaceuticals in the environment. Annu Rev Env Resour 35:57–75
8. Richardson SD, Ternes TA (2014) Water analysis: emerging contaminants and current issues. Anal Chem 86(6):2813–2848
9. Castillo GC, Vila IC, Neild E (2000) Ecotoxicity assessment of metals and wastewater using multitrophic assays. Environ Toxicol 15(5):370–375
10. Bolong N, Ismail AF, Salim MR, Matsuura T (2009) A review of the effects of emerging contaminants in wastewater and options for their removal. Desalination 239(1–3):229–246
11. Cuklev F, Kristiansson E, Fick J, Asker N, Förlin L, Larsson DJ (2011) Diclofenac in fish: blood plasma levels similar to human therapeutic levels affect global hepatic gene expression. Environ Toxicol Chem 30(9):2126–2134
12. Galus M, Jeyaranjaan J, Smith E, Li H, Metcalfe C, Wilson JY (2013) Chronic effects of exposure to a pharmaceutical mixture and municipal wastewater in zebrafish. Aquat Toxicol 132:212–222
13. Kostić J, Kolarević S, Kračun-Kolarević M, Aborgiba M, Gačić Z, Paunović M, Višnjić-Jeftić Ž, Rašković B, Poleksić V, Lenhardt M, Vuković-Gačić B (2017) The impact of multiple

stressors on the biomarkers response in gills and liver of freshwater breams during different seasons. Sci Total Environ 601:1670–1681
14. International Commission for the Protection of the Danube River (2015) Danube river basin management plan. ICPDR, Vienna. https://www.icpdr.org/main/sites/default/files/nodes/documents/drbmp-update2015-annexes.pdf
15. Rastall AC, Neziri A, Vukovic Z, Jung C, Mijovic S, Hollert H, Nikcevic S, Erdinger L (2004) The identification of readily bioavailable pollutants in Lake Shkodra/Skadar using semipermeable membrane devices (SPMDs), bioassays and chemical analysis. Environ Sci Pollut Res 11 (4):240
16. Stesevic D, Feiler U, Sundic D, Mijovic S, Erdinger L, Seiler TB, Heininger P, Hollert H (2007) Application of a new sediment contact test with Myriophyllum aquaticum and of the aquatic Lemna test to assess the sediment quality of Lake Skadar. J Soil Sediment 7(5):342–349
17. Perović A, Perović S, Erdinger L, Hollert H (2012) Assessment of the genotoxic potential of Lake Skadar sediments using Ames test and comet assay on the fish cell line RTL-W1. Arch Biol Sci 64(1):249–256
18. Perović S, Perović A, Erdinger L, Hollert H (2013) Assessment of the mutagenic potential of Skadar Lake sediments using the Salmonella microsomal assay. Arch Biol Sci 65(3):1189–1194
19. Perović A, Perović S, Seiler TB, Hollert H (2013) In vitro cytotoxic and teratogenic potential of sediment extracts from Skadar Lake using fish cells RTL-W1 and Danio rerio embryos. Arch Boil Sci 65(4):1539–1546
20. Perović S, Sljukic B, Šrut M, Perović A, Klobučar GI (2019) Evaluation of DNA damage in haemolymph of freshwater mussels *Unio pictorum* from Lake Skadar. Biologia:1–6
21. Kastratović V, Krivokapić S, Đurović D, Blagojević N (2013) Seasonal changes in metal accumulation and distribution in the organs of *Phragmites australis* (common reed) from Lake Skadar, Montenegro. J Serb Chem Soc 78(8):1241
22. Kastratović V, Bigović M, Jaćimović Ž, Kosović M, Đurović D, Krivokapić S (2018) Levels and distribution of cobalt and nickel in the aquatic macrophytes found in Skadar Lake, Montenegro. Environ Sci Pollut Res 25(27):26823–26830
23. Petrović D, Jančić D, Furdek M, Mikac N, Krivokapić S (2016) Aquatic plant *Trapa natans* L. as bioindicator of trace metal contamination in a freshwater lake (Skadar Lake, Montenegro). Acta Bot Croat 75(2):236–243
24. Mesi A, Kopliku D (2013) Toxicity bio-monitoring of Shkodra Lake surface water using a higher plant assay. Acad J Interdiscipl Stud 2(8):133
25. Rakocevic J, Sukovic D, Maric D (2018) Distribution and relationships of eleven trace elements in muscle of six fish species from Skadar Lake (Montenegro). Turk J Fish Aquat Sci 18 (5):647–657
26. Vukašinović-Pešić V, Blagojević N, Vukanović S, Savić A, Pešić V (2017) Heavy metal concentrations in different tissues of the snail Viviparusmamillatus (Küster, 1852) from lacustrine and riverine environments in Montenegro. Turk J Fish Aquat Sci 17(3):557–563
27. Vukašinović-Pešić V, Blagojević N (2018) Metal pollution: evidences from plants, aquatic invertebrates and fish from Lake Skadar. In: The Skadar/Shkodra Lake environment. Springer, Cham, pp 141–151
28. Kolarević S, Kračun-Kolarević M, Jovanović J, Ilić M, Paunović M, Kostić-Vuković J, Martinović R, Jokanović S, Joksimović D, Pešić V, Kirschner A, Linke R, Ixenmaier S, Farnleitner A, Savio D, Reischer G, Tomić N, Vuković-Gačić B (2019) Microbiological water quality of rivers in Montenegro. In: Pešić V, Paunović M, Kostianoy A (eds) The rivers of Montenegro. The handbook of environmental chemistry. Springer, Cham
29. Tomović S (2008) Ciljevi vodoprivrednog razvoja Crne Gore. Vodoprivreda 40:127–137
30. EPA (2013) Report on the state of the environment in Montenegro based on indicators. Environmental Protection Agency of Montenegro, Podgorica
31. EPA (2017) Report on the state of the environment in Montenegro based on indicators. Environmental Protection Agency of Montenegro, Podgorica

32. Van der Oost R, Beyer J, Vermeulen NP (2003) Fish bioaccumulation and biomarkers in environmental risk assessment: a review. Environ Toxicol Pharmacol 13(2):57–149
33. Smit MGD, Bechmann RK, Hendriks AJ, Skadsheim A, Larsen BK, Baussant T, Bamber S, Sanni S (2009) Relating biomarkers to whole-organism effects using species sensitivity distributions: a pilot study for marine species exposed to oil. Environ Toxicol Chem 28:1104–1109
34. Hook SE, Gallagher EP, Batley GE (2014) The role of biomarkers in the assessment of aquatic ecosystem health. Integr Environ Assess Manag 10(3):327–341
35. Broeg K, Westernhagen HV, Zander S, Korting W, Koehler A (2005) The "bioeffect assessment index" (BAI), a concept for the quantification of effects of marine pollution by an integrated biomarker approach. Mar Pollut Bull 50:495–503
36. Timbrell JA (1998) Biomarkers in toxicology. Toxicology 129(1):1–12
37. Mussali-Galante P, Tovar-Sánchez E, Valverde M, del Castillo ER (2013) Biomarkers of exposure for assessing environmental metal pollution: from molecules to ecosystems. Rev Int Contam Ambient 29(1):117–140
38. EU (2000) Directive 2000/60/EC of the European Parliament and of the Council establishing a framework for the Community action in the field of water policy (EU Water Framework Directive). Off J EU 327:1–72
39. Nagel R, Isberner K (1998) Testing of chemicals with fish – a critical evaluation of tests with special regard to zebrafish. In: Fish ecotoxicology. Birkhäuser, Basel, pp 337–352
40. Keiter S, Rastall A, Kosmehl T, Wurm K, Erdinger L, Braunbeck T, Hollert H (2006) Ecotoxicological assessment of sediment, suspended matter and water samples in the Upper Danube River. A pilot study in search for the causes for the decline of fish catches (12 pp). Environ Sci Pollut Res 13(5):308–319
41. Lammer E, Carr GJ, Wendler K, Rawlings JM, Belanger SE, Braunbeck T (2009) Is the fish embryo toxicity test (FET) with the zebrafish (*Danio rerio*) a potential alternative for the fish acute toxicity test. Comp Biochem Physiol C 149:196–209
42. Braunbeck T, Kais B, Lammer E, Otte J, Schneider K, Stengel D, Strecker R (2015) The fish embryo test (FET): origin, applications, and future. Environ Sci Pollut Res 22 (21):16247–16261
43. Schulte C, Nagel R (1994) Testing acute toxicity in the embryo of zebrafish, Brachydanio rerio, as an alternative to the acute fish test: preliminary results. Atla 22(1):12–19
44. ISO (2007) Water quality – determination of the acute toxicity of waste water to zebrafish eggs (*Danio rario*). ISO 15088:2007
45. EU (2010) Directive 2010/63/EU of the European Parliament and of the Council of 22 September 2010 on the protection of animals used for scientific purposes. Off J EU 276:33–79
46. Dai YJ, Jia YF, Chen N, Bian WP, Li QK, Ma YB, Chen YL, Pei DS (2014) Zebrafish as a model system to study toxicology. Environ Toxicol Chem 33:11–17
47. Babić T, Dinić J, Stojković Burić S, Hadžić S, Pešić M, Radojković D, Divac Rankov A (2018) Comparative toxicity evaluation of targeted anticancer therapeutics in embryonic zebrafish and sea urchin models. Acta Biol Hung 69:395–410
48. Jha AN (2008) Ecotoxicological applications and significance of the comet assay. Mutagenesis 23(3):207–221
49. Kolarević S, Milovanović D, Kračun-Kolarević M, Kostić J, Sunjog K, Martinović R, Đorđević J, Novaković I, Sladić D, Vuković-Gačić B (2017) Evaluation of genotoxic potential of avarol, avarone, and its methoxy and methylamino derivatives in prokaryotic and eukaryotic test models. Drug Chem Toxicol:1–10
50. Đorđević J, Kolarević S, Jovanović J, Kostić J, Novaković I, Sladić D, Vuković-Gačić B (2018) Evaluation of genotoxic potential of tert-butylquinone and its derivatives in prokaryotic and eukaryotic test models. Drug Chem Toxicol. https://doi.org/10.1080/01480545.2018.1514043
51. Martinović R, Kolarević S, Kračun-Kolarević M, Kostić J, Marković S, Gačić Z, Kljajić Z, Vuković-Gačić B (2015) Genotoxic potential and heart rate disorders in the Mediterranean

mussel *Mytilus galloprovincialis* exposed to Superdispersant-25 and dispersed diesel oil. Mar Environ Res 108:83–90
52. Martinović R, Kolarević S, Kračun-Kolarević M, Kostić J, Jokanović S, Gačić Z, Vuković-Gačić B (2016) Comparative assessment of cardiac activity and DNA damage in haemocytes of the Mediterranean mussel *Mytilus galloprovincialis* in exposure to tributyltin chloride. Environ Toxicol Pharmacol 47:165–174. Environmental Sciences: 91/229; IF: 2.313 (2016)
53. Gačić Z, Kolarević S, Sunjog K, Kračun-Kolarević M, Paunović M, Knežević-Vukčević J, Vuković-Gačić B (2014) The impact of *in vivo* and *in vitro*exposure to base analogue 5-FU on the level of DNA damage in haemocytes of freshwater mussels *Unio pictorum* and *Unio tumidus*. Environ Pollut 191:145–150
54. Kolarević S, Knežević-Vukčević J, Paunović M, Kračun M, Vasiljević B, Tomović J, Vuković-Gačić B, Gačić Z (2013) Monitoring of DNA damage in haemocytes of freshwater mussel *Sinanodonta woodiana* sampled from the Velika Morava River in Serbia with the comet assay. Chemosphere 93:243–251
55. Vuković-Gačić B, Kolarević S, Sunjog K, Tomović J, Knežević-Vukčević J, Paunović M, Gačić Z (2014) Comparative study of the genotoxic response of freshwater mussels *Unio tumidus* and *Unio pictorum* to environmental stress. Hydrobiologia 735:221–231
56. Kolarević S, Kračun-Kolarević M, Kostić J, Slobodnik J, Liška I, Gačić Z, Paunović M, Knežević-Vukčević J, Vuković-Gačić B (2016) Assessment of the genotoxic potential along the Danube River by application of the comet assay on haemocytes of freshwater mussels: the joint Danube survey 3. Sci Total Environ 540:377–385
57. Kolarević S, Gačić Z, Kostić J, Sunjog K, Kračun-Kolarević M, Paunović M, Knežević-Vukčević J, Vuković-Gačić B (2016) Impact of common cytostatics on DNA damage in freshwater mussels *Unio pictorum* and *Unio tumidus*. CLEAN Soil Air Water 44:1471–1476
58. Sunjog K, Gačić Z, Kolarević S, Višnjić-Jeftić Z, Jarić I, Knežević-Vukcević J, Vuković-Gačić B, Lenhardt M (2012) Heavy мetal accumulation and the genotoxicity in barbel (*Barbus barbus*) as indicators of the Danube river pollution. Sci World J. https://doi.org/10.1100/2012/351074
59. Sunjog K, Kolarević S, Kračun-Kolarević M, Gačić Z, Skorić S, Đikanović V, Lenhardt M, Vuković-Gačić B (2014) Variability in DNA damage of chub (*Squalius cephalus* L.) blood, gill and liver cells during the annual cycle. Environ Toxicol Pharmacol 37:967–974
60. Deutschmann B, Kolarevic S, Brack W, Kaisarevic S, Kostic J, Kracun-Kolarevic M, Liska I, Paunovic M, Seiler T-B, Shao Y, Sipos S, Slobodnik J, Teodorovic I, Vukovic-Gacic B, Hollert H (2016) Longitudinal profile of the genotoxic potential of the River Danube on erythrocytes of wild common bleak (*Alburnus alburnus*) assessed using the comet and micronucleus assay. Sci Total Environ 573:1441–1449
61. Kolarević S, Aborgiba M, Kračun-Kolarević M, Kostić J, Simonović P, Simić V, Milošković A, Reischer G, Farnleitner A, Gačić Z, Milačič R, Zuliani T, Vidmar J, Pergal M, Piria M, Paunović M, Vuković-Gačić B (2016) Evaluation of genotoxic pressure along the Sava River. PLoS One 11:e0162450. https://doi.org/10.1371/journal.pone.0162450
62. Kostić J, Kolarević S, Kračun-Kolarević M, Aborgiba M, Gačić Z, Lenhardt M, Vuković-Gačić B (2016) Genotoxicity assessment of the Danube River using tissues of freshwater bream (*Abramis brama*). Environ Sci Pollut Res 23:20783–20795
63. Kračun-Kolarević M, Kolarević S, Jovanović J, Marković V, Ilić M, Simonović P, Simić V, Gačić Z, Diamantini E, Stella E, Petrović M, Majone B, Bellin A, Paunović M, Vuković-Gačić B (2016) Evaluation of genotoxic potential throughout the upper and middle stretches of Adige river basin. Sci Total Environ 571:1383–1391. Environmental Sciences: 22/229; IF: 4.9 (2016)
64. Jovanović J, Kolarević S, Milošković A, Radojković N, Simić V, Dojčinović B, Kračun-Kolarević M, Paunović M, Kostić J, Sunjog K, Timilijić J, Djordjević J, Gačić Z, Žegura B, Vuković-Gačić B (2017) Evaluation of genotoxic potential in the Velika Morava River basin *in vitro* and *in situ*. Sci Total Environ 621:1289–1299

65. Matić D, Vlahović M, Kolarević S, Mataruga VP, Ilijin L, Mrdaković M, Vuković-Gačić B (2016) Genotoxic effects of cadmium and influence on fitness components of *Lymantria dispar* caterpillars. Environ Pollut 218:1270–1277
66. Kračun-Kolarević M, Kolarević S, Atanacković A, Marković V, Gačić Z, Paunović M, Vuković-Gačić B (2015) Effects of 5-fluorouracil, etoposide and CdCl2 in aquatic Oligochaeta *Limnodrilus udekemianus* Claparede (Tubificidae) measured by comet assay. Water Air Soil Pollut 226:1–9
67. Aborgiba M, Kostić J, Kolarević S, Kračun-Kolarević M, Elbahi S, Knežević-Vukčević J, Lenhardt M, Paunović M, Gačić Z, Vuković-Gačić B (2016) Flooding modifies the genotoxic effects of pollution on a worm, a mussel and two fish species from the Sava River. Sci Total Environ 540:358–367
68. Scholz S, Fischer S, Gündel U, Küster E, Luckenbach T, Voelker D (2008) The zebrafish embryo model in environmental risk assessment – applications beyond acute toxicity testing. Environ Sci Pollut Res 15(5):394–404
69. Jarvis RB, Knowles JF (2003) DNA damage in zebrafish larvae induced by exposure to low-dose rate γ-radiation: detection by the alkaline comet assay. Mutat Res 541(1–2):63–69
70. Kosmehl T, Hallare AV, Reifferscheid G, Manz W, Braunbeck T, Hollert H (2006) A novel contact assay for testing genotoxicity of chemicals and whole sediments in zebrafish embryos. Environ Toxicol Chem 25(8):2097–2106
71. Kosmehl T, Krebs F, Manz W, Braunbeck T, Hollert H (2007) Differentiation between bioavailable and total hazard potential of sediment-induced DNA fragmentation as measured by the comet assay with zebrafish embryos. J Soil Sediment 7(6):377–387
72. Kosmehl T, Hallare AV, Braunbeck T, Hollert H (2008) DNA damage induced by genotoxicants in zebrafish (Danio rerio) embryos after contact exposure to freeze-dried sediment and sediment extracts from Laguna Lake (The Philippines) as measured by the comet assay. Mutat Res 650(1):1–14
73. Eleršek T, Plazar J, Filipič M (2013) A method for the assessment of DNA damage in individual, one day old, zebrafish embryo (Danio rerio), without prior cell isolation. Toxicol In Vitro 27(8):2156–2159
74. Hollert H, Keiter S, König N, Rudolf M, Ulrich M, Braunbeck T (2003) A new sediment contact assay to assess particle-bound pollutants using zebrafish (Danio rerio) embryos. J Soil Sediment 3(3):197
75. Levan A (1938) The effect of colchicine on root mitoses in Allium. ~. Hereditas 24:471–486
76. Fiskesjo G (1985) The Allium test as a standard in environmental monitoring. Hereditas 102:99–112
77. Fiskesjo G (1985) Allium test on river water from brain and Saxdn before and after closure of a chemical factory. Ambio 14(2):99–I03
78. Bosio S, Laughinghouse IV HD (2012) Bioindicator of genotoxicity: the *allium cepa* test. Environ Contam. https://doi.org/10.5772/31371
79. Rank J, Nielsen MH (1994) Evaluation of the Allium anaphase-telophase test in relation to genotoxicity screening of industrial wastewater. Mutat Res 312:17–24
80. Chauhan LKS, Saxena PN, Gupta SK (1999) Cytogenetic effects of cypermethrin and fenvalerate on the root meristem cells of *Allium cepa*. Environ Exp Bot 42:181–189
81. Leme DM, Marin-Morales MA (2009) Allium cepa test in environmental monitoring: a review on its application. Mutat Res 682(1):71–81
82. Roberto MM, Jamal CM, Malaspina O, Marin-Morales MA (2016) Antigenotoxicity and antimutagenicity of ethanolic extracts of Brazilian green propolis and its main botanical source determined by the *Allium cepa* test system. Genet Mol Biol 39(2):257–269. https://doi.org/10.1590/1678-4685-gmb-2015-0130
83. Vujošević M, Anđelković S, Savić G, Blagojević J (2008) Genotoxicity screening of the river Rasina in Serbia using the Allium anaphase–telophase test. Environ Monit Assess 147 (1–3):75–81

84. Radić S, Stipaničev D, Vujčić V, Rajčić MM, Širac S, Pevalek-Kozlina B (2010) The evaluation of surface and wastewater genotoxicity using the Allium cepa test. Sci Total Environ 408 (5):1228–1233
85. Ma T-H, Xu Z, Xu C, McConnell H, ValtierraRabago E, Adriana Arreola G, Zhang H (1995) The improved Allium/Vicia root tip micronucleus assay for clastogenicity of environmental pollutants. Mutat Res 334(2):185–195. https://doi.org/10.1016/0165-1161(95)90010-1
86. Dane F, Ekici N, Aktas YK (2006) The effect of waste water on root growth and mitosis in onion (Allium cepa) root apical meristem. Asian J Plant Sci 5(2):331–334
87. Abdissa Y, Tekalign T, Pant LM (2011) Growth, bulb yield and quality of onion (Allium cepa L.) as influenced by nitrogen and phosphorus fertilization on vertisol I. growth attributes, biomass production and bulb yield. Afr J Agric Res 6(14):3252–3258

Microbiological Water Quality of Rivers in Montenegro

Stoimir Kolarević, Margareta Kračun-Kolarević, Jovana Jovanović, Marija Ilić, Momir Paunović, Jovana Kostić-Vuković, Rajko Martinović, Sandra Jokanović, Danijela Joksimović, Vladimir Pešić, Alexander K. T. Kirschner, Rita Linke, Simone Ixenmaier, Andreas Farnleitner, Domenico Savio, Georg Reischer, Nevenka Tomić, and Branka Vuković-Gačić

S. Kolarević (✉) and B. Vuković-Gačić
Faculty of Biology, Center for Genotoxicology and Ecogenotoxicology, University of Belgrade, Belgrade, Serbia
e-mail: stoimirk@bio.bg.ac.rs

M. Kračun-Kolarević, J. Jovanović, M. Ilić, and M. Paunović
Institute for Biological Research "Siniša Stanković", National Institute of Republic of Serbia, University of Belgrade, Belgrade, Serbia
e-mail: margareta.kracun@ibiss.bg.ac.rs

J. Kostić-Vuković
Institute for Multidisciplinary Research, University of Belgrade, Belgrade, Serbia
e-mail: jkostic@imsi.rs

R. Martinović, S. Jokanović, and D. Joksimović
Institute of Marine Biology, University of Montenegro, Kotor, Montenegro
e-mail: rajko.mar@ac.me

V. Pešić
Faculty of Science and Mathematics, University of Montenegro, Podgorica, Montenegro
e-mail: vladopesic@gmail.com

A. K. T. Kirschner
Institute for Hygiene and Applied Immunology, Water Microbiology, Medical University of Vienna, Vienna, Austria

Interuniversity Cooperation Center Water and Health (ICC), Vienna, Austria

Karl Landsteiner University of Health Sciences, Krems an der Donau, Austria
e-mail: Alexander.Kirschner@meduniwien.ac.at; https://www.waterandhealth.at

R. Linke, S. Ixenmaier, and G. Reischer
Interuniversity Cooperation Center Water and Health (ICC), Vienna, Austria

Research Group Environmental Microbiology and Molecular Diagnostics, Institute of Chemical, Environmental and Bioscience Engineering, TU Wien, Vienna, Austria
e-mail: rita.linke@tuwien.ac.at; https://www.waterandhealth.at

Vladimir Pešić, Momir Paunović, and Andrey G. Kostianoy (eds.), *The Rivers of Montenegro*, Hdb Env Chem (2020) 93: 135–156, DOI 10.1007/698_2019_420,

Contents

Abstract The literature data on the microbiological water quality of the surface and groundwaters in Montenegro is very scarce. Therefore, this chapter aims to provide an insight in the microbiological water quality of rivers in Montenegro by compiling the data obtained in period 2009–2018 in national monitoring program and the data collected within the Montenegro survey in 2019 with an emphasis on the hotspots of faecal pollution and possible sources of pollution. Despite the high risk that poor implementation of wastewater treatment might represent for Montenegro, the obtained dataset indicates that overall microbiological water quality of rivers in Montenegro is quite good. About 80% of the investigated sites have bathing water quality according to Montenegrin legislation. Only four sites were identified as hotspots of faecal pollution where the human source of pollution prevails.

Keywords Faecal indicator bacteria, Microbial source tracking, Microbiology, Montenegro, Rivers

---

A. Farnleitner and D. Savio
Interuniversity Cooperation Center Water and Health (ICC), Vienna, Austria

Karl Landsteiner University of Health Sciences, Krems an der Donau, Austria

Research Group Environmental Microbiology and Molecular Diagnostics, Institute of Chemical, Environmental and Bioscience Engineering, TU Wien, Vienna, Austria
e-mail: andreas.farnleitner@tuwien.ac.at; https://www.waterandhealth.at

N. Tomić
Institute of Hydrometeorology and Seismology – Montenegro, Podgorica, Montenegro
e-mail: nena.tomic@meteo.co.me

1 Introduction

Montenegro is characterized by watersheds rich with water, even compared to worldwide standards [1]. About 95.3% of rivers in Montenegro are formed at the territory of this country. There are several important currents which drain in two directions: toward the Black Sea and toward the Adriatic Sea. Sizeable portion of Montenegro is made of continental karst as part of the Dinaric karst aquifer system which extends from Italy through Slovenia, Croatia, Bosnia and Herzegovina, Montenegro, and Albania. There is a complex link of terrestrial and lentic systems throughout the streams, rivers, and the Adriatic Sea. Therefore, knowledge on microbiological quality of water is of great importance as the pollution in any of these compartments can have serious consequence on numerous essential ecosystem services such as drinking water production, water exploitation for irrigation, recreation etc. [2, 3]. Water represents an important resource of Montenegro for all of the indicated ecosystem services. Recreational activities on surface waters such as rafting, kayaking, canyoning, and fly-fishing are significant aspect of tourism [4]. Karst aquifers are a major source for drinking water production in Montenegro [5]. These highly fragile ecosystems and the exploitation of their resources or inappropriate land uses give rise to environmental problems (water pollution, subsidence, flooding, changes in the subterranean environment, etc.). This issue goes beyond the borders considering that changes in such complex water system can be reflected at the level of the whole watershed. Surprisingly, the literature data on the microbiological water quality of the surface and groundwater in Montenegro is still very scarce. The most reliable information for this aspect of water quality in Montenegro can be obtained from the national monitoring program. Therefore, this chapter aims to provide an overview on the national monitoring data obtained in period 2009–2018, an overview of the status quo based on the data collected in 2019 within the Montenegro Survey with an emphasis on the hotspots of faecal pollution and possible sources of pollution.

2 Watersheds in Montenegro

2.1 *The Black Sea Basin*

The area of the Black Sea watershed expands on 7,260 km^2 or 52.5% the territory of Montenegro. The major drainage channels are rivers Ibar (tributary of the Zapadna Morava River), Lim, Tara, Piva, and Ćehotina (tributaries of the Drina River).

The Ibar River springs at the mountain Hajla. The major tributaries are Županica, Limnička, Ibarac, Grahovska, Bukovačka, Baltička, and Baćka. The area of the Basin covers 413.6 km^2. *The Lim River* is the most important river in Montenegro from a hydrological aspect. It originates from the Plavsko Lake and receives numerous tributaries: Murinska River, Zlorečica, Đurička, Rženička, Velička,

Kraštica, Trebička, Ševarinska, Bistrica, Šekularska, Kaluđerska, Brzava, Ljubovida, Dapsićka, Lješnica, Bjelopoljska Lješnica, and Bjelopoljska Bistrica. The area of the Lim Basin within the Montenegro territory is 2,880 km^2. *The Tara River* spring is situated beneath the Maglić Kariman. The major tributaries are Opasanica, Drcka, Pčinja, Plašnica, Štitarica, Ravnjak, Ljutica, Skrbuša, Svinjača, Jezerštica, Rudnjača, Bjelojevićka, and Selačka. The Tara Basin covers an approximate area of 2,040 km^2. *The Piva River Basin* is composed of several rivers. Upper stretch of the river, starting from springs at the Durmitor up to Šavnik is called Bukovica. From the confluence of Bijela in Šavnik to the confluence of Komarnica, it is known as Pridvorica. Further on, it is called Komarnica until the confluence with river Sinjaci, where it becomes Piva. In Šćepan Polje, Piva and Tara confluence are forming Drina. The area of the Piva Basin is about 1,784 km^2.

The Ćehotina River springs beneath the mountain Stožer. The major tributaries are Korička, Maočnica, Vezišnica, and Voloder. The area of the Basin is approximately 809.8 km^2.

2.2 *The Adriatic Sea Basin*

The area of watershed drained to the Adriatic Sea covers around 6,560 km^2 or 47.5% of the territory. A major drainage channel in the Adriatic Basin is Morača together with the Skadar Lake and the Bojana River. *The Morača River* Basin covers an area of 2,628 km^2. The most significant tributaries are Zeta, Cijevna, and Ribnica. *Zeta* as the most significant tributary of Morača covers an area of approximately 1,216 km^2. Morača fills the Skadar Lake, the largest lake in the Balkan Peninsula with a surface area that seasonally fluctuates between 370 and 600 km^2 [6]. The Lake is shared between the countries Montenegro and Albania. Besides Morača, important tributaries from the Montenegrin side are Rijeka Crnojevića (12.3 km long) and Plavnica. The Lake is drained to the Adriatic Sea by Bojana (41 km long).

3 Wastewaters as Significant Pressure on Surface Waters in Montenegro

When discussing the ecological and chemical status of the river basins worldwide, wastewaters represent one of the most significant pressures. Untreated wastewaters lead to the deterioration of ecosystem quality due to high input of organic load in the recipient. The situation is quite worrying as wastewaters represent a significant health hazard as sources of numerous pathogens. Various ways of surface water exploitation lead to the direct contact of host and pathogen which might have enormous consequences such as disease outbreaks. The level at which the wastewaters are processed before their release in a recipient differ significantly among the countries as a consequence of differences in economic, cultural, and educational standards. Situation is even more complex when discussing transboundary basins, shared by multiple countries, which is the case with the Danube River Basin (DRB). There is an evident difference in water quality between the EU and non-EU countries within the basin. In our previous work we have indicated these differences in faecal contamination along the whole Danube [7–9] and the most significant tributaries of Danube in the Balkan region such as Sava, Tisza, and Velika Morava [2, 10–14]. Area of Montenegro represents an important part of the Danube River Basin, and knowledge on the pressures in this area is of great significance for filling the gaps at the whole basin level. The Danube River Basin Management Plan updated in 2015 [15] indicates that about 55% of wastewaters in Montenegro are collected but not treated, while 38% of the generated load was neither collected nor treated. The statistics are generated based on the data from 7 agglomerations $\geq$2,000 person equivalents (PE) in Montenegro (year 2011–2012). Based on the data from the Ministry of Economic Development dating from 2008, Montenegro has about 17 municipalities with >10,000 PE, out of which 9 are situated at the river banks (363,377 PE which is about 59% of total PE) and 6 at the coastline (145,847 PE which is about 24% of total PE).

4 Microbiological Pollution of Surface and Groundwaters

Water pollution by faecal material, which is the source of many pathogens, represents a high health risk for both humans and animals [16]. Intensive growth of human population and activities has led to pollution of water resources by biological pollutants such as viruses and bacteria. Microbiological pollution may originate from the point and nonpoint sources of pollution. Point sources include the discharge of improperly treated and untreated communal wastewater and livestock enterprises, while nonpoint sources include surface runoff from urban and agricultural surfaces and outflow from septic tanks [7]. The impact of both kind of sources can be multiple, such as deterioration in quality of the surface waterbodies and groundwater resources. The second point is of special interest in Montenegro as groundwater resources from mountainous karst aquifers play an important role in public water supplies throughout the world [17]. Assessment of the sanitary aspect of water quality is therefore necessary for undertaking measures for more efficient management of natural resources and implementation of the wastewater treatment plants.

Faecal pollution of anthropogenic origin represents an important issue for all aspects of water usage because the presence of untreated faecal waste may lead to various health issues in exposed humans [18, 19]. When compared to pathogenic microorganisms which may be present in faecal material, bacteria that comprise the normal gut microbiota are present in much higher concentrations, and thus are easily detectable. If they are not present in the water sample, it may be concluded that pathogenic bacteria are absent from the water as well [20]. Therefore, instead of detection and isolation of all potential enteric pathogens present in water which is expensive and time consuming, the focus is placed on the more approachable bacterial groups in means of cultivation and isolation that are used as indicators of the presence of enteric pathogens [21].

Indicator bacteria which are widely used in the assessment of the sanitary aspect of water quality belong to the group of coliforms. Coliform bacteria, primarily *Escherichia coli*, and intestinal enterococci are considered the basis of faecal pollution monitoring [8, 22].

Coliform bacteria are present in the environment, as well as in human and animal waste. Although the term "coliform" does not have a taxonomic value, it is used to describe different genera within the family Enterobacteriaceae which ferment lactose. Coliforms usually do not cause disease, but their presence in water points to the possible presence of pathogens.

Within the group of coliforms, there is a division into three groups of indicator microorganisms: total coliforms, faecal (thermotolerant) coliforms, and *E. coli*. The presence of each group points to a different risk level. If the total coliforms are present in the water, the sample is further examined to the presence of faecal coliforms and *E. coli* as better indicator of the faecal contamination [23]. In addition, it is recognized that there is a strong correlation between the *E. coli* levels and both pathogenic organisms and gastrointestinal illnesses [24].

The group of total coliforms includes aerobic and facultative anaerobic, gram-negative rod-shaped bacteria, which do not form endospores, and ferment lactose with the production of gas and acid during 48 h at 35 $\pm$ 2°C with the help of an enzyme β-galactosidase [25].

Faecal coliforms are a subgroup of total coliforms which is directly related to the faecal pollution originating from the homeotherms. These bacteria possess all the characteristics of total coliforms, but they have the ability to grow and ferment lactose at 44.5°C during 48 h of incubation. In thermotolerant coliforms, the physiological basis of growth at elevated temperatures is interpreted as an adaptation of proteins to maintain stability at temperatures that prevail in the intestinal tract of animals [26].

The only species within the family of Enterobacteriaceae, *E. coli*, is almost always of faecal origin and is considered the most precise indicator of faecal contamination and the potential presence of pathogens [27]. Besides, *E. coli* is the most common of all coliforms in the intestinal flora of warm-blooded animals [21].

Enterococci, especially *Enterococcus faecalis*, are normally present in human and animal faecal material, and thus are also used as indicators of the faecal pollution presence. They belong to the genus *Enterococcus* which includes about 30 types of gram-positive cocci that do not form endospores. The optimal temperature for growth for the majority of species is 35–37°C. They are classified as facultative anaerobes. *E. faecalis* and *Enterococcus faecium* are the most prevalent species cultured from humans, accounting for more than 90% of clinical isolates. Surface waters are not a natural habitat for enterococci and their presence in this environment is considered to be the result of faecal pollution [28].

5 Faecal Indicator Bacteria in Montenegrin National Legislation on Water Quality and Compliance with EU Directives

5.1 Microbiological Indicators in Classification and Categorization of Surface and Groundwaters

In Montenegro, monitoring and assessment of microbiological water quality is regulated by national legislative through the Law on Waters which is linked to the Regulation on classification and categorization of surface and groundwaters (Official Gazette of the Republic of Montenegro 27/07). The Regulation uses a trilateral system for water classification based on the purpose of water usage: (a) drinking water production and food industries, (b) cultivation of fish and mussels, and (c) bathing water. Microbiological indicators are included in each part of trilateral systems for classification purposes. Tables 1, 2, and 3 show limit values for classes of water quality established based on the bacterial numbers. The national legislation is in general in compliance with EU regulations in case of the bathing water quality [29] thus according to Montenegrin legislation, the values listed are limit values of single measurements.

Table 1 Microbiological indicators for quality for water used for drinking water production

Class mark	A	A1	A2	A3
Total coliforms/100 mL	10	10	500	5,000
Faecal coliforms/100 mL	10	20	2,000	20,000
Intestinal enterococci/100 mL	<LoD	20	1,000	10,000

A water which can be used for drinking in its current form, *A1* water which requires only simple methods for purification and disinfection before drinking, *A2* water which can be used for drinking after advanced purification steps, *A3* water which requires intensive treatment before usage for drinking (Official Gazette of the Republic of Montenegro 27/07)

Table 2 Microbiological indicators for water quality used for cultivation of fish and mussels

Class mark	S	Š	C
Total coliforms/100 mL	2,000	100	10,000
Faecal coliforms/100 mL	–	<300	–

S suitable for cultivation of salmonids, *Š* suitable for cultivation of mussels, *C* suitable for cultivation of fish species with lesser commercial value (cyprinids) (Official Gazette of the Republic of Montenegro 27/07)

Table 3 Microbiological indicators for water quality used for bathing

Class mark	K1	K2
Intestinal enterococci/100 mL	200	330
E. coli/100 mL	500	900

K1 good quality, *K2* sufficient quality (Official Gazette of the Republic of Montenegro 27/07)

The Regulation also prescribes the reference methods for the assessment of number of the microbiological indicator bacteria. For total and faecal coliforms, the reference method is membrane filtration and cultivation at Tergitol lactose agar or Endo agar at 37 and 44°C for total coliforms and faecal coliforms, respectively. Intestinal enterococci are also assessed by membrane filtration followed by cultivation on sodium azide agar at 37°C.

5.2 *Microbiological Indicators in Assessment of Wastewaters Quality*

As the area of Montenegro was characterized as a vulnerable part of the DRB based on the fact that the majority of wastewater is released in recipients without any treatment [15], the sanitary quality of wastewaters must be monitored. The Regulation on the methods and conditions for measuring the quantities of wastewater discharged into the receiver (Official Gazette of the Republic of Montenegro 24/10) includes microbiological indicators of pollution and prescribes reference methods for intestinal enterococci (MPN method SRPS EN ISO 7899-1:2009 or membrane filtration method SRPS EN ISO 7899-2:2010) and faecal coliforms (MPN method SRPS EN ISO 9308-3:2009 or membrane filtration method SRPS EN ISO 9308-1:2010). The quality of wastewaters in Montenegro is the subject of the Regulation on quality and sanitary and technical conditions for discharging wastewater into the recipient and public sewerage, the method and procedure of testing the quality of wastewater, the minimum number of tests and the content of the report on the established quality of wastewater (Official Gazette of the Republic of Montenegro 45/08, 9/10, 26/12, 52/12 and 59/13). However, it is interesting that the Regulation does not include microbiological indicators and their maximum allowable concentrations in discharged wastewaters.

6 Dataset Obtained Within the National Monitoring Program in Period 2009–2018

Systematic assessment of the water quality in Montenegro is performed by the Department of Water Quality of the Institute of Hydrometeorology and Seismology – Montenegro in compliance with the program for assessment of quantitative and qualitative characteristics of water. Map of the sampling stations is available in the study of Vukašinović Pešić et al. [30]. The program has a seasonal basis; four samples are obtained per monitoring station annually.

To give an overview on the surface water quality in Montenegro, we have analyzed the monitoring data available for the period 2009–2018 for 37 sites. A total of 1,599 samples taken over 10 years were processed, and the data are summarized in Fig. 1a in the light of the classification proposed by Kavka et al.

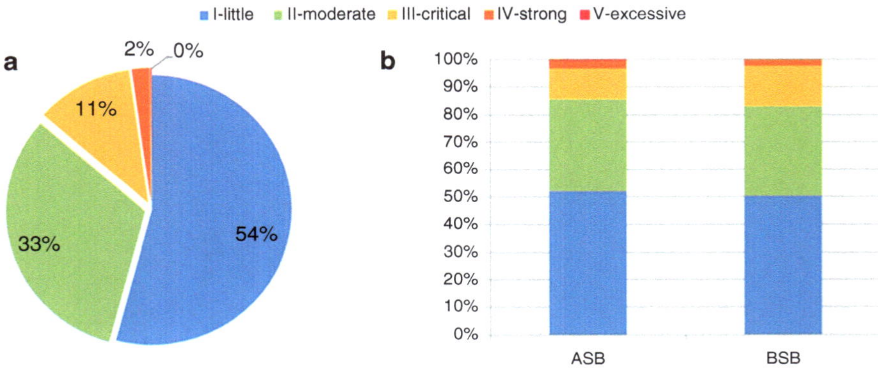

Fig. 1 Summary of monitoring data for indicators of faecal pollution (faecal coliform bacteria) obtained in period 2009–2018 for 37 monitoring sites in Montenegro (**a**); comparative indication of the distribution of the samples in classes based on the level of pollution in the Adriatic Sea Basin (ASB) and Black Sea Basin (BSB) (**b**)

[31]. In this system, five classes of faecal pollution were defined such that classes I and II are below and quality classes III, IV, and V exceed the faecal pollution limit values for good bathing water quality according to Montenegrin legislation [8]. We decided to use this classification system to make the obtained data comparable with the rest of the data available for Danube River Basin. Analysis show that 54% of all samples were in class I, 33% in class II, 11% in class III, 2% in class IV, and only 0.13% of all samples were in class V. Based on the analyzed dataset, one can say that the overall quality of surface water in Montenegro is quite good considering that above 80% of all samples had good or sufficient bathing water quality based on Montenegrin legislation. When comparing the datasets obtained for the Adriatic Sea Basin (ASB) with Black Sea Basin (BSB) data, a similar pattern in distribution of the samples based on water quality was observed (Fig. 1b).

6.1 The Adriatic Sea Basin

In ASB, monitoring data were available for six sites at Morača, four sites at Zeta, two sites at Cijevna, and per one site at Bojana and Rijeka Crnojevića. Additionally, three sites related to the Skadar Lake were included in the analyses (Fig. 2). The most affected site at Morača, and for all rivers in ASB as well, was the site downstream of the wastewater outlet in Podgorica. At this site about 90% of the collected samples indicated critical or even higher level of pollution. The impact was evident on the sites downstream as well, the site *Grbavci*, with 50% of the samples critically polluted and the site *Vukovci* with 25% of the samples critically polluted. The impact of Morača on Skadar was not so evident; only 2 out of 158 samples indicated critical pollution. Still about 26% of the samples were moderately polluted indicating presence of anthropogenic influence. The study of Šundić and Radujković [32]

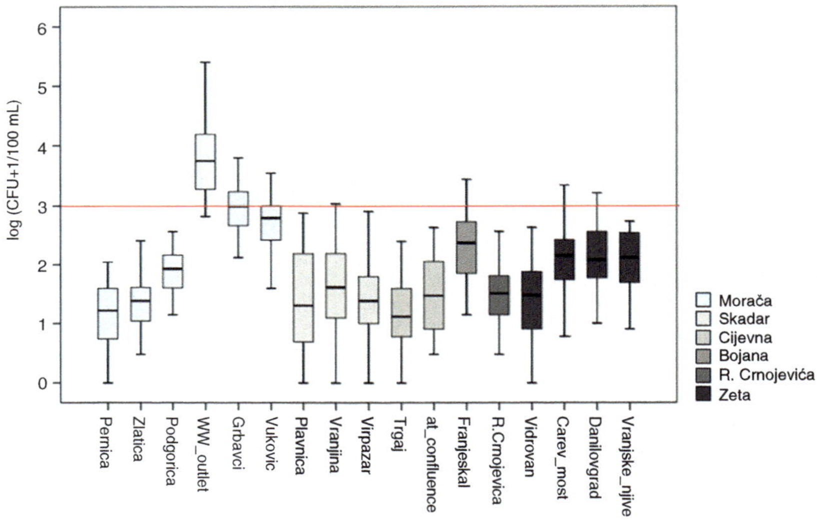

Fig. 2 Summary of monitoring data for indicators of faecal pollution (faecal coliform bacteria) obtained in period 2009–2018 for monitoring sites belonging to the Adriatic Sea Basin (ASB) in Montenegro; Box 25–75% range, Whisker minimum, median, and maximum; horizontal red line represents a border between the moderate and critical level of pollution

indicates that the Skadar Lake receives a significant amount of wastewaters from Podgorica (45,706 m^3), Danilovgrad (1,048 m^3), Nikšić (34,748 m^3), and Cetinje (1,516 m^3) on a daily basis. Sporadic pollution events were noticed at Rijeka Crnojevića and Zeta (sites *Duklov most* and *Danilovgrad*) with 5–10% of the samples showing critical pollution level. This should not be neglected as the karstic springs and reservoirs belonging to the area of Nikšić and Cetinje are the most vulnerable to pollution coming from the surface [5].

6.2 The Black Sea Basin

In BSB, monitoring data were available for six sites at Lim and six sites at Tara, five sites in the Ćehotina Basin, two sites at Ibar and one site at Piva (Fig. 3). The most affected site in the BSB was the site at Ćehotina situated *downstream Pljevlja* where approximately 80% of the samples indicated critical or even higher level of pollution. The impact was evident further downstream (*downstream Vezišnica*) with 50% of the samples polluted critically or above. These sites are commonly recognized as the most polluted in Montenegro [33–35]. Gradac and Vezišnica also had more than 25% of samples that were higher than critically polluted. Critical level of pollution was also recorded in samples collected at several different sites at Lim. The highest level of pollution was detected *downstream Bijelo Polje* where 50% of collected

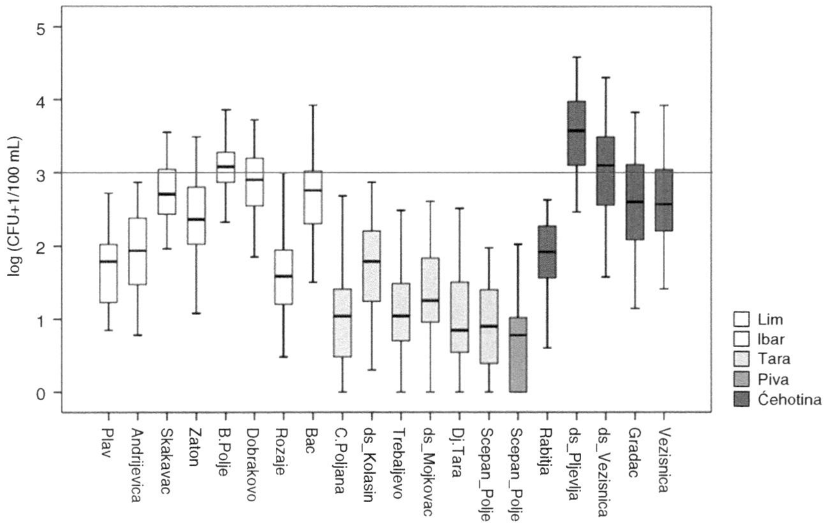

Fig. 3 Summary of monitoring data for indicators of faecal pollution (faecal coliform bacteria) obtained in period 2009–2018 for monitoring sites belonging to Black Sea Basin (BSB) in Montenegro; Box 25–75% range, Whisker minimum, median, and maximum; horizontal red line represents a border between the moderate and critical level of pollution

samples were critically polluted. At Ibar, sporadic pollution events were detected at the site *Bać*, where 25% of the samples were critically polluted. At Tara, all samples had good or sufficient bathing water quality according to national legislation.

Vukašinović Pešić et al. [30] indicated that water quality has increasing trend from 2012 based on the Water Quality Index (WQI). Considering that assessment of WQI implements number of coliform bacteria we have investigated annual trends in concentration of faecal coliform. For most of the sampling sites, including the most polluted ones, a declining trend in faecal coliforms can be observed (Fig. 4).

7 Montenegro Microbiological Survey

To obtain the most recent dataset on the microbiological quality of the surface waters in Montenegro and to identify the hotspots of faecal pollution and source of pollution as well, a comprehensive sampling survey was performed in a short period of time (4 days) in May 2019. Survey included the two-step selection processes in the assessment of water quality that comprised of the assessment of the level of faecal pollution in the first step, and the microbial source tracking analyses in the critically polluted samples in the second step.

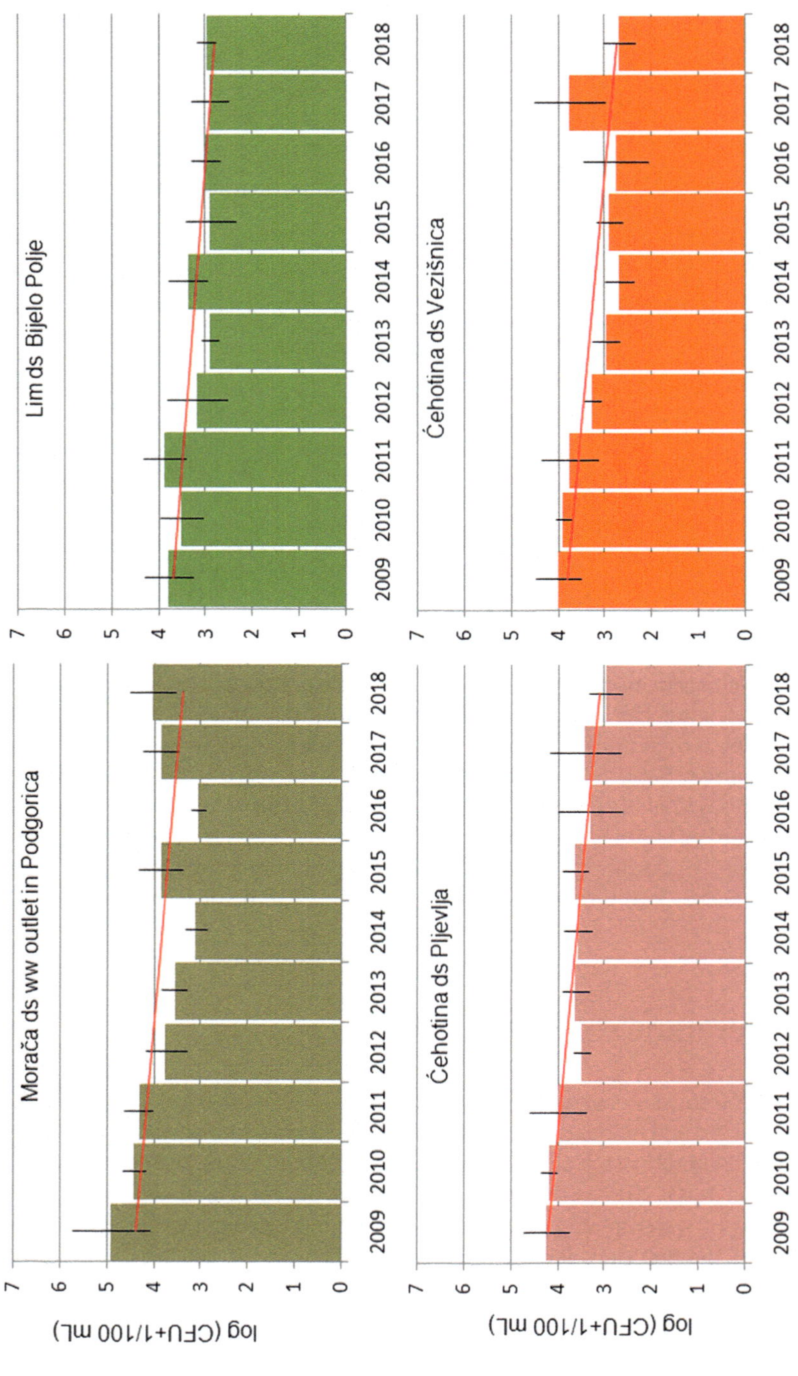

Fig. 4 Annual trends in concentration of faecal coliforms for the most polluted sites in Montenegro in period 2009–2018; trend line is labeled with red

7.1 Samples and Sites

Within the Montenegro survey, total of 25 sites were analyzed of which 22 are included in the national monitoring while an additional three were included as interesting tributaries that were not monitored before. An overview of sampling sites is indicated in Fig. 5.

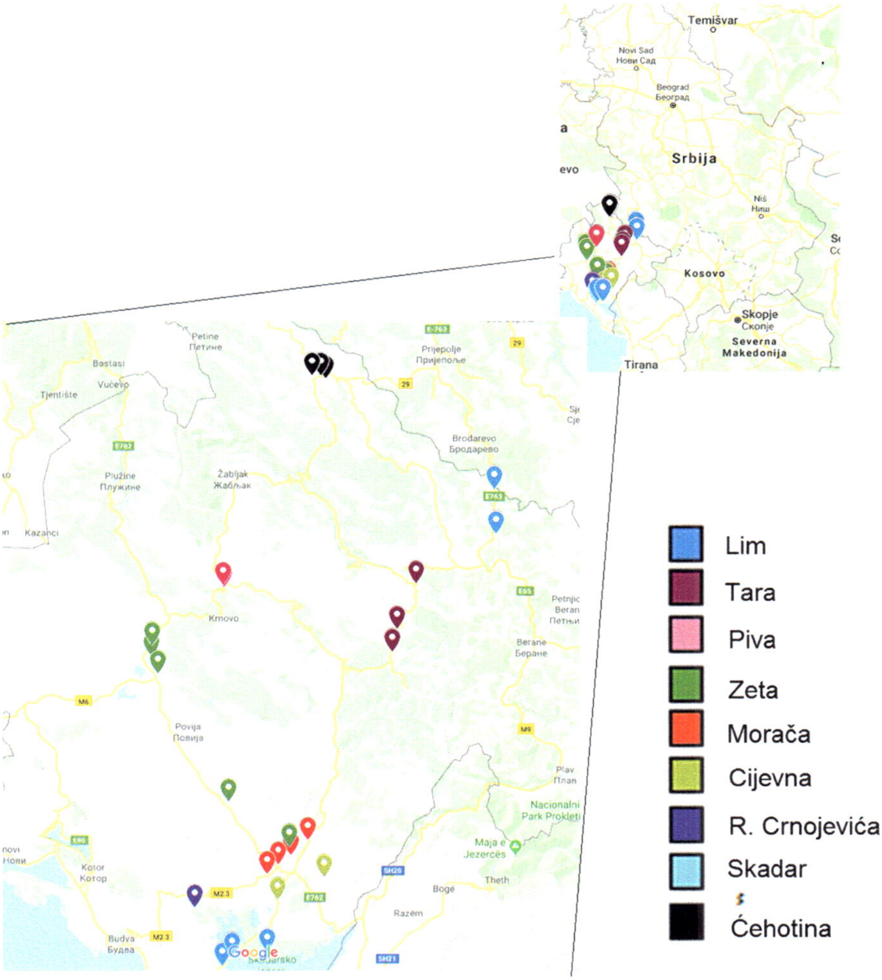

Fig. 5 Map of the sampling sites at which the water was collected for microbiological analyses within the Montenegro survey 2019 (maps were designed using Google Maps editor)

7.2 Methodology

Water samples were collected in 500 mL sterile glass bottles approximately 30 cm below the water surface. Samples were stored in dark at 4°C and transferred to the laboratory of the Institute for Marine Biology, Kotor, where the samples were further processed (within 6 h of sampling).

Detection of Total Coliforms, *E. coli* and Enterococci

Quantification of total coliforms (TC), *E. coli* (EC) and intestinal enterococci (IE) was performed by Colilert Quanti-Tray 2000 system, which provides a Most Probable Number (MPN) result, based on a color change/fluorescence in 97 wells. For both coliform bacteria and enterococci one dilution was analyzed (1:10 and 1:5, respectively). Powdered reagents Colilert-18 and E-enterolert were used for cultivation of coliforms and enterococci, respectively, following the instructions of manufacturer. For coliforms, trays were incubated at 37°C, for at least 18 h. For enterococci, incubation was performed at 44°C for at least 24 h. For assessment of the water quality based on total coliforms, *E. coli* and enterococci our classification scheme developed for the Danube River was used [7, 31].

7.3 Results

The results are summarized in Fig. 6. Based on the concentration of *E. coli* in analyzed samples, 40% of the sites were little polluted, 36% moderately, 16% critically, and per 4% strong and excessively polluted. When comparing the data obtained within the Montenegro survey 2019 and national monitoring dataset, significant correlation can be observed ($R^2 = 0.724$) as shown in Fig. 7.

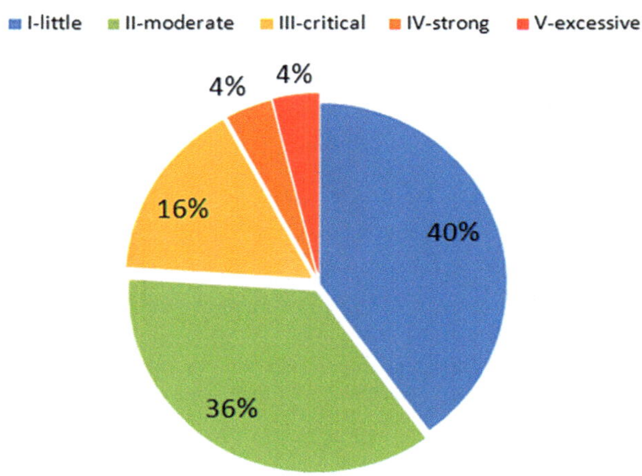

Fig. 6 Summary of data for indicators of faecal pollution (*E. coli*) obtained within the Montenegro survey in 2019

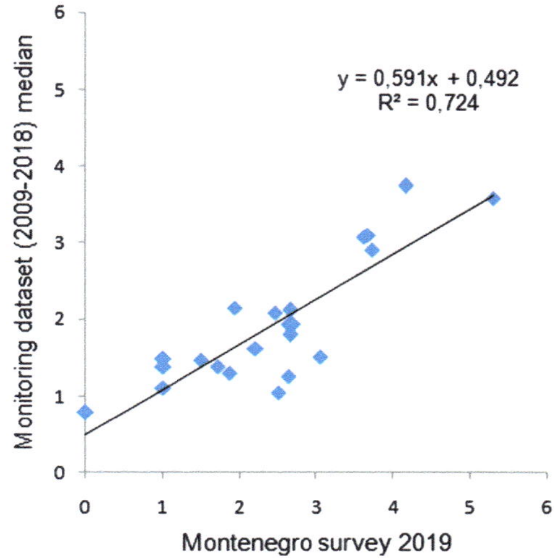

Fig. 7 Correlation between the data obtained within Montenegro survey 2019 (log (MPN + 1/100 mL)) and data from national monitoring (median log (CFU + 1/100 mL))

7.3.1 The Adriatic Sea Basin

In ASB, samples were collected at four sites at Morača, five sites at Zeta, two sites at Cijevna, one site at Rijeka Crnojevića, and three sites related to Skadar (Fig. 8). The most affected site at Morača, and for all rivers in ASB as well, was the site impacted by the wastewater outlet in Podgorica. At this site IE and EC indicated strong pollution which was not surprising considering that 93% of the samples from the annual dataset obtained from this site were heavily polluted. Critical level of pollution was recorded at *R. Crnojevića* according to EC and IE which is in compliance with annual data which indicated sporadic pollution events at this site. The best water quality was detected in upper parts of Zeta (*Vidrovan*), Cijevna (*Trgaj*), and Morača (*Zlatica*, upstream of Podgorica).

7.3.2 The Black Sea Basin

In BSB, samples were collected at three sites at Tara, three at Ćehotina, and per two sites at Lim and Piva rivers (Fig. 9). The most affected site in the BSB was the site at Ćehotina, situated *downstream of Pljevlja*, where EC and IE indicated an excessive level of pollution. The impact was evident on the site situated downstream (*downstream Vezišnica*) where critical level of pollution was recorded. Based on the fact that annual dataset indicated persistent pollution at these sites (83 and 59% of critically contaminated samples respectively) they were identified as hotspots of pollution. Critical level of pollution was also recorded at the river Lim *downstream of the town Bijelo Polje*. Impact was evident further downstream at the site

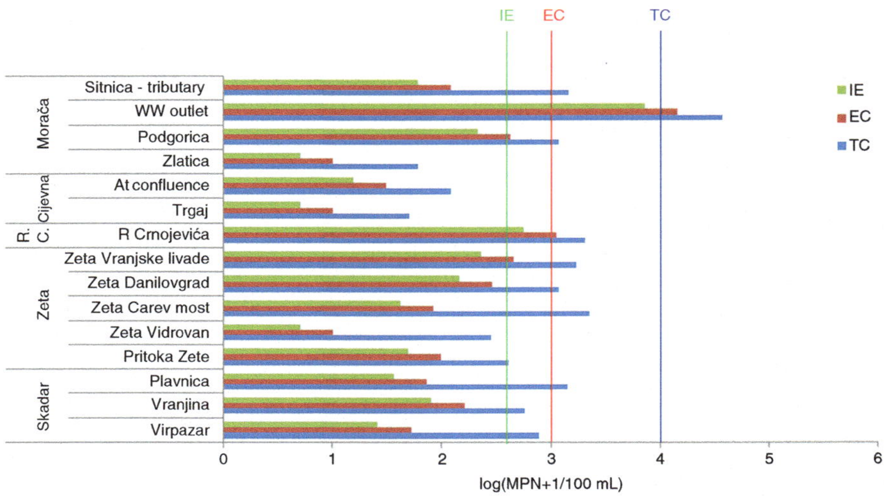

Fig. 8 Data on total coliforms (TC), *E. coli* (EC) and intestinal enterococci (IE) for sampling sites at rivers belonging to the Adriatic Sea Basin obtained within the Montenegro survey 2019. Values are presented as log (MPN + 1/100 mL). Vertical lines represent borders between class II and III for IE (green), EC (red), and TC (blue) as indication of critical pollution

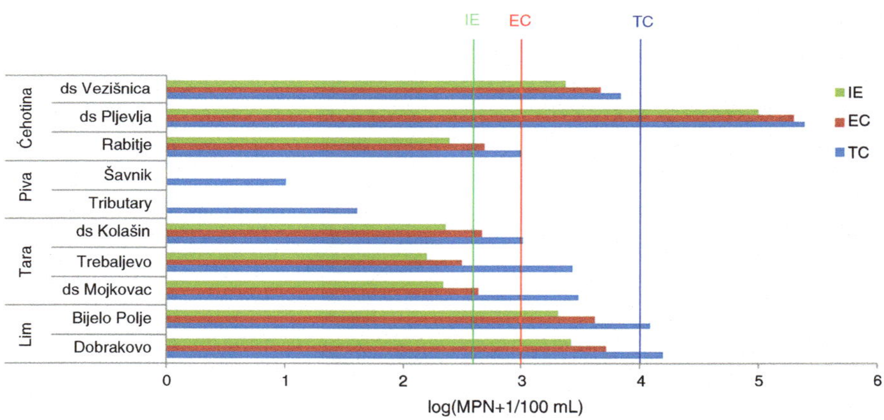

Fig. 9 Data on total coliforms (TC), *E. coli* (EC) and intestinal enterococci (IE) for sampling sites at rivers belonging to the Black Sea Basin obtained within the Montenegro survey 2019. Values are presented as log (MPN + 1/100 mL). Vertical lines represent borders between class II and III for IE (green), EC (red), and TC (blue) as indication of critical pollution

Dobrakovo. In the annual dataset, impact of Bijelo Polje was evident in 59% of the samples and therefore this site is also marked as a hotspot. All three sites at the Tara River showed moderate level of pollution. The Piva River had the best water quality; IE and EC were below the level of detection. The long-term monitoring dataset indicated occasional pollution events at the river Ibar but this river was excluded from the Survey in 2019 due to unfavorable hydrological and meteorological conditions in the basin of this river in the period of sampling.

7.4 Microbial Source Tracking

The concept that the origin of faecal pollution can be traced using microbiological, genotypic, phenotypic, and chemical methods has been termed microbial source tracking [36]. Quantitative PCR (qPCR)-based assays for the analysis of general-, human-, wastewater-, or animal-associated genetic Bacteroidetes faecal markers have gained increasing popularity in the field of faecal pollution analysis and microbial source tracking (MST) during recent years [37]. Bacteroidetes are one of the dominating bacterial groups in human and animal faecal excreta (representing up to 30% of the biomass) which show remarkable host adaptations [7]. The human-associated faecal marker BacHum [38] and modified version of the HF183II [39] could be used to track human-associated faecal pollution, ruminant-associated pollution can be traced by BacR qPCR assay [40], while Pig2Bac qPCR assay [41] is the method for detecting pig-associated sources. Considering that ruminants and pigs are major livestock commonly cultivated in Montenegro, we wanted to investigate prevalence of the human-associated pollution at the most affected sites.

7.4.1 Sample Preparation and DNA Extraction

Taking together the data available from the monitoring database and the data obtained within the Montenegro survey conducted in 2019, only four sites were identified as hotspots of faecal pollution. From these sites, 100 mL water samples collected from the different surface waters were frozen at −20°C immediately upon sampling and transferred to the laboratory. After thawing of the samples, (bacterial) biomass was filtered onto 0.2 μm polycarbonate Millipore filters which were subsequently frozen at −20°C. A clean filter was frozen and stored alongside the sample filters as filter control. DNA was extracted by a mechanical, bead-beating-based DNA extraction protocol with chemical/phenol-chloroform-based isolation of DNA [16, 37, 42]. DNA was eluted in 100 μL of Tris buffer.

7.4.2 Microbial Faecal Source Tracking Assays

The human-associated genetic faecal markers BacHum and HF183II were enumerated by quantitative PCR (qPCR) allowing the detection and quantification of presumable human-associated faecal pollution. The qPCR assays targeting ruminant-associated (BacR qPCR assay) and pig-associated genetic markers

(Pig2Bac qPCR assay) were included to test for possible animal faecal pollution. All qPCR assays were adapted to run on the Rotor-Gene Q qPCR-thermocycler with the Rotor-Gene Multiplex PCR master mix (Qiagen Inc.). Quantification was achieved by parallel quantification of a dilution series of standard plasmids of known concentration. Results are expressed as marker equivalents (ME) per 100 mL taking into account the filtration volume as well as the dilution factor used for the respective DNA extract [14].

7.4.3 Results

The human-associated genetic faecal markers BacHum and HF183II were detected in each sample of the four selected sites exhibiting considerable faecal pollution level, while the ruminant-associated BacR and the pig-associated Pig-2-Bac markers were below the level of detection for all sites. This suggests prevalence of human-associated sources of faecal pollution at these sites. Results are summarized in the Table 4. Trends in concentrations of ME/100 mL corresponded to those of *E. coli*. The site with the highest concentration of human-associated genetic faecal markers was the site *ds Pljevlja at Ćehotina* which could obviously be identified as the biggest hotspot of faecal pollution based on the obtained microbiological and molecular-biological data.

Table 4 Summarized data from the monitoring database, Montenegro survey 2019, and MST analyses for four sites identified as hotspots for faecal pollution in Montenegro

	Monitoring database 2009–2018	Montenegro survey 2019	MST (log of ME + 1/100 mL)				
Site	Critically polluted samples (%)	log (MPN *E. coli* + 1/ 100 mL)	Allbac	HF183II	BacHum	BacR	Pig2Bac
Lim ds Bijelo Polje	59	3.61	5.34	3.03	3.48	0.00	0.00
Morača ds WW outlet	93	4.16	5.45	3.22	3.69	0.00	0.00
Ćehotina ds Pljevlja	83	5.3	8.26	6.31	6.57	0.00	0.00
Ćehotina ds Vezišnica	59	3.66	4.81	3.09	2.92	0.00	0.00

8 Conclusions

In this chapter we have compiled data from 10 years of national monitoring (2009–2018) which provided an insight into the persistence of pollution at the selected sites with data from the Montenegro survey performed in 2019 as a status quo overview and the data from MST analyses, which provided indication on prevalence of pollution source. Despite the high risk that poor implementation of wastewater treatment might represent for Montenegro, the obtained dataset indicates that overall microbiological water quality of rivers in Montenegro is quite good. About 80% of the investigated sites had good or sufficient bathing water quality according to Montenegrin legislation. Only four sites were identified as hotspots of faecal pollution based on the high frequency of the critically polluted samples in the long-term monitoring data. This observation was additionally confirmed with the dataset of the Montenegro survey in 2019. MST data indicated prevalence of human-associated faecal pollution at the sites. Based on the analyzed data, the impact of wastewaters of the towns Podgorica, Bijelo Polje, and Pljevlja situated at the riverbanks of Morača, Lim, and Ćehotina is more than evident.

Acknowledgments This study was performed within the framework of the bilateral project of Republic of Serbia and Republic of Austria No. 451-03-02141/2017-09/30. We would like to acknowledge Luka Mitrović, Head of Institute of Hydrometeorology and Seismology – Montenegro, and our colleagues Željka Đurišić and Kumrija Šestović for provided monitoring dataset 2009–2018. The authors are grateful to Luka Gačić who provided improvements to our English.

References

1. DIKTAS (2012) Protection and sustainable use of the Dinaric Karst Transboundary Aquifer System. Montenegro country report
2. Kolarević S, Knežević-Vukčević J, Paunović M, Tomovic J, Gačić Z, Vuković-Gačić B (2011) Antropogenic impact on water quality of the River Danube in Serbia: microbiological analysis and genotoxicity monitoring. Arch Biol Sci 63:1209–1217
3. Savio D, Sinclair L, Ijaz UZ, Parajka J, Reischer GH, Stadler P, Blaschke A, Bloschl G, Mach R, Kirschner A, Farnleitner AH, Eller A (2015) Bacterial diversity along a 2600 km river continuum. Environ Microbiol 17:4994–5007
4. Bulatović J, Rajović G (2017) Some aspects of the sport tourism with the review of Montenegro: overview. Tour Educ Stud Pract 4:18–28
5. Government of Montenegro, Ministry of Agriculture and Rural Development (2017) Strategija upravljanja Vodama Crne Gore, p 178
6. Pešić V, Karaman GS, Kostianoy AG (2018) Introduction. In: Pešić V, Karaman G, Kostianoy A (eds) The Skadar/Shkodra Lake environment. The handbook of environmental chemistry, vol 80. Springer, Cham, pp 1–10
7. Kirschner AKT, Kavka G, Velimirov B, Mach RL, Sommer R, Farnleitner AH (2009) Microbiological water quality along the Danube River: integrating data from two whole-river surveys and a transnational monitoring network. Water Res 43:3673–3684

8. Kirschner AK, Kavka G, Reischer GH, Sommer R, Blaschke AP, Stevenson M, Vierheilig J, Mach RL, Farnleitner AH (2014) Microbiological water quality of the Danube River: status quo and future perspectives. In: Liska I (ed) The Danube River Basin. Springer, Berlin, Heidelberg, pp 439–468
9. Kirschner AKT, Reischer GH, Jakwerth S, Savio D, Ixenmaier S, Toth E, Sommer R, Mach R, Linke R, Eiler A, Kolarevic S, Farnleitner AH (2017) Multiparametric monitoring of microbial faecal pollution reveals the dominance of human contamination along the whole Danube River. Water Res 124:543–555
10. Kolarević S, Knežević-Vukčević J, Paunović M, Tomović J, Gačić Z, Vuković-Gačić B (2011) The anthropogenic impact on water quality of the river Danube in Serbia: microbiological analysis and genotoxicity monitoring. Arch Biol Sci 63:1209–1217
11. Kolarević S, Knežević-Vukčević J, Paunović M, Gačić Z, Vuković-Gačić B (2011) Assessment of the microbiological quality of the river Tisa in Serbia. Water Res Manag 1:57–61
12. Kolarević S, Knežević-Vukčević J, Paunović M, Vasiljević B, Kračun M, Gačić Z, Vuković-Gačić B (2012) Seasonal variations of microbiological parameters of water quality of the Velika Morava river. Arch Biol Sci 64:1017–1027
13. Kapetanović D, Smrzlić IV, Valić D, Teskeredžić E, Kolarević S, Sunjog K, Tomović J, Kračun-Kolarević M, Knežević-Vukčević J, Paunović M, Gačić Z, Vuković-Gačić B (2015) Microbial characterisation of the Sava river. In: Milačič R, Ščančar J, Paunović M (eds) The Sava river. Springer, Berlin, Heidelberg, pp 201–228
14. Vrzel J, Vuković-Gačić B, Kolarević S, Gačić Z, Kračun-Kolarević M, Kostić J, Aborgiba M, Farnleitner A, Reischer G, Linke R, Paunović M, Ogrinc N (2016) Determination of the sources of nitrate and the microbiological sources of pollution in the Sava River Basin. Sci Total Environ 573:1460–1471
15. International Commission for the Protection of the Danube River (2015) Danube river basin management plan. ICPDR, Vienna. https://www.icpdr.org/main/sites/default/files/nodes/documents/drbmp-update2015-annexes.pdf
16. Reischer GH, Haider JM, Sommer R, Stadler H, Keiblinger KM, Hornek R, Zerobin W, Mach RL, Farnleitner AH (2008) Quantitative microbial faecal source tracking with sampling guided by hydrological catchment dynamics. Environ Microbiol 10:2598–2608
17. Ford DC, Williams PW (2013) Introduction to karst. In: Karst geomorphology and hydrology. Wiley, Chichester
18. Budnick GE, Howard RT, Mayo DR (1996) Evaluation of Enterolert for enumeration of enterococci in recreational waters. Appl Environ Microbiol 62:3881–3884
19. Sinton LW, Donnison AM, Hastie CM (1993) Faecal streptococci as faecal pollution indicators: a review. Part II: sanitary significance, survival, and use. N Z J Mar Freshw Res 27:117–137
20. WHO (1997) Guidelines for drinking-water quality. Surveillance and control of community supplies, vol 3. 2nd edn. World Health Organization, Geneva
21. Rompre AP, Servais J, Baudart M, Roubin R, Laurent P (2002) Detection and enumeration of coliforms in drinking water: current methods and emerging approaches. J Microbiol Methods 49:31–54
22. Byamukama D, Mach RL, Kansiime F, Manafi M, Farnleitner AH (2005) Discrimination efficacy of fecal pollution detection in different aquatic habitats of a high-altitude tropical country, using presumptive coliforms, *Escherichia coli* and *Clostridium perfringens* spores. Appl Environ Microbiol 71:65–71
23. Evanson M, Ambrose RF (2006) Sources and growth dynamics of fecal indicator bacteria in a coastal wetland system and potential impacts to adjacent waters. Water Res 40:475–486
24. Raina PS, Pollari FL, Teare GF, Goss MJ, Barry DA, Wilson JB (1999) The relationship between *E. coli* indicator bacteria in well-water and gastrointestinal illnesses in rural families. Can J Public Health 90:172–175
25. Bitton G (2005) Microbial indicators of fecal contamination: application to microbial source tracking. Report submitted to the Florida Stormwater Association 719, pp 4–6

26. Kavka G, Poetsch E (2002) Microbiology. In: Joint Danube Survey – technical report of the International Commission for the Protection of the Danube River, pp 138–150
27. Bonde GJ (1977) Bacterial indication of water pollution. In: Advances in aquatic microbiology. Academic Press, London, pp 273–364
28. Cabral JP (2010) Water microbiology. Bacterial pathogens and water. Int J Environ Res Public Health 7:3657–3703
29. European Commission (2006) Directive 2006/7/EC of the European Parliament and of the Council of 15 February 2006 concerning the management of bathing water quality and repealing Directive 76/160/EEC
30. Vukašinović Pešić V, Blagojević N, Savić A, Tomić N, Pešić V (2019) The change in the water chemistry of the rivers of Montenegro over a 10-year period. In: Pešić V, Paunović M, Kostianoy A (eds) The rivers of Montenegro. The handbook of environmental chemistry. Springer, Cham
31. Kavka GG, Kasimir GD, Farnleitner AH (2006) Microbiological water quality of the River Danube (km 2581–km 15): longitudinal variation of pollution as determined by standard parameters. In: Proceedings of the 36th international conference of IAD. Austrian Committee Danube Research, Vienna, pp 415–421
32. Šundić D, Radujković B (2012) Pollution of the Skadar Lake. NGO Green Home. http://www.greenhome.co.me/fajlovi/greenhome/attach_fajlovi/eng/main-pages/2013/12/pdf/Pollution_of_the_Skadar_Lake.pdf
33. EPA (2013) Report on the state of the environment in Montenegro based on indicators. Environmental Protection Agency of Montenegro, Podgorica
34. EPA (2017) Report on the state of the environment in Montenegro based on indicators. Environmental Protection Agency of Montenegro, Podgorica
35. Tomović S (2008) Ciljevi vodoprivrednog razvoja Crne Gore. Vodoprivreda 40:127–137
36. Scott TM, Rose JB, Jenkins TM, Farrah SR, Lukasik J (2002) Microbial source tracking: current methodology and future directions. Appl Environ Microbiol 68:5796–5803
37. Mayer E, Reischer G, Ixenmaier S, Derx J, Blaschke A, Edbon J, Linke R, Egle L, Ahmed W, Blanch A, Byamuka D, Savill M, Mushi D, Cristobal H, Edge T, Schade M, Aslan A, Brooks Y, Sommer R, Masago Y, Sato M, Taylor H, Rose J, Wuertz S, Shanks O, Piringer H, Mach R, Savio D, Zessner M, Farnleitner A (2018) Global distribution of human-associated fecal genetic markers in reference samples from six continents. Environ Sci Technol 52:5076–5084
38. Kildare BJ, Leutenegger CM, McSwain BS, Bambic DG, Rajal VB, Wuertz S (2007) 16S rRNA-based assays for quantitative detection of universal, human-, cow-, and dog-specific fecal Bacteroidales: a Bayesian approach. Water Res 41:3701–3715
39. Green HC, Haugland RA, Varma M, Millen HT, Borchardt MA, Field KG, Walters WA, Knight R, Kelty CA, Shanks OC (2014) Improved HF183 quantitative real-time PCR assay for characterization of human fecal pollution in ambient surface water samples. Appl Environ Microbiol 80:3086–3094
40. Reischer GH, Kasper DC, Steinborn R, Mach RL, Farnleitner AH (2006) Quantitative PCR method for sensitive detection of ruminant fecal pollution in freshwater and evaluation of this method in alpine karstic regions. Appl Environ Microbiol 72:5610–5614
41. Mieszkin S, Furet JP, Corthier G, Gourmelon M (2009) Estimation of pig fecal contamination in a river catchment by real-time PCR using two pig-specific Bacteroidales 16S rRNA genetic markers. Appl Environ Microbiol 75:3045–3054
42. Griffiths RI, Whiteley AS, O'Donnell AG, Bailey MJ (2000) Rapid method for coextraction of DNA and RNA from natural environments for analysis of ribosomal DNA-and rRNA-based microbial community composition. Appl Environ Microbiol 66:5488–5491

The Biodiversity and Biogeographical Characteristics of the River Basins of Montenegro

Vladimir Pešić, Michał Grabowski, Sead Hadžiablahović, Drago Marić, and Momir Paunović

Contents

V. Pešić (✉) and D. Marić
Department of Biology, University of Montenegro, Podgorica, Montenegro
e-mail: vladopesic@gmail.com; dragomrc@yahoo.com

M. Grabowski
Department of Invertebrate Zoology and Hydrobiology, University of Lodz, Łódź, Poland
e-mail: michal.grabowski@biol.uni.lodz.pl

S. Hadžiablahović
Nature and Environment Protection Agency of Montenegro, Podgorica, Montenegro

Faculty of Philosophy, University of Montenegro, Podgorica, Montenegro
e-mail: seadh@t-com.me

M. Paunović
Institute for Biological Research "Siniša Stanković", University of Belgrade, Belgrade, Serbia
e-mail: mpaunovi@ibiss.bg.ac.rs

Vladimir Pešić, Momir Paunović, and Andrey G. Kostianoy (eds.), *The Rivers of Montenegro*, Hdb Env Chem (2020) 93: 157–200, DOI 10.1007/698_2019_414,

Abstract The territory of Montenegro includes three freshwater ecoregional units: the Dalmatian, the Southeastern Adriatic, and the Dniester-Lower Danube Ecoregions. The boundaries of the Dalmatian and Southeastern Adriatic Ecoregions in Montenegro are not clear and require revision. Contrary to the accepted FEOW global freshwater ecoregion system, which places the eastern border of the Dalmatian Ecoregion near the town of Budva, we have accepted that the eastern border of the Dalmatian Ecoregion ends or overlaps with the eastern boundary of the Trebišnjica river basin (Lake Bileć). The northwestern boundaries of the Southeastern Adriatic Ecoregion overlap with the distribution border of the fish *Pelasgus minutus*. We emphasize that not only those fish that are traditionally used to model ecoregional delineations of freshwater ecosystems but also other groups of organisms, such as underground fauna, have proven to be good biogeographical indicators, especially in the karst areas where surface running waters are scarce and drains underground.

The karst had a decisive influence on the hydrographic isolation and vicariance, which has led to high numbers of so-called local endemics, whose distribution has been useful in understanding the biogeographic patterns of river basins in Montenegro. In this chapter, we provide an overview of the diversity and ecology issues related to the aquatic biota of the main river basins in Montenegro and identify the main gaps in our knowledge on these unique ecosystems.

Keywords Adriatic basin, Danube basin, Diversity, Ecoregional delineation, Endemism, Montenegro, Rivers

1 Introduction

Due to its complex geological and paleogeographical history, its climatic conditions and its pedological and orographic characteristics, Montenegro is characterized by an extraordinary rich and diverse aquatic biota that inhabits various types of running waters. Most rivers run through narrow valleys which are isolated from each other that lead to geographical isolation and consequently to allopatric speciation and vicariant events resulting in a high endemism among local biota. This endemism, which is particularly impressive in the southern parts of Montenegro, has caught the attention of a large number of researchers for almost two centuries (see [1] for an overview), who were mainly focused on systematics and taxonomy issues. On the other hand, biogeography of the Montenegrin river basins, including a regional-scale review of the biota inhabiting these ecosystems, has not been the focus of research. The biogeographical regionalization of Montenegrin freshwater systems, which is often considered as the first step of water resource protection plan [2], was generally based on broader biogeographic freshwater divisions. In the ecoregion concept proposed by J. Illies [3], Montenegro was assigned to only one ecoregion – the Dinaric Western Balkans (Ecoregion 5). This region seems to be too large to be used as an a priori class to describe all of the various aquatic community types and to provide a framework for regional managements of biodiversity and aquatic resources. Nevertheless, despite criticism of Illies' classification (see [4]), this system today is still widely used for implementing the European Union Water Framework Directive 2000/60/EC [5].

Since 2008, the ecoregion concept adopted by the Freshwater Ecoregions of the World (FEOW) [6] has become a widely used tool for describing and evaluating the diversity and ecological and evolutionary patterns of aquatic biota. This concept is primarily based on freshwater fish species [7] and in recent years has become popular especially among conservation biologists. Montenegro, with a surface of 13,812 km^2, comprises parts of the three FEOW ecoregional units (Fig. 1; see below under *Ecoregional delineations*).

Despite the general benefit that the FEOW concept has brought to the biogeographical characterization of Montenegrin freshwater, a comprehensive analysis of the achievements of the proposed system for the main river basins is still lacking. In this chapter we provide a brief overview of the knowledge on some well-known groups of aquatic biota in Montenegro, which are recognized as useful biogeographic indicators. Recently published data has demonstrated that some of the aquatic biotic taxa that cannot readily disperse over watershed barriers, such as fish, aquatic snails, and troglobiotic fauna can be used in redefining the ecoregional boundaries in Montenegro. Moreover, we offer a brief review of the main river basins in Montenegro with respect to their biodiversity and endemism. This will help to identify the main gaps in our knowledge on the distribution and diversity of the main aquatic biotic groups inhabiting running water ecosystems but also encourage the better protection and management of threatened water resources in Montenegro.

2 The (Uncertain) Paleogeography of the Hydrological Basins of Montenegro

2.1 The Geological Setting

From the geotectonic point of view, the territory of Montenegro, together with all the other parts of the Balkan Peninsula and most of the Mediterranean Region, belongs to the Alpine belt (Alpides), also referred to as Neo-Europe. The latter name reflects the fact that this highly mountainous part of the continent was consolidated as the last one, through the accretion of Gondwana-derived continental microplates. It emerged from the late Paleozoic-Mesozoic-Tertiary

Fig. 1 The freshwater ecoregion delineations of Montenegro according to the FEOW [6]. (1) The Dniester-Lower Danube Ecoregion, (2) the Southeastern Adriatic Ecoregion, (3) and the Dalmatian Ecoregion

ocean of Tethys, predominantly during Cretaceous-Cenozoic orogenic events, mainly those of Alpidic/Apine orogeny [8]. The Alpidic orogeny has not yet been completed; active tectonic processes in the area, reflected, *inter alia*, by frequent earthquakes, are still ongoing, as the Mediterranean Sea is not closed [9].

The mountain range known as the Dinarides, constituting the southeast continuation of the Alps and extending longitudinally along the Adriatic coast, covers the whole Montenegro's territory. The highest mountains of the Dinarides are, generally, situated along the watershed between the Black Sea (Danube) and the Adriatic basins. Among them, only the highest groups of peaks rising well above 2,000 m above sea level, such as the Durmitor and Prokletije mountains in Montenegro, have Alpine characteristics. The main ridges comprising the watershed between the Black, Adriatic, and Aegean Seas extend toward the east and northeast. The watersheds possibly reflect the Alpidic orogeny collision zones of the tectonic plates shifting from the southwest (the Italian Adriatic plate) and from the southeast (the Aegean-Macedonian-Albanian plate) [10].

The Dinarides in Montenegro belong to two major geomorphological units: the Inner and the Outer Dinarides [11] (Fig. 2). The boundary between the two units in Montenegro runs along the line between Sarajevo in Bosnia and Herzegovina, through Kolašin and Skadar/Shkodra to Elbasan in Albania [10]. The Inner Dinarides, composed predominantly of Paleozoic and Mesozoic limestones covered by Neogene lacustrine deposits, flysch, and ophiolites, are located in northern Montenegro. They include some karst areas, such as the Pešter Plateau, on the border between Montenegro and Serbia, with the longest (c.a. 17 km) cave in Montenegro – Pećina nad Vražjim Firovima (Đalovića Pećina). The major river systems in the Inner Dinarides, i.e., the Ćehotina, the Piva, the Tara, and the Lim in Montenegro, belong to the Danube catchment area and, effectively, to the Black Sea basin. The Outer Dinarides are characterized by the domination of thick layers of Triassic, Jurassic, and Cretaceous limestone. The highest mountain areas in Montenegro, i.e., Durmitor and Prokletije, reaching way above 2,000 m above sea level, are found here, but also the largest plain area, i.e., the Zeta Plain, with an average elevation of c.a. 40 m above sea level, as well as the major basin of Lake Skadar. The, so far, deepest (−775 m) cave in Montenegro, Jama na Vjetrenim brdima, is located in Durmitor. This area is also known as the High Karst (*Visoki Krš*) and has a well-developed subterranean hydrological network,

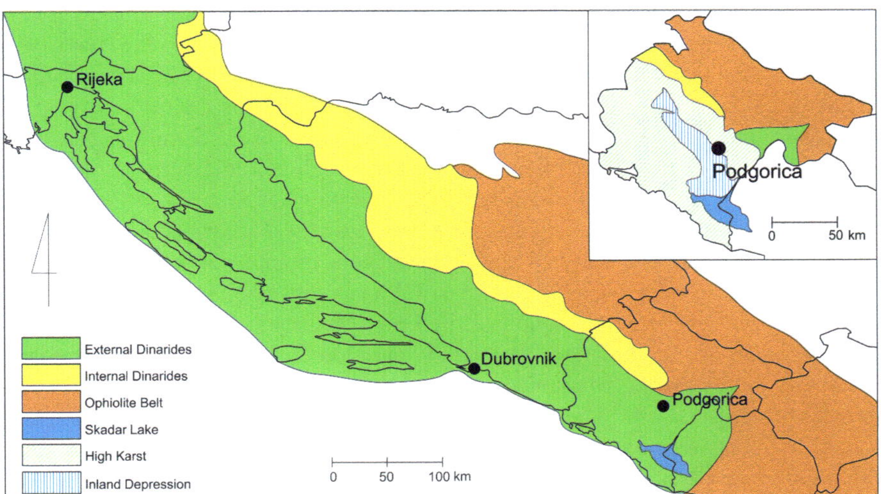

Fig. 2 Geomorphological regionalization of the Dinarides mountain range (modified after [12]). Inset: geomorphological regionalization of Montenegro (modified after [13])

alongside the presence of karstic spring systems, often with very high discharge rates, exceeding 10 m^3/s. The Morača river system with its main affluent the Zeta as well as the Crnojevića drain the High Karst into Lake Skadar. The lake discharges its waters through the River Bojana/Buna into the Adriatic. A number of small coastal rivers and streams drain the Primorje (Littoral) Mountains directly into the sea. Some of the coastal springs discharge into the Adriatic below sea level, e.g., in Boka Kotorska Bay.

2.2 *Uncertain Paleogeography*

Generally, as in the case of most sedimentary rocks in southern Europe, the formations in the Dinarides and in the northern Hellenides accumulated in the vast, relatively stable shallow area of the ancient Tethys Ocean that originated over 250 million years ago and which, since the Jurassic Age, ca. 150 million years ago, separated Laurasia from Gondwana [14]. The collision of the African and Eurasian continental plates leads to the gradual closure of Tethys and to the formation of the semi-closed Proto-Mediterranean basin. In the Cretaceous period, the Dinaric area started to emerge from the waters of the Tethys, and from as early as the Palaeogene, it was a large uplifted terrain bordered to the south by the also uplifting Hellenids and, to the west, by the present Adriatic basin. Many of the river valleys of the Adriatic tributaries in the Dinarides and the Hellenides, including at least some of the present drainages, may have originated in the Paleogene, some 30 million years ago, and passed through a lacustrine phase until some 10 million years ago, when the presently known rivers finally formed [15]. Such a hypothesis on the ancient origin of local drainages is concordant with the estimated time of divergence of the cyprinid fish *Telestes montenegrinus*, endemic to the rivers of the Skadar basin, from its closest relative, *T. muticellus*, widely distributed in the Apennine Peninsula and northern Adriatic drainages and also with the divergence times of other Balkan species of *Telestes* (from ca. 16 to ca. 5 million years ago) [16]. However, it must be remembered that the high tectonic activity of the area resulted in apparent alterations to the local hydrological systems, at least several times in their history. The schedule and nature of these changes as well as their impact on the biogeography of the aquatic biota remain almost unstudied. The closure of Tethys finalized at ca. 5.96 million years ago, in the late Miocene, resulting in a complete desiccation of the Proto-Mediterranean Sea. This relatively short episode of predominantly terrestrial conditions with some hypersaline water bodies is known as the Messinian salinity crisis. The extensive karstification and erosion processes in the Dinarides, and thus the formation of the High Karst, are often hypothesized to be a result of the Miocene regression of the Proto-Mediterranean and concomitant climatic changes [17]. After what were probably a few intrusions of brackish/fresh waters from the Paratethys Sea, the Proto-Mediterranean basin was refilled by the Atlantic waters at ca. 5.33 million years ago, during the so-called Zanclean Flood, marking the beginning of the Pliocene and of the present Mediterranean Sea [18–21]. The aridization of the climate started from as early as the Miocene and intensified greatly during the Pleistocene glaciations, further modifying the hydrology in the Balkans and in the entirety of southern Europe, leading to a periodical depletion of the riverine network, with isolated spring systems and lakes that could serve as a microrefugia for aquatic biota [22–24]. The origins of the Lake Skadar basin and the associated system of springs are also quite obscure. However, it is clear that the springs, with numerous faunal endemics, are millions of years older than the large and shallow lake that we know today and that came to existence only within the last 1,200 years, largely due to local tectonic activity (for a review and discussion see [25]).

In conclusion, due to its turbulent geological history over the last 30 million years, the Dinaric area as well as the whole Balkan Peninsula has been undergoing enormous and dynamic landscape change, resulting in great complexity. The paleoenvironmental reconstructions and also the present-day physiographic patterns show significant habitat diversity and patchiness for both terrestrial and aquatic ecosystems [26]. The formation of local hydrological networks has been a very complex process, with the temporary existence of large lake systems, the fragmentation of drainages, frequent river captures, and progressing karstification. At present, the hydrological network is composed of only a few larger and hundreds of smaller drainages with relatively unstable watersheds [27]. Such habitat patchiness has been a stimulus for a plethora of allopatric, peripatric, and parapatric speciation events in the area (see [24, 28–30]). In consequence, the Mediterranean region, and the Western Balkans, in particular, is recognized as one of the most important present-day hot spots of biodiversity and endemism globally and a model system for studies of biogeography and evolution of numerous organisms [31–34].

3 The Diversity and Biogeographic Characteristics of the Riparian Vegetation and Aquatic Biota of the Running Waters of Montenegro

3.1 Riparian Vegetation

Riverine riparian vegetation includes vegetation alongside rivers that is influenced by the adjacent water body. The transitional character of the riparian zone is characterized by the existence of distinctive hydrological, soil, and biotic conditions, which results in the presence of several distinct riparian plant communities, some of which are recognized as habitat types of European importance by the EU and require protection measures [35]. Different forms of anthropogenic activity, especially the excessive exploitation of sand and gravel, as well as large infrastructure facilities (such as the construction of highways, bridges, and tourist facilities), affect a change in the composition as well as the disappearance of riparian vegetation, sometimes even in areas that are strictly protected. The rich riparian and aquatic vegetation with abundant charophyte populations at the confluence of the River Plavnica completely vanished as a result of the construction of tourist facilities in the protected area of the Lake Skadar National Park [36].

Riparian vegetation along the banks of large rivers in Montenegro includes different types of forest vegetation, most of them included in the Catalogue of Habitat Types of EU Importance of Montenegro [37].

91E0 Alluvial forests with *Alnus glutinosa* and *Fraxinus excelsior* (*Alno-Padion, Alnion incanae, Salicion albae*). Forests (*Salicetum albo-fragilis*) dominated with the white willow (*Salix alba*) and the brittle willow (*Salix fragilis*) occur along the river- and lake banks at altitudes from 50 to 700 m asl. Communities with black alder (*Alnus glutinosa*) occupy rivers at altitudes from 50 to 900 m asl, while riparian forest (*Oxali-Alnetum incanae*) dominated by gray alder (*Alnus incana*) can be found in the sub-Mediterranean region at elevations between 800 and 1,100 m. According to Petrović et al. [37], this type of habitat is present along the banks of the Zeta, the Morača, the Lim, the Tara, the Ćehotina, and the Ibar Rivers, as well as the tributaries of the latter river close to Rožaje.

91F0 Riparian mixed forests of *Quercus robur*, *Ulmus laevis*, and *Ulmus minor*, *Fraxinus excelsior*, or *Fraxinus angustifolia*, along the great rivers (*Ulmenion minoris*). This type of habitat could be developed along the coast of the Lim River, on recent alluvial deposits [37].

92A0 *Salix alba* and *Populus alba* galleries. This type of habitat is present along the banks of the rivers in the Skadar Lake area (such as the Gostiljska Rijeka) and the Bojana/Buna River.

92D0 Southern riparian galleries and thickets (*Nerio-Tamaricetea* and *Securinegion tinctoriae*). This type of habitat (*Vitex agnus-castus* community and fragments of *Tamarix dalmatica* community) is present on the banks of some intermittent rivers such as the Cijevna/Cemi River, especially in their lower reaches, as well as on the mouths of some rivers such as the Rivers Morača and Bojana/Buna.

The Catalogue of Habitat Types of EU Importance of Montenegro [37] comprises also eight habitat types in the section Freshwater Habitats – Running Water Present in Montenegro:

3220 Alpine rivers and the herbaceous vegetation along their banks – communities of herbaceous plants (e.g., *Calamagrostis pseudophragmites* and *Epilobium dodonaei*) are present on the pebbly banks of many mountain rivers in Montenegro [37, 38].

3230 Alpine rivers and their ligneous vegetation with *Myricaria germanica* – communities of low shrubs dominated by European grain (*Myricaria germanica*) and the scaly river willow (*Salix* spp.) are recorded on gravel deposits rich in fine silt of the Rivers Lim and Tara and their tributaries [37, 38].

3240 Alpine rivers and their ligneous vegetation with *Salix elaeagnos* – communities dominated by the gray willow *Salix elaeagnos* developed on various types of alluvial deposits, mainly on the limestone substrate dominated by gravel and sand, were found along the banks of mountain rivers such as the Lim, the Tara, the Komarnica, the Bukovica, the Tušina, the Bijela, the Pridvorica, the Bistrica, Bjelojevićka Rijeka, and the Mojkovačka Bistrica [37].

3250 Constantly flowing Mediterranean rivers with *Glaucium flavum* – communities dominated by *Glaucium flavum* and *Oenothera biennis* are present in the valleys of the River Bojana/Buna and other smaller rivers in the Mediterranean and sub-Mediterranean part of Montenegro [37].

3260 Water courses of plain to montane levels with the *Ranunculion fluitantis* and *Callitricho-Batrachion* vegetation – communities with submerged or flotant vegetation of waterfalls (*Ranunculion fluitantis* and *Callitricho-Batrachion*) and aquatic mosses are probably common in Montenegrin rivers, but the distribution data are scarce [37]. A well-developed community dominated by *Ranunculus trichophyllus* is present along the River Zeta. On the Gostiljska Rijeka (a northern tributary of Lake Skadar), this type of habitat is represented by the *Ranunculus trichophyllus* community and *Hippuris vulgaris* var. *fluviatilis* community.

3270 Rivers with muddy banks with *Chenopodion rubri* p.p. and *Bidention* p.p. vegetation – 1-year pioneering nitrophilic vegetation of *Chenopodion rubriques* p.p. and *Bidention* p.p. develop on the muddy banks of the rivers of the lowland and mountainous regions. According to Petrović et al. [37], this type of habitat could be developed along the coasts of the largest Montenegrin rivers that have a slow water flow, such as the Rivers Zeta and Bojana/Buna, and the lower parts of the Morača and Lim Rivers, as well as around Lake Skadar.

3280 Constantly flowing Mediterranean rivers with *Paspalo-Agrostidion* species and hanging curtains of *Salix* and *Populus alba* – communities dominated by *Paspalum paspalodes* are present on the muddy banks of perrenial rivers (the Zeta, the Crmnica, Rijeka Crnojevića, and the Bojana/Buna) in the Mediterranean and sub-Mediterranean part of Montenegro [37, 39].

3290 Intermittently flowing Mediterranean rivers of the *Paspalo-Agrostidion* – communities (with characteristic species *Paspalum paspalodes*, *Polygonum amphibium)* inhabit the muddy and nitrified shores of intermittent rivers in the Mediterranean and sub-Mediterranean part of Montenegro [37].

Vegetation of freshwater springs, shorelines, and swamps in Montenegro is represented by several communities belonging to the following classes: *Montio-Cardaminetea*, *Littorelletea uniflorae*, *Isoëto-Nanojuncetea*, and *Phragmito-Magnocaricetea*. The class *Montio-Cardaminetea* that comprises vegetation developing in cold and well-oxygenated springs is presented with different communities from the *Cardamino-Montion* alliance. The classes *Littorelletea uniflorae* and *Isoëto-Nanojuncetea* are listed as "3130 Oligotrophic to mesotrophic-standing waters with vegetation of the *Littorelletea uniflorae* and/or of the *Isoëto-Nanojuncetea*" habitat type in Catalogue of Habitat Types of EU Importance [37]. This type of vegetation is characteristic for the banks of oligotrophic to mesotrophic lakes, ponds, and pools but can be found along the banks of rivers such as the Crnojevića Rijeka and the Gostiljska Rijeka.

The class *Phragmito-Magnocaricetea* (reed swamp, sedge bed, and herbland vegetation of freshwater or brackish water bodies and streams) can also be found along the banks of rivers (such as the Zeta, the Lim, the Matica, Gostiljska Rijeka, Rijeka Crnojevića, the Karatuna) where it is represented with communities belonging to the following alliances: *Phragmition communis*, *Glycerio-Sparganion*, *Carici-Rumicion hydrolapathi*, and *Eleocharito palustris-Sagittarion sagittifoliae*.

The class *Molinio-Arrhenetheretea* is represented by communities from the alliances *Calthion palustris* (present along the banks of Gostiljska Rijeka) and *Filipendulo-Petasition*, the latter belonging to "6430 Hydrophilous tall herb fringe communities of plains and of the montane to alpine" habitat type, present on the banks of the Biogradska Rijeka (Fig. 3).

3.2 Algae

Algae, which represent important components of running water assemblages, are generally dominated by diatoms (Bacillariophyceae), Chlorophyta (green algae), and Cyanobacteria (blue-green algae). Petković [40] studied the autumn aspect of the algae assemblages of five rivers in Montenegro (the Rivers Morača, Zeta, Lim, Piva, and Tara) and reported 150 species belonging to diatoms, green algae, blue-green algae, Pyrrophyta (dinoflagellates), Euglenophyta (euglenoids), and Rhodophyta (red algae). Diatoms dominated in taxa number and abundance and were mainly represented by the periphyton species attached to a wide variety of substratum.

Fig. 3 The mouth of the River Biogradska rijeka into Lake Biogradsko is located in a forest reserve of 1,600 ha, which preserves one of the last remaining "primeval" old-growth beech forests (*Lamio orvalae-Fagetum*) in Europe. A well-developed community dominated by *Petasites hibrydus* (Butterbur) belonging to the "6430 Hydrophilous tall herb fringe communities of *Filipendulo-Petasition* alliance" habitat type is present here. Photo S. Hadžiablahović

The study showed greater species richness in the slowly flowing and lowland reaches of the larger rivers (e.g., the River Lim, 66; the Tara, 64; and the Morača, 53 spp.) than on the steep reaches (e.g., the River Piva with 33 species). The highest number (94 taxa) was recorded for the River Zeta [40]. The study by Petković [40] reveals a small difference in the algae composition between lotic and lentic habitats. The algae assemblages in lentic habitats were dominated by diatoms, green algae (Chlorococcales), and desmid algae (Zygnematophyceae), while in lotic habits, the dominant types were diatoms and the chrysophyte *Hydrurus foetidus*, a species common in mountain streams in early spring and also in lowland rivers [41].

In general, the algae assemblages of running waters were dominated by benthic algae, but the larger and deeper lowland rivers sustain a phytoplankton component. Petković [40] mentioned that the phytoplankton community of the lentic and deeper parts of the Zeta and the Morača was dominated by green algae (Chlorococcales) and to a lesser extent by blue-green algae, chrysophytes, and red algae.

As emphasized by many studies, the development of algal assemblages in the river environment is affected by numerous factors, of which temperature and the availability of nutrients and lights are recognized as the main driver of changes in their biomass and taxonomic composition [42]. Moreover, the spatial and temporal pattern of algal assemblages depends on anthropogenic activity, which makes them a good indicator for the ecological monitoring of rivers [42, 43]. Diatoms in particular have been routinely used for the monitoring of surface water in Europe [5]. Recently, Rakočević [44] proved the applicability of epilithic diatoms in the ecological assessment of Montenegrin mountain rivers.

3.3 Aquatic Macrophytes

Macrophytes, which include large algae (e.g., Charophytes), bryophytes, and vascular plants, are recognized as an important group for assessing the water quality of surface water. Literature on the macrophytes of the running waters of Montenegro is scant, and most studies have focused on larger water bodies, such as Lake Skadar/Shkodra, while the running water systems have not yet been studied sufficiently. Hadžiablahović [35] gives an overview of the current state of the research on the vegetation and diversity of the vascular plants of Lake Skadar, while Blaženčić et al. [36] provides a systematic overview of Charophyta in the Lake Skadar basin.

The occurrence of macrophytes depends on a number of factors such as water flow velocity, catchment geology, and the substrate composition [42]. Haslam [45] lists three main factors that diminish river vegetation: the higher water velocity of rivers at high altitude, river straightening at low or medium altitudes, and pollution. In the middle and lower reaches of the rivers where the water flow is slow enough (and often creates a system of secondary channels and oxbows), macrophytes can become very abundant, often showing the dominance of those species that prefer higher trophic states.

The freshwater aquatic vegetation of the classes *Lemnetea* and *Potamogetonetea* is especially well developed on the small tributaries of Lake Skadar and in the River Zeta. The freshwater aquatic vegetation of the class *Potamogetonetea* of the River Zeta is represented by several communities (i.e., *Nymphaeo albae-Nupharetum luteae*, *Potamogetonetum denso-nodosi*, *Potamogetonetum lucentis*, *Potamogetonetum pectinati*, *Potamogetonetum perfoliati*, *Myriophylletum spicati*, and *Ceratophylletum demersi*) belonging to the habitat type "3150 Natural eutrophic lakes with *Magnopotamion* and *Hydrocharition* type vegetation." Semiaquatic plant communities or their fragments from the *Phragmito-Magnocaricetea* class (i.e., *Phragmites australis* comm., *Scirpus lacustris* comm., *Myosotis scorpioides* comm., *Mentha aquatica* comm., *Oenanthe aquatica* comm., *Veronica beccabunga* comm., *Cyperus longus* comm., *Paspalum paspalodes* comm., and so on) are also present along the watercourse. The faster stretches of the River Zeta, which have a sandy-gravel bottom, are predominated populated by *Myriophylletum spicati* and *Potamogetonetum pectinati* communities. Bushes of the willow moss *Fontinalis antipyretica* are also very abundant here.

The class *Lemnetea* is represented by several communities (such as *Lemno minoris-Riccietum fluitantis*, *Lemno-Utricularietum*, *Hydrocharitetum morsus-ranae*) that can be found on some scattered places in the Rivers Zeta and Bojana/Buna.

A *Potamogetono pectinati-Myriophylletum spicati* community is predominated in the aquatic vegetation of the River Karatuna (Fig. 4). Apart from the dominant species, an important role in forming this community is also played by *Ceratophyllum demersum*, *Potamogeton perfoliatus*, *P. lucens*, and *Sparganium emersum*. The community of charophyte species *Nitella gracilis* belonging to the habitat type "3140 Hard oligo-mesotrophic waters with benthic vegetation of *Chara* spp." is also present. In the lower part of the River Karatuna, the aquatic vegetation is represented by several communities (i.e., *Nymphaeo albae-Nupharetum luteae*, *Potamogetonetum perfoliati*, and *Ceratophylletum demersi*) or their fragments that belong to habitat type "3150 Natural eutrophic lakes with *Magnopotamion* and *Hydrocharition* type vegetation." At Gostiljska Rijeka, the latter habitat type is represented by the communities *Potamogetonetum denso-nodosi*, *Sparganio emersi-Potamogetonetum pectinati*, *Potamogetono pectinati-Myriophylletum spicati*, and *Nupharetum pumilae*.

3.4 Macroinvertebrates

In terms of diversity and abundance, macroinvertebrates have often been emphasized as the most important animal assemblage of running waters [46]. Macroinvertebrate assemblages include two groups with different life cycles, specifically fully aquatic taxa on the one hand, and aquatic insects with a terrestrial dispersal stage on the other. The differences in the life cycle of these two groups

Fig. 4 The River Karatuna, an outflowing stream of Lake Malo Blato, formed from the Rivers Biševina and Šegrtnica, which connect Lake Skadar and Malo Blato. Photo S. Hadžiablahović

can affect the diversity of indexing macrobenthic assemblages, presumably as a result of the sampling time [47] which requires a multiple sampling survey in order to give an appropriate estimation of overall macroinvertebrate richness.

Aquatic invertebrates are still insufficiently studied in Montenegro, and, with some exceptions (such as the study of the life cycle of endemic *Homoeogammarus scutarensis* in [48]), there is a lack of detailed ecological studies. Most studies have focused on faunistic or taxonomic research. There are a large number of endemic species, and many of them are still awaiting full description. This is especially true for the biotic groups that inhabit more isolated habitats such as springs and groundwater. The greatest level of endemism were found among freshwater hydrobioid snails – about 82% of the snails inhabiting the Lake Skadar/Shkodra basin are point endemics [49]. Recent literature has provided more information on the diversity of aquatic mollusk species inhabiting Montenegro, especially its southern part [49–52].

The annelid fauna of Montenegro has also been intensively studied, and thus far 113 species have been reported, of which 82 species and subspecies belong to Oligochaeta [53], 29 species and subspecies are Hirudinea [54, 55], and 2 species are Brachiobdellida [56]. The most abundant Oligochaeta species in the rivers of the Danube basin is *Nais elinguis* which in the River Ibar reach an abundance of 35,518.36 ind. m^{-2}. *Nais barbata* is the most dominant species in the rivers of the Adriatic Sea basin with a peak density of 6,367.34 ind. m^{-2} reported for the River Mrtvica [53]. Water mites (Hydrachnidia) are represented by 190 species [57, 58], diversified to occupy different types of running waters, even in intermittent conditions [59]. The endemism of both groups, the Oligochaeta and the water mites, is greatest in the southern part of Montenegro (see [46] for an overview).

Among the crustaceans, the greatest number of endemic species is also found in the southern part of Montenegro. Twelve of the thirty-four malacostracan species that have been recorded so far from the Lake Skadar basin are endemics [46]. Of the crustaceans that inhabit running waters, the endemic amphipod *Homoeogammarus scutarensis* [60] and freshwater shrimps [61] represent especially interesting biogeographic indicators. Approximately 90 stygobiotic species are known from the aquatic subterranean environment of the Lake Skadar/Shkodra basin, with a high number of endemics especially among crustaceans (Copepoda, Amphipoda, and Ostracoda) and gastropods (for an overview of the troglobiotic aquatic fauna of the Lake Skadar basin, see [62]).

Insect communities dominate, both in terms of diversity and abundance, in the macroinvertebrate assemblages of running waters. Water flow velocity and changes in the substrate composition (often as a result of organic or inorganic pollution) are often identified as the dominant environmental parameters shaping the structure of macroinvertebrate assemblages in running waters.

Dragonflies (Odonata) are among the best studied insect groups with 67 species that have been registered so far in Montenegro (for an overview of dragonfly diversity and research see [46]). Several insect orders, such as Ephemeroptera, Plecoptera, and Trichoptera (EPT) and Diptera (Chironomidae), are frequently used in the routine monitoring activities related to freshwater ecosystems in assessing water quality status according to the WFD.

Chironomid midges represent an abundant, diverse, and widespread group of invertebrates in all types of running waters in Montenegro [46]. The current number of non-biting midges reported from Montenegro [63] is obviously underestimated, and it is likely that further studies and the application of molecular techniques will likely increase the number of species found [46]. The largest number of mayflies (Ephemeroptera) and stoneflies (Plecoptera) has been registered for the Lake Skadar basin (for an overview of diversity of these groups see [46]). On the other hand, the greatest number of caddisflies (Trichoptera) has been registered in the northern part of Montenegro, especially in the Durmitor mountains, which hosts around 150 species [64]. There are many endemics among the aquatic invertebrates, and today they are exposed to various anthropogenic factors that increase the threat of extinction [50]. Unfortunately, the Red Data Book for Montenegro is not yet available, and most of these species are still not protected at the national or international level [46].

3.5 Hyporheic Fauna

Interstitial habitats exist at a depth of several centimeters or decimeters beneath streams and rivers. Regardless of the fact that this habitat is more accessible for studying than other types of underground water habitats [62], interstitial (hyporheic) fauna has only been poorly studied in Montenegro, and the available data remains highly dispersed. The first attempt to systematize the published data was done recently for the aquatic subterranean environment of the Lake Skadar Basin [62]. Of the approximately 90 stygobiotic species occurring in the subterranean environment of the Lake Skadar basin, 28 have been reported from interstitial waters. Hydrachnidia (water mites) with 12 species and Copepoda (8 species) predominated [62]. The above-mentioned survey on the interstitial fauna of the Lake Skadar Basin reveals a lack of marked endemicity versus cave water environments, which are richer in endemics [62]. Among interstitial species, only *Stygohydracarus karanovici* (known from the interstitial habitats of Lake Plav and Lake Skadar, and the spring of the River Biogradska) is endemic to Montenegro [62].

3.6 Fish

Freshwater fish are among the most commonly used organisms for assessing water quality, following the Water Framework Directive. In addition, fish are often used as a model for biogeographical research and ecoregion delimitations (see below). The freshwater fish fauna of Montenegro currently consists of 89 species [65]. Among them, at least 14 freshwater fish have not been recorded for a long time, and thus the current number is lower and tallies 75 species, of which 62 species are native and the other 13 are introduced [65]. Two of them, i.e., *Salmo zetensis* (soft-muzzled trout) and *Barbatula zetensis* (the Zeta stone loach), are endemic to Montenegro [65]. The freshwater ichthyofauna of the Lake Skadar basin (in Montenegro and Albania) includes eight endemic species, with one of them, *Chondrostoma scodrense* (the Skadar nase), assumed to be extinct [66].

The ichthyofauna of the inland waters of Montenegro can be clearly divided into two groups that correspond to the Danube (Black Sea) and the Adriatic Sea basin, respectively. In the Adriatic Sea basin, there are considerably more autochthonous species (43 spp.) than in the Black Sea basin (19 spp.), presumably due to the presence of the brackish habitats of the River Bojana/Buna and the small coastal tributaries [65]. The autochthonism of some species, for example, the presence of the Black Sea salmon (*Salmo labrax*) in the upper part of the River Zeta is questionable [65]. Moreover, there are more introduced (13 spp.) and translocated (7 spp.) species in the Adriatic than in the Danube basin [65]. Introduction and translocation affects the actual number of fish species that inhabit Montenegrin river basins, which requires long-term research and continuous updating with regularly published checklists. The best studied is the ichthyofauna of the Lake Skadar/Shkodra basin, which has a long tradition of research (for an overview the diversity of and conservation of freshwater fish of the Lake Skadar basin see [66]).

The biogeographic characteristics of the ecoregions and the main river basins in Montenegro with regards to those species that are particularly useful in ecoregional delimitations are discussed below.

3.7 Amphibians and Reptiles

The basin of Lake Skadar is the best studied in terms of reptiles and amphibians and represents one of the hotspots of diversity of these groups in the Balkans [67]. Recent surveys reveal the numbers of 15 amphibians and 36 reptile species inhabiting this region (for an overview the diversity of and conservation of amphibians and reptiles of the Lake Skadar basin see [67]). Among them, one species, *Pelophylax shqipericus* (the Albanian Water Frog), listed by the UICN as endangered, is a local endemic. This lowland species, which is restricted to swampy habitats around Lake Skadar/Shkodra, has recently been recorded from a swampy area along the Rivers Matica and Sušica between Podgorica and Danilovgrad [68]. Other species of conservation importance include *Emys orbicularis* and *Triturus macedonicus* which could become endangered in the future [67]. This is also the case for *Mauremys rivulata* which habitats in the coastal region may become threatened by intensive urbanization. In the mountainous region, the destruction of breeding sites and the introduction of fish threaten the populations of Alpine newts (*Ichthyosaura alpestris).* Alpine salamander (*Salamandra atra*) inhabits the very high mountains (it was first detected on Mount Bogićevica at 2,100–2,200 m above sea level) and has a restricted distribution in the Lake Skadar basin [67]. Apart from intensive human pressure, the highly fragile populations of the latter species are also threatened by ongoing climate change [67].

3.8 Birds

Montenegro is characterized by a diverse and rich avifauna, and with 230 breeding bird species, it is in fact richer than that of most neighboring countries (see [69] for an overview). Among them there are 45 species found in Annex I of the EU's Birds Directive and 26 species listed by BirdLife International as Species of European Conservation Concern [69].

The rivers themselves and associated water bodies provide habitats for many birds. For example, many water birds that nest in the swampy part of Lake Skadar, such as *Microcarbo pygmaeus* (Pygmy Cormorants), *Egretta garzetta (*Little White Egrets), *Ardea alba* (Great White Egrets), *Plegadis falcinellus* (Glossy Ibises), *Ardeola ralloides* (Squacco), and *Nyctycorax nyctycorax* (Night Herons), visit the River Morača to rest and feed [69, 70]. The latter species, Night Herons, were found also to nest in the tall trees along the upper part of the River Zeta [69]. The eroded banks of the Lim, the Zeta, and the Sitnica are suitable places for nesting populations of *Alcedo atthis* (Common Kingfishers) [69]. We should emphasize the importance of canyons in preserving populations of many birds. The river canyons of the Morača, the Mrtvica, Mala Rijeka, the Tara, the Sušica, the Bukovica, and the Komarnica are important nesting places for *Aquila chrysaetos* (the Golden Eagle). The more thermophilous Cijevna/Cemi river canyon is the place where a population of *Bubo bubo* (Eagle Owls) and colonies of *Hirundo rupestris* (Eurasian Crag Martins) live [69].

3.9 Mammals

The semiaquatic mammal fauna of Montenegro is poorly investigated. Two species that deserve special attention are *Castor fiber* (the European beaver), often considered a keystone "ecosystem engineer" in running waters, and *Lutra lutra* (the Eurasian otter) which is a well-known "signal species" for aquatic biodiversity issues [71]. The European beaver in the past was widely distributed in Montenegro and neighboring countries but became extinct probably in the second half of the nineteenth century, as a result of overhunting [72]. As with other Balkan countries, the written sources on the existence of beavers in Montenegro are lacking, and therefore it is difficult to give the exact beaver extinction date for Montenegro [72, 73]. Recently, the beaver has been successfully reintroduced in neighboring countries (Croatia and Serbia), so its future presence in Montenegro (if not already returned) is also likely.

Most information on the presence of *Lutra lutra* in Montenegro relates to Lake Skadar and its tributaries including the River Bojana/Buna [74]. The otter is an apex predator in both riverine and wetland systems. The diet of the otter of Lake Skadar has not been studied, but some studies reveal that the general otter diet is dominated by *Anguilla anguilla* [75]. This could be a concern, as the European eel is an endangered fish in the Lake Skadar basin [66].

4 The Ecoregional Delineations of Montenegrin Fresh Waters

Abell et al. [7] have stated that ecoregion delineation "benefits from the best available data describing species and systems ecology, but can proceed with imperfect information [76]." The FEOW ecoregion boundaries are based on expert judgment and therefore strongly depend on the availability and quality of data applicable to a selected model (mainly fish) as well as knowledge of the level of research on ecological and evolutionary patterns in the studied area. This knowledge can considerably vary from region to region, which has resulted in the current ecoregional regionalization often encountering controversy. It seems reasonable to expect that the proposed ecoregional regionalization cannot be applied to all groups of organisms. Many groups of organisms have shown significant differences in distribution patterns, which depends on a number of factors, including differences in their mobility (poor or active dispersers) and their life cycle (e.g., taxa that are wholly aquatic or taxa dispersed terrestrially by flying). In most similar studies aimed at ecoregional delineations of freshwater ecosystems, fish were used as the model, but some other groups of organisms such as snails [15] and underground fauna [62, 77] have proved to be good biogeographical indicators. Contemporary studies of the phylogeographical patterns of selected groups that inhabit the Lake Skadar basin (for an overview see [25]) have contributed to a better understanding of the distribution and evolution of particular groups, which may warrant revisions of the ecoregional freshwater division of Montenegro.

The territory of Montenegro includes three freshwater ecoregional units: the Dalmatian, the Southeastern Adriatic, and the Dniester-Lower Danube Ecoregions [6]. The boundaries of the latter ecoregion in Montenegro coincide with the border between the Danube and the Adriatic Sea basins (Fig. 1). Of the 110 native species that are endemic to the Dniester-Lower Danube ecoregion, only *Cobitis elongata* (the Balkan loach), *Gobio obtusirostris* (the Balkan gudgeon), *Phoxinus csikii*, and *Hucho hucho* (the huchen) are present in the waters of Montenegro [65]. The boundary of the distribution of some aquatic taxa coincides well with the boundary between the two basins. An illustrative example is the springsnail genus *Bithynella* (Fig. 5d) widespread in the Danube basin, but with the exception of one locality (the spring at "Dujovo Točilo" near Lijeva Rijeka not far from the hydrological drainage boundary of the two basins), not present in the Adriatic Sea basin [78]. The Drin brook lamprey *Eudontomyzon stankokaramani* (Fig. 5f), with ammocoetes found in detritus-rich sands or clay sediments, inhabit the piedmont and montane zones of streams in Montenegro that belong to the Drim River drainage [79].

Based on the FEOW ecoregion system, the main part of the Adriatic coast of Montenegro belongs to the Southeastern Adriatic Ecoregion. A smaller part of the Adriatic Sea basin in Montenegro, which includes certain small tributaries in the Bay of Kotor and the Lake Bileć basin, belongs to the Dalmatian Ecoregion whose eastern border lies near the town of Budva (Fig. 1). Among the fish species that belong to the genera (i.e., *Delminichthys*, *Aulopyge*, and *Phoxinellus*) which are endemic to the latter ecoregion, only *Delminichthys ghetaldii* (the Popovo minnow) found in Lake Bileć is present in the Montenegrin part assigned to the FEOW's Dalmatian Ecoregion [65].

The Southeastern Adriatic Ecoregion in Montenegro includes the southern part of the Adriatic Sea basin, i.e., the Lake Skadar/Shkodra basin and the River Bojana/Buna, as well as the River Željeznica, a small tributary running into the Adriatic Sea. Moreover, a small unnamed stream (under Mount Čakor) in the municipality of Plav that runs into the River Pećka Bistrica (Fig. 1) and the River Drim also belong to Southeastern Adriatic Ecoregion [7].

This Southeastern Adriatic Ecoregion does not have any endemic genera but hosts more than 30 endemic species. There is a large number of endemic species shared by the three large lakes (Lake Prespa, Lake Ohrid, and Lake Skadar) that belong to this ecoregion. The Montenegrin part of the Southeastern Adriatic Ecoregion hosts two species, *Pelasgus minutus* (the Ohrid minnow, Fig. 5g) and *Pachychilon pictum* (the Albanian roach) of the two genera, respectively, that are not present in the Dalmatian Ecoregion [65]. Moreover, the Montenegrin part hosts two species *Salmo zetensis* and *Barbatula zetensis* which are endemic to Montenegro and 14 species that are endemic to the Southeastern Adriatic Ecoregion [65].

Fig. 5 Representatives of animal taxa from Montenegrin freshwater ecosystems: (**a**) shrimp (*Atyaephyra vladoi*), (**b**) leech *Hirudo verbana*, (**c**) adults of stonefly *Isoperla pesici*, (**d**) springsnail *Bithynella luteola*, (**e**) water mite *Mideopsis roztoczensis*, (**f**) the Drin brook lamprey *Eudontomyzon stankokaramani*, (**g**) the Ohrid minnow *Pelasgus minutus*. Photographs by M. Grabowski (**a**, **b**), V. Pešić (**c**), P. Glöer (**d**), I. Pozojević (**e**), S. Zogaris (**e**, **f**)

One especially interesting biogeographic indicator is the fish *Pelasgus minutus*, present in Morača and Bojana and in several small coastal tributaries that directly enter the Adriatic Sea (the River Jaška, Jankove vode, the River Morinjska). West of the River Sutorina (from the Morinj springs) located near the administrative border of Montenegro, *Pelasgus minutus* has not been registered. According to Marić [65], *Phoxinus karsticus* originally described from the endorheic-system of the River Trebišnjica seems to be the only common fish species found in the Dalmatian Ecoregion and the Adriatic Sea basin in Montenegro.

The biogeographic affiliation of the Trebišnjica basin and the boundary between its drainage and other parts of the Adriatic drainage in Montenegro is difficult to establish, because there are very few surface waters in the border area. This relatively large area in the western part of Montenegro has almost no surface waterbodies, with the exception of Lake Bileć. The only surface stream is a short River Nudol

(Nudoljska), while the rest of the surface waters drains underground. Among the freshwater fish species which are endemic to the Dalmatian Ecoregion, only *Delminichthys ghetaldi*, *Alburnus neretvae* (the Neretva bleak), and *Squalius svallize* (the Neretva chub) are present in the Montenegrin part of the Trebišnjica river basin [65].

The cluster analysis of freshwater fish native to Montenegrin hydrographic basins grouped the Trebišnjica river basin with the Coastal rivers (excl. The River Bojana/Buna) region (Fig. 6).

According to Marić [65], the eastern border of the Dalmatian Ecoregion ends or overlaps with the eastern boundary of the Trebišnjica river basin (Lake Bileć) and the distribution border of *Delminichthys ghetaldi*, *Alburnus neretvae*, and *Squalius svallize*. On the other hand, the western, or northwestern, boundaries of the South-eastern Adriatic Ecoregion overlap with the distribution border of *Pelasgus minutus* (Fig. 7).

The abovementioned ecoregional delineation finds confirmation in the biogeographical pattern of the troglobiotic fauna. Pešić et al. [62] stated that the southeastern border of the distribution area of some holodinaric species or groups, supposed to cover the entire region of Dinaric Karst [80], coincided with the northwest administrative border of Montenegro [77]. Characteristic examples include the European cave salamander *Proteus anguinus*, the cave shrimp *Troglocaris* agg. *anophthalmus*, the polychaete *Marifugia cavatica*, the hydrozoan *Velkovrhia enigmatica*, and the bivalvian *Mytilopsis kusceri*. Recent phylogenetic studies have confirmed that most of these species have an approximately similar phylogeographic pattern [81, 82] indicating the common biotic break of holodinaric taxa in the northwest region of Montenegro. This is probably a result of the geological and hydrological history of the area, most often associated with the presence of prominent barriers to species dispersal.

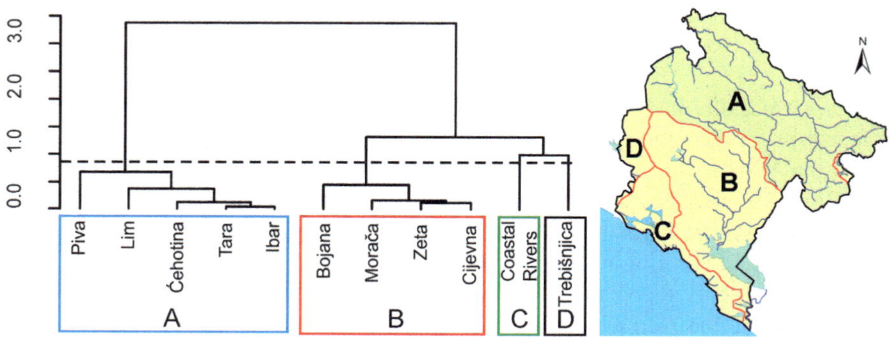

Fig. 6 Cluster analysis (Ward's method) with the presence/absence of native freshwater fishes (61 species) in 11 hydrographic basins. The arbitrarily defined cutoff at 9% creates four groupings: (A) the "Danube basin," (B) the "Skadar Lake basin," (C) the "Coastal rivers (excl. the River Bojana/Buna)," and (D) the "Trebišnjica basin"

Fig. 7 The revised freshwater ecoregion delineations of Montenegro. (1) The Dniester-Lower Danube Ecoregion, (2) the Southeastern Adriatic Ecoregion, (3) the Dalmatian Ecoregion

5 The Diversity of Aquatic Biota in the Main River Basins of Montenegro

Montenegro can be clearly divided in two regions: (1) the Mediterranean and sub-Mediterranean region and (2) the continental part of Montenegro. These two regions correspond to Mediterranean and Alpine biogeographical regions, respectively [83], whose borders in Montenegro, running approximately parallel to shore of Adriatic Sea, divide the country into two almost equal parts. The biogeographical boundaries coincide with the border between the Adriatic and Black Sea basins in Montenegro.

5.1 The Black Sea Basin

The Black Sea basin in Montenegro covers 7,545 km^2 [84]. The basin is predominantly mountainous and characterized by the rocks of the southeastern Dinarides. The basin has a "continental" type climate, with relatively cool and humid summers and long and harsh winters [85]. The rivers of the Black Sea basin are generally longer and are characterized by more water than those of the Adriatic Sea basin. Moreover, the river network is denser, while the share of intermittent rivers and ephemeral streams (IRES) in the river network is lower than in the Adriatic Sea basin. The Black Sea basin has five main river basins whose biogeographical characteristics are described below.

5.1.1 The Tara Basin

At 141 km long, the River Tara is the longest river in Montenegro. From the confluence of the two rivers, the Opasanica and the Veruša, that form the Tara to the town of Kolašin, its riverbed is very wide and flat with rather steep sides, but after the town of Mojkovac, the Tara changes direction, running through a canyon valley. The hydrological complexity of the Tara basin (1,915 km^2 [84]) that includes numerous springs, waterfalls, and the deepest river canyon in Europe [86], sustains very rich and diverse aquatic biota, including several endemic species. However, only a few groups of aquatic biota have been thoroughly studied so far.

The taxonomic composition of the benthic algae and meiofauna assemblages of the River Tara and its tributaries have been studied by Petković and Petković [87]. Among algae, diatoms dominated with 143 species, followed by blue-green algae (42 spp.), Chlorophyceae (31 spp.), and Conjugatophyceae (28 spp.). In the summer period, there is an evident increase in the number of periphyton species belonging to Chlorophyceae and Conjugatophyceae whose number decreases downstream, while the number of Bacillariophyceae and Chlorophyceae in the lower course of the river increases [87]. The diversity and abundance of algae assemblages increase from winter to summer [44, 87], with a slight decline in the autumn [87]. Rotatoria predominated in the meiofauna with 47 recorded species, followed by protozoans (15 spp.), microcrustaceans (Cladocera and Copepoda with 8 and 7 taxa, respectively), and Gastrotricha with one species [87].

The macroinvertebrate fauna is insufficiently studied, and the only available ecological data are those from a monitoring survey of the Bar-Boljare highway development activities around the River Tara. The monitoring that started in January

2019 was carried out at the sites of Mateševo and Rovačko Trebaljevo. The highest density of macroinvertebrate assemblages at Rovačko Trebaljevo was revealed in February and March, being 5,247 ind. m^{-2} and 3,971 ind. m^{-2}, respectively. The density at Mateševo was much lower, which is a consequence of the negative ecological impact associated with highway development activities in the immediate vicinity of this site (see [88] for the controversy surrounding highway construction over the River Tara). The low density at Trebaljevo was recorded in April (517 ind. m^{-2}) what is in the correlation with the high monthly discharge (flow) of the river in this period, which has the effect of reducing the invertebrate abundance. Ephemeroptera dominated in all months except February and March, when Trichopterans predominated.

As can be seen at Fig. 8, the share of EPT taxa and Chironomidae, respectively, at Rovačko Trebaljevo showed the opposite trend. The study also revealed a significant difference when lentic assemblages were analyzed. Ephemeropteran (particularly *Ephemera danica*) dominance in the lentic part was observed in most of the studied months except March–April, when Chironomids dominates. In August aquatic bugs and *Radix labiata* were abundant in the lentic zone. The highest density was found in January (2,167 ind. m^{-2}) and March (2,486 ind. m^{-2}) and the lowest at 573 ind. m^{-2} and 397 ind. m^{-2} in April and May, respectively.

The autochthonous ichthyofauna of the Tara basin is represented by eight species *Hucho hucho*, *Salmo labrax*, *Thymallus thymallus*, *Cottus gobio*, *Barbus balcanicus*, *Chondrostoma nasus*, *Squalius cephalus*, and *Phoxinus csikii* [65].

The Tara basin is the *locus typicus* for several endemic caddisfly species: *Plectrocnemia mojkovacensis* (Bistrica stream), *Polycentropus devetaki* (Mojkovac), *Rhyacophila balcanica* (Bjelasica), *R. diakoftensis* (the River Tara), *R. nurga* (Crkvičko Polje near Rudine), *Drusus gombos*, and *Potamophylax ureges* (both described from a forest stream at Dobrilovina [89]). The springsnail genus *Bithynella* is represented by three species, *B. dispersa*, *B. luteola*, and *B. taraensis*, that inhabit

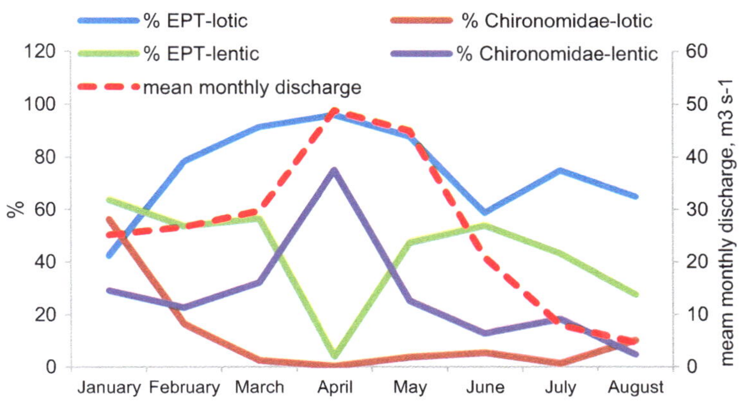

Fig. 8 The relationship between the mean monthly discharge and the share of EPT taxa and Chironomidae, respectively, in the lotic and lentic areas of Tara River at Rovačko Trebaljevo

springs in the Tara basin [78]. *Paladilhiopsis tarae* is known from the groundwater underlying the Tara Valley [90]. The leech *Dina minuoculata* [54] and the stonefly *Besdolus illyricus* [91] inhabit small mountain streams shaded by the forest.

5.1.2 The Lim Basin

The River Lim is the most important tributary of the River Drina and has the river basin surface of 2,291 km^2 [84]. The total length of the river is 220 km, of which 123 km is in Montenegro. The total decline of the river is around 40%, but the upper part near Berane is much steeper (8–10 m/km) than the lower part (2–4 m/km), resulting in a higher average water velocity (of up to 2 m/s) in the upper part [86]. Downstream of the town of Berane, the River Lim has a significantly wider riverbed, in some places being over 100 m [92].

The first study of the benthic algae assemblages of the River Lim and its tributaries was conducted in the 1970s [40]. In total, 66 algae species were recorded at the outflow of the River Lim from Lake Plav [40]. Diatoms predominated with 48 species (73%), while the share of other groups was lower. Apart from the other species, the red algae *Batrachospermum vagum* and aquatic hyphomycete *Tetracladium marchalianum* (as "*Asterotrix raphidioides*") were found [40]. The occurrence of the latter species was also reported in the Morača, the Zeta [40], and the Piva [93].

The invertebrate fauna of the River Lim remains poorly studied. Marković et al. [94] reported 66 macroinvertebrate taxa in the Serbian part of the river. Šundić and Radujković [53] reported 21 species of Oligochaeta in the Montenegrin part of the River Lim. The other macroinvertebrate groups remain poorly studied, and baseline species inventory work is required.

The Lim basin hosts 18 autochthonous fish species: *Hucho hucho*, *Salmo labrax*, *Thymallus thymallus*, *Alburnoides bipunctatus*, *Alburnus alburnus*, *Cottus gobio*, *Lota lota*, *Esox lucius*, *Cobitis elongate*, *Sabanejewia balcanica*, *Rutilus virgo*, *Barbatula barbatula*, *Barbus balcanicus*, *Barbus barbus*, *Chondrostoma nasus*, *Squalius cephalus*, *Telestes rysela*, and *Phoxinus csikii* [65]. *Salmo labrax* dominated in the fish assemblages of the Lim basin, while *Barbus balcanicus* and *Squalius cephalus* dominate in the river itself [95]. In terms of noteworthy fish species, we should mention huchen, which in the River Lim attain their greatest age and size [96]. *Acherontacarus fonticolus* is an endemic water mite known only from a spring near Đurđevi Stupovi Monastery in Berane [57]. Two species of the caddisfly genus *Drusus* (*D. siveci* and *D. krusniki*), inhabiting spring habitats, and the first order of mountain streams were originally described from the Gnjili Potok stream [89].

5.1.3 The Ćehotina Basin

The drainage basin of the River Ćehotina is one of the smallest in Montenegro (1,113 km^2 as per [84]). The total length of the river is 125 km, of which 100 km is in Montenegro. The different geological composition and structure of the land, dominated by clastic rock (schist, eruptive, and lacustrine deposits) in the main part of the basin, makes this basin different from the others in Montenegro.

The pollution that comes primarily from the Pljevlja thermal power plant, from the air, from the washing of the land, and from water flows (notably from the streams of Paleški potok and Vezišnica), which lead the pollutants directly into the river, is thus an important factor that affects the taxonomic composition, abundance, and growth of the aquatic fauna of the Ćehotina. The saprobic level of the Ćehotina was found to range from oligosaprobity-β-mesosaprobity to α-mesopsaprobity at most sites along the watercourse, including also in its tributary, the Vezišnica stream [53].

The inflow of pollutants (chemicals but also ash and slag) from the thermal power plant causes a drastic change in the water chemistry, often rapidly increasing the pH of the water. This affects invertebrate and fish communities below the pollutants effluent discharge, causing a high mortality rate among the local fish and invertebrate fauna. One such example occurred on July 5, 2019, when about 800 kg of fish died as a result of the illegal release of chemicals into the water of the River Ćehotina. Despite the devastating impact on the local fauna, it seems that the process of recolonization by *Gammarus balcanicus* from the nearby non-affected sites is proceeding rapidly. One study carried out on the Ćehotina (just a few kilometers from the site when the pollutants were released) a week after the mass death of fish and other organisms revealed the presence of *Gammarus balcanicus* in large numbers. Some individual mayflies, chironomids, *Agabus* beetle, and even the leech *Dina lineata* were found.

With the exception of Oligochaeta which are represented by 28 recorded species [53], most of the other macroinvertebrate groups in the River Ćehotina and its basin are still insufficiently studied. The most abundant Oligochaeta species in the Ćehotina is *Limnodrilus hoffmeisteri*, followed by *Stylodrilus heringianus*, *Tubifex ignotus*, *Stylodrilus lemani*, and *Chaetogaster diastrophus* [53]. Plociennik and Pešić [63] showed that the chironomid assemblages of the Ćehotina are formed by taxa typical of cold, rocky-bottomed, fast-flowing streams, mainly with orthocladiinae from the *Cricotopus* and *Orthocladius* groups.

The autochthonous ichthyofauna of the Ćehotina basin includes 12 species: *Hucho hucho*, *Salmo labrax*, *Thymallus thymallus*, *Alburnoides bipunctatus*, *Cottus gobio*, *Barbatula barbatula*, *Barbus balcanicus*, *Barbus barbus*, *Chondrostoma nasus*, *Squalius cephalus*, *Telestes rysela*, and *Phoxinus csikii* [65].

5.1.4 The Piva Basin

The River Piva has a river basin surface of 1,731 km^2 [84]. The name of the Piva refers to the part of the river that is formed by the merging of the River Komarnica and the River Sinjac, which is considered to be a branch of the Piva, whose source is the "Pivsko oko" spring. In 1976, downstream of the confluence with the River Mratinjska, the Mratinje Hydroelectric Power Plant, a 220-m-high dam, was constructed in the canyon valley. As a result, an artificial lake with a length of 42 km and a volume of 880,000,000 m^3 was formed [97], flooding part of the Piva canyon. The newly formed lake partially entered into the valleys of the River Komarnica, the River Vrbnica, and Mratinjski Creek, whose watercourses supply the lake with most of its water.

The first hydrobiological studies of the Piva and the Komarnica were conducted in the 1970s [93, 98] before the dam was constructed. The algal communities were represented by 124 taxa [93]. Diatoms predominated with 66 recorded taxa. *Planothidium lanceolatum* (as *Achnanthes laceolata* in [93]), *A. gibberula*, *Diatoma vulgare*, *Iconella spiralis*, *Meridion circulare*, *Melosira varians*, *Cympatopleura elliptica*, *Ulnaria ulna* (as *Synedra ulna* in [93]), *Campylodiscus noricus*, *Gyrosigma acuminatum*, *Rhoicosphenia abbreviata* (as *R. curvata* in [93]), *Surirella librile* (as *Cymatopleura solea* in [93]), *Gomphonema acuminatum* (as *G. acuminatum* var. *coronatum* in [93]), and *Nitzschia sigmoidea* were the most common taxa at that time. About 42% of the diatom species reported before the construction of the dam were indicators of oligotrophic conditions [93]. The share of the species which where indicators of crenobiotic oligotrophy was also significant and included *Amphora ovalis*, *Aulacoseira distans* (as *Melosira distans* in [93]), *Hannaea arcus* (as *Ceratoneis arcus* in [93]), *Denticula elegans*, *Eunotia vanheurckii* (as *Eunotia faba* in [93]), *E. praerupta*, *Fragilariforma bicapitata* (as *Fragilaria bicapitata* in [93]), *Staurosirella pinnata* (as *Fragilaria pinnata* in [93]), *Gomphonema angustatum*, *Staurosirella martyi* (as *Opephora martyi* in [93]), *Rhopalodia gibba* (as *Pinnularia gibba* in [93]), *Surrirela didyma*, and *Tabellaria flocculosa* [93]. The phytoplankton community was quite stable throughout the year, with a lower abundance of diatoms during a rainy period of the year, i.e., from October to early May [93]. Apart from the other species, the chrysophytan *Hydrurus foetidus* and *Characiopsis pyriformis* [93], red algae *Batrachospermum gelatinosum* (reported as *B.moniliforme* in [93]) and *Bangia atropurpurea* [40] were present.

The meiofauna community of the River Piva and its tributaries before the dam construction consisted of 62 species [93]. Rotatorians predominated with 40 species, followed by Protozoa with 16 species, while Copepoda and Cladocera were represented by 3 species each [93]. The most dominant taxa were the copepods *Eucyclops serrulatus* and *Bryocamptus pygmaeus*; the rotatorians *Notholca labis*, *N. squamula*, and *Trichotria tetractis*; and the protozoon *Vorticella campanula* [93].

The limnocrene spring "Pivsko Oko" was investigated in the 1970s before the Mratinje dam was completed [98]. No typical phytoplankton species were found, and the algal assemblages were predominantly diatoms. The abundant occurrence of *Diatoma vulgare* was recorded, as well as scattered specimens of *Iconella spiralis* which is an indicator of oligotrophy. A low number of filamentous charophyte green algae *Spirogyra* and desmid *Closterium*, as well as the well-developed branching filaments of *Cladophora glomerata* and gelatinous sheaths of cyanophite *Sphaeronostoc* [98], were also recorded. Zooplankton was represented by rare specimens of the copepod *Eucyclops serrulatus*, the cladoceran *Chydorus sphaericus*, the protozoan *Arcella conica*, and a more diverse number of Rotatoria, including *Colurella adriatica*, *Cephalodella gibba*, *Lecane hamata*, and *Notholca labis* [98].

After the formation of the Piva reservoir, the composition of the algae, meiofauna, and macroinvertebrates significantly changed in relation to the state found before the construction of the dam [99, 100]. For example, in the first year of the formation of the accumulation, 22 taxa were found in phytoplankton [99]. Diatoms (*Synedra*), Chrysophyta, and Volvocales (palmeloid stages) predominated. As can be seen from Fig. 9, there is an evident change in the development of the phytoplankton community, which is generally a characteristic of all accumulation, especially of a canyon type such as the Piva reservoir [99].

In 1976, when the accumulation was formed, the zooplankton community had 76 species. Rotatoria dominated with 44 species, while the 3 other groups, Cladocera, Copepoda, and Protozoa, had the same species numbers (10 spp.) [99].

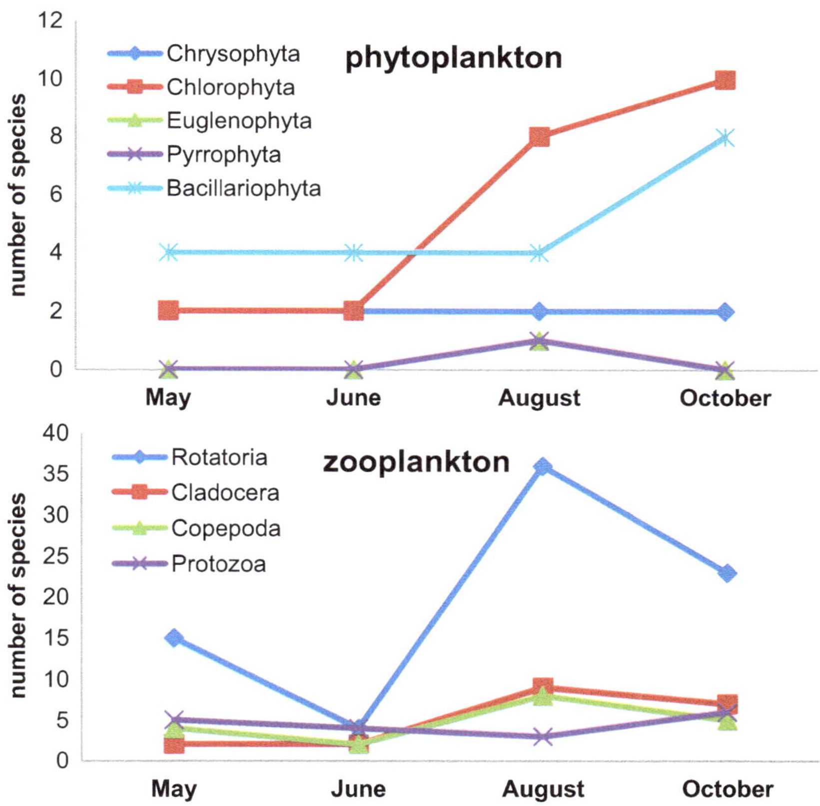

Fig. 9 Formation of the phytoplankton and zooplankton communities in the first year, 1976, when the Piva reservoir formed (from [99])

In terms of abundance, Cladocera dominated in August, but in October at most of the studied localities, Rotatoria became more abundant [99]. A lower number of species was registered in the first months with a peak in diversity in August and a fall in October (see Fig. 9). Within 5 years (1976–1980) of the formation of the accumulation, the zooplankton richness in the Piva reservoir increased to 172 species [100]. Rotatoria were the most diverse with 98 species, followed by Protozoa (31 spp.), Cladocera (24 spp.), and Copepoda (19 spp.). In the term of abundance, the zooplankton community after 5 years was dominated by Rotatoria (57% of the total zooplankton abundance) and to a lesser extent by Protozoans (18%) [100].

The changes in the faunal composition apply also to ichthyofauna. Before the dam was completed, 12 fish species were known from the Piva basin [95]. After the dam construction, only three autochthonous fish species (the European grayling, the Danubian basin brown trout, and the minnow) were found in the Piva accumulation [95], while the other species were not observed. The present-day autochthonous ichthyofauna of the Piva basin includes four species: *Salmo labrax*, *Thymallus thymallus*, *Cottus gobio*, and *Phoxinus csikii* [65].

Many of the rivers in the Piva basin that are today threatened by the building of small hydropower plants (see [101] for a discussion) host a number of endemic and endangered species. The endemics includes the hydraenid beetles *Ochthebius pretneri* known from the stream at Skakavac near Šavnik and *Hydraena biltoni* described from the canyon of the River Komarnica and the mayfly species *Rhitrogena marinkovici* known from the lower part of the River Bukovica [102].

5.1.5 The Ibar Basin

The River Ibar in the northeast part of the country has a basin surface area in Montenegro of only 433 km^2 [84]. Its watercourse in Montenegro is only 34 km long, originating from the spring at "Vrelo Ibra" on the northern slope of Mt. Hajla. The waters of the Ibar empty into the Zapadna Morava, while the waters of the other four main river of the Danube basin in Montenegro flow into the River Drina. This has affected the differences in the composition of selected biotic groups, including the ichthyofauna of these waters; for example, the fish genus *Telestes* is not present in the Morava basin [65].

The macroinvertebrate fauna of the River Ibar is insufficiently studied [53, 103]. Tubić et al. [104] reported 57 macroinvertebrate taxa in the Serbian part of the river. The only available ecological data are those from a study carried by Skenderović et al. [103] in summer 2017 at four sites: the Vrelo Ibra spring, site L2-bridge, Besnik, and Bać. *Gammarus fossarum* and *Nemoura cinerea* dominated in the spring section. At site L2, apart from the species of *Gammarus* and hydrobid snails, trichopterans (*Philopotamus montanus*, *Stenophylax latipennis*) were most common. Below the town of Rožaje, the macroinvertebrate abundance decreased. Trichoptera predominated with the larvae of *Hydropsyche* and *Rhyacophila*, while the share of gammarid amphipods decreased at Besnik, and at the lowermost point of Bać, they completely disappeared [103]. On the other hand, the share of oligochaetes increased which indicate organic pollution [103]. The Oligochaeta assemblage of the River Ibar is most similar to those of the River Lim, but it has the highest density (40,493.78 ind. m^{-2}) among all the rivers of the Black Sea basin [53]. Among the 27 Oligochaeta species that were recorded in the Ibar, the species of the genus *Nais* (*N. elinguis*, *N. barbata*, *N. communis*, and *N. pseudobtusa*) predominated [53].

The autochthonous ichthyofauna of the Ibar basin includes ten species: *Hucho hucho*, *Salmo labrax*, *Thymallus thymallus*, *Alburnoides bipunctatus*, *Cottus gobio*, *Barbatula barbatula*, *Barbus balcanicus*, *Chondrostoma nasus*, *Squalius cephalus*, and *Phoxinus csikii* [65].

5.2 The Adriatic Sea Basin

The Adriatic Sea basin in Montenegro covers 6,268 km^2 [84]. It can be divided into three areas: the Bojana/Buna river basin or the Lake Skadar/Shkodra basin, the Trebišnjica river basin, and the Montenegrin Coastal Area, which includes several small rivers directly flowing into the sea. The diversity and biogeographical characteristics of the aquatic biota that inhabits those basins are briefly described below.

5.2.1 The Rivers in the Region of the Montenegrin Coast

Most of the Montenegrin Coast area is formed of dry-land calcareous mountainous landscapes, having scarce running water ecosystems. However, this area hosts numerous siphonic lakes and pools in the underground karst [105] that often appear on the surface as strong karstic springs, mainly in the proximity of the seashore, as well as under the sea. Most rivers are short and descend abruptly into the sea after a short stretch. The largest part of the Boka Kotorska (Boka Bay) hinterland is part of the Montenegrin Coast region (including Grahovsko polje and Njeguši) where, along with numerous intermittent springs and rivers (the River Grahovo and the streams of Grabovica and Ceklinski potok), a few perennial water courses are present, in which fish can be found [65].

The largest river in the vicinity of Boka Kotorska Bay is the 7-km-long River Sutorina, whose lower part has been channelized. Most of the other watercourses around Boka Kotorska (e.g., the Meljine, the Repaj, the Pijavica, and the Jošica) are short streams of a torrential nature, as a result of the strong influence of fresh waters during the rainy season. The hydrographic network in the southern part of the Adriatic basin in Montenegro is better developed, and there are several smaller rivers (including the Međurečka Rijeka, the Miđanska (Vladimirska), and the Rastiška) and a larger one, the River Bojana/Buna, which forms the border with Albania. Most of the smaller rivers in the region of the Montenegrin Coast dry up in the summer, which affects the composition of their aquatic biota.

Due to the small size of most of the rivers, the average species richness is low. Moreover, the endemicity of epigean fauna is not high, presumably due to the intermittent nature of many of the watercourses. The aquatic biota remain still insufficiently studied, but there are a number of rare (e.g., *Homoeogammarus thoni*, *Echinogammarus pungens*, and *Atyaephyra desmaresti*) or even endemic (the hydraenid *Ochthebius olicinium* and the caddisfly *Agapetus kampos*) species which have been reported. One interesting biogeographic indicator is the monotypic gastropod genus *Antibaria*, distributed from the spring at Bostanj in Pelinovo, near the town of Tivat, to the Bunar spring near the town of Stari Bar [106]. *Antibaria notata* is also known from a few springs near Virpazar on Lake Skadar, but always from middle elevations, never descending to the shore of the lake [106]. This species has been listed as extinct [107], but it's presence at multiple sites, most of which dry in the summer, qualified it as being of Least Concern, although it may be more accurate to qualify it as Near Threatened [108].

According to Marić [65], the autochthonous ichthyofauna of the region (without the River Bojana/Buna) is represented by 14 species: *Anguilla anguilla*, *Platichthys flesus*, *Atherina boyeri*, *Chelon aurata*, *Chelon labrosus*, *Chelon ramada*, *Chelon saliens*, *Syngnathus abaster*, *Syngnathus acus*, *Mugil cephalus*, *Aphanius fasciatus*, *Pomatoschistus marmoratus*, *Dicentrarchus labrax*, and *Pelasgus minutus*. Of these, one particularly interesting biogeographical indicator is the Ohrid minnow (*Pelasgus minutus*).

This region sustains the Balkan terrapin (*Mauremys rivulata*) and a number of endemic hydrobiid snails that inhabit the springs of this area [52, 106] including at least one hydrobiid genus *Stygobium* endemic to this region [106].

The River Bojana/Buna The River Bojana/Buna is an outflow of Lake Skadar/Shkodra, which is 42 km long with an average flow of 304 m^3/s [109].

The aquatic vegetation of the Bojana was studied by Fanneli et al. [39]. The floating/submerged macrophytes included two belts, one outer belt in deep waters with dominant *Potamogeton pectinatus* and an inner belt in the shallower waters with dominant *Potamogeton nodosus* and locally, in disturbed sites, *Elodea canadensis*. Among the helophytes, close to the riverbank, four belts were distinguished: in deeper waters the rather sparse *Scirpus lacustris*, followed by *Iris pseudacorus*, *Typha angustifolia*, and finally *Phragmites australis* [39].

The autochthonous ichthyofauna of the Bojana includes 38 species [65] and has similar characteristics to that of Lake Skadar [66]. The terminal part of the River Bojana represent an estuary inhabited by complex communities composed of those species that are able to tolerate a wide range of salinity. Among them there are *Syngnathus abaster*, *S. acus*, *Aphanius fasciatus*, *Pelasgus minutus*, and five species of mullets [66].

The River Bojana/Buna is a possible corridor for the introduction of the alien *Dikerogammarus villosus* (the killer shrimp) into Lake Skadar [46], as well as of the invasive *Dreissena polymorpha* (the zebra mussel) which is already present in Lake Šas and the River Bojana [49]. Therefore, the continuous monitoring of the River Bojana communities is highly recommended.

5.2.2 The Rivers in the Lake Skadar Basin

The rivers in the central part of Montenegro belong to the Lake Skadar basin, whose surface area is 5,631 km^2 [109]. The biggest tributary of Lake Skadar is the River Morača, which provides about 63% of the lake's water.

This region of the Lake Skadar/Shkodra basin is characterized by a high number of endemic freshwater invertebrates, including endemic aquatic gastropods, insects (e.g., Plecoptera and Coleoptera, among other groups), water mites and various crustacean taxa with the endemic freshwater shrimp (*Atyaephyra vladoi*). Moreover, the Lake Skadar basin hosts an endemic amphibian (*Pelophylax shqipericus*) and 59 freshwater fish species, of which 45 are native to the watershed and 15 introduced, including 7 recent species which are endemic [66]. The recent book *The Skadar/Shkodra Lake Environment* [1] provides a comprehensive overview of the diversity, endemism, and various threats to the aquatic biota inhabiting this unique basin.

The River Morača (Fig. 10) The first comprehensive study of the benthic algae was conducted in the 1970s [40]. In total, 66 species were registered. Diatoms predominated with 49 species, but both the Chlorophyta (22 spp.) and the Cyanophyta (16 spp.) were also diverse [40].

The macroinvertebrate assemblages of the upper and middle course of the River Morača were studied in the 1970s at the following sites: Morača-upper, Morača-Monastir, Morača-Međuriječje, Morača-middle, and Morača-Bioče [110] (Fig. 11). The highest density of macroinvertebrate assemblage was found at

Fig. 10 The canyon of the Morača River. Photo by S. Popović

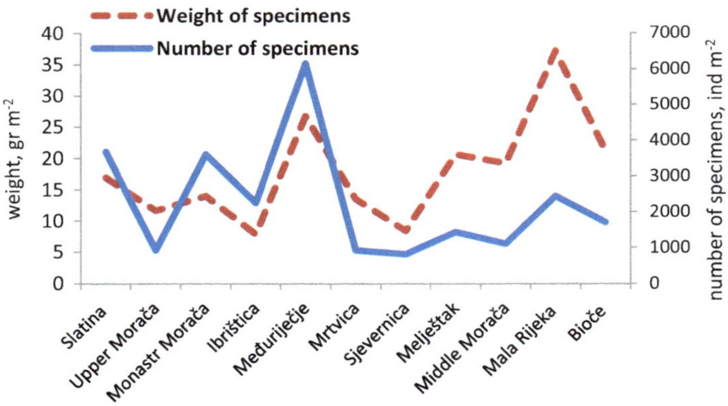

Fig. 11 The biomass and abundance of macroinvertebrate assemblages of Morača River and its tributaries. Modified from [110]

Međurječje (6,150 ind. m^{-2}). At the two other sites that belong to the upper course, Upper Morača and Monastir Morača, the densities were lower 3,671 ind. m^{-2} and 3,608 ind. m^{-2}, respectively. The indicator taxa of the upper part of the River Morača were the caddisfly species *Micrasema minimum*, *Hydropsyche instabilis*, and *Rhyacophila* sp. The most dominant group at the Upper Morača sites was Trichoptera (particularly *Hydropsyche instabilis*), which contributed almost 73% of the biomass [110]. Ephemeroptera was also abundant here, while the share held by other groups like Mollusca, Oligochaeta, coleopterans, dipterans, and plecopterans each did not exceed 5% of the zoobenthos biomass [110]. At Monastir Morača, the share of caddisflies dominated here by *Micrasema minimum* (2,337 ind. m^{-2}) dropped, while the participation of Plecoptera which contribute approximately one quarter of the macroinvertebrate biomass increased. Downstream, at Međurječje, *Micrasema minimum* almost doubled its density (4,576 ind. m^{-2}) forming 74% of the assemblage abundance at this site [110].

In the middle course, the composition of the macroinvertebrate assemblage changed. The density of macroinvertebrate assemblage in the middle course decreases to 1,093 ind. m^{-2} at the Middle Morača site and 1,697 ind. m^{-2} at Bioče. In terms of biomass, Plecoptera was the most dominant group, contributing up to 51% at the Middle Morača site [110]. *Perla marginata* predominated in the biomass at the Middle Morača site, while *Isoperla grammatica* predominated at Bioče [110]. Trichopterans, and especially mayflies, were also abundant at Bioče in all seasons [111]. *Rhitrogena diaphana*, *Heptagenia* spp., *Baetis alpinus*, and *Ecdyonurus aurantiacus* (reported as *E. fluminum* in [112]) were the most abundant mayfly species at the latter site [112]. At the subsequent site of Rogami, the abundance of *Rhitrogena diaphana* and *Heptagenia* dropped, while the abundance of *Ecdyonurus* species increased [112].

Chironomids were low in density and biomass at most of the studied sites along the upper and middle course of the River Morača [110]. *Nilotanypus dubius* and *Diamesa* are common chironomids in the upper part of Morača River (see [63] for a fauna survey on the chironomid midges of the River Morača). At the Bioče site, the occurrence of chironomid midges increased. Here *Synorthocladius semivirens* and *Eukiefferilla* species *(clypeata*, agg. *fittkaui*, and *ikleyensis*-type) were abundant, but *Cricotopus* agg. *cylindraceaus*, *Tvetenia*, and *Polypedilum convictum* were also found [63].

A study of the macroinvertebrate assemblages of the lower course of the River Morača, below the city of Podgorica, was conducted in the 1970s at three sites: Kolovrat, Botun, and Vukovci [111]. Ephemeroptera predominated at all three sites in all seasons, except in the summer, when chironomida larvae predominated [111]. The highest proportion of mayfly larvae was observed at Kolovrat, where they in the summer formed almost 96.2% of the total assemblage abundance. Here, *Rhitrogena diaphana* predominated, while the proportion of *Heptagenia* species significantly decreased. *Baetis alpinus* disappeared, while the proportion of *B. fuscatus* increased [112]. At the lower sites, Botun and Vukovci, the occurrence of Heptageniidae and Baetidae species decreased [112]. *Ephemerella ignita* dominated at Vukovci and *Ephemera danica* at Botun, while at Vukovci, *Caenis macrura* were also abundant [112].

The highest density of chironomid larvae was observed at Botun in the summer when they formed 79% of the total assemblage abundance [111]. Oligochaeta also reached their maximum density here, indicating strong human impact at this site. The abundance of Gastropoda was greatest in the summer at the site of Kolovrat, while the amphipod gammarids reached their maximum abundance at Vukovci in the spring [111]. Similarly, the abundance of aquatic beetles (Coleoptera) and bugs (Heteroptera) increased in the summer, reaching their highest densities at Kolovrat and Vukovci, respectively [111].

The autochthonous ichthyofauna of the River Morača is represented by 24 species: *Alosa* sp., *Salmo farioides*, *Anguilla anguilla*, *Salmo marmoratus*, *Alburnoides ohridanus*, *Alburnus scoranza*, *Gasterosteus gymnurus*, *Knipowitschia montenegrina*, *Ninnigobius canestrinii*, *Ninnigobius montenegrensis*, *Cobitis ohridana*, *Barbatula zetensis*, *Rhodeus amarus*, *Rutilus prespensis*, *Barbus rebeli*, *Pelasgus minutus*, *Scardinius knezevici*, *Chondrostoma ohridanum*, *Squalius platyceps*, *Gobio skadarensis*, *Telestes montenigrinus*, *Pachychilon pictum*, *Phoxinus karsticus*, and *Gobio obtusirostris* [65]. Four species, *Salmo farioides*, *Barbus rebeli*, *Telestes montenegrinus*, and *Phoxinus karsticus* occur in the upper part of the river [66]. *Thymallus thymallus* was translocated to the River Morača from the Danube basin [66].

The underground fauna of the Morača River is poorly known. There is evidence that the water of Morača River sinking in the underground between Bioče and Zlatica (Podgorica) appear on some springs (the Iverak and the Stanjevića Rupa) downstream near the confluence with the Zeta River (Sekulić personal communication). The thanatocoenosis of the Iverak spring compromise a number of endemic subterranean species (*Zeteana ljiljanae*, *Plagigeyeria lukai*, *Iverakia hausdorfi*, *Bythiospeum demattiai*), including two monotypic genera *Zeteana* and *Iverakia*, known only from the latter spring system [51]. It is likely that these species are characteristic inhabitants of the underground water of the Morača River.

The Morača and its tributaries are also inhabited by a number of epigean endemic and subendemic taxa. These include the water mites *Atractides longisetus* described from the upper part of the River Mrtvica [113] and *Torrenticola lukai* described from the Mrtvica River but also reported from the upper part of Tara River and its tributaries [58], two mayflies, *Serratella ikonomovi* reported from the River Sjevernica but also from the River Cijevna [114] and *Siphlonurus abraxas* described from the Morača near Bioče and further reported from the stream of Rujela near Tuzi [115]. Most of the endemic and subendemic species, i.e., the stoneflies *Taeniopteryx auberti*, *Brachyptera tristis*, *Capnioneura balkanica balkanica*, *Leucatra jahoriensis*, *L. malcor*, *Protonemoura auberti*, *P. aestiva*, and *Isoperla pesici* (Fig. 5c), are restricted to mountain spring habitats [46].

The Tributaries of the Upper Part of the Morača In the 1970s, a comprehensive study was carried out on the macroinvertebrate assemblages of several tributaries of the River Morača: the Mrtvica, the Slatina, Mala Rijeka, the Melještak, the Sjevernica, and the Ibrištica [110]. Mayfly larvae, especially species of the genera *Rhitrogena*, *Ephemerella*, and *Baetis* predominated in the Melještak, the Mrtvica, and Mala Rijeka, contributing over 50% of the total biomass [110]. Jacobi [110] revealed the highest biomass (approx. 37 g. m^{-2}) at Mala Rijeka, due to the higher share of stoneflies larvae (*Isoperla grammatica*) in that tributary. Trichopterans, especially *Hydropsyche instabilis* and *Rhyacophila pascoei*, were predominant in the Sjevernica, contributing to more than 65% of the total biomass [110]. Caddisfly larvae also dominated in the Slatina and the Ibrištica, but their share here was lower and did not exceed 50% of the total biomass [110]. The majority of caddisfly species were collected in the tributaries of the River Morača (32 spp.), while in the river itself, only 16 species were collected [116].

Of the other groups, it is worth mentioning that the River Mrtvica, which harbored a rich Oligochaeta fauna of 21 recorded species, has the highest density of oligochaetan assemblage (9,910.10 ind. m^{-2}) among the rivers of the Adriatic Sea basin [53].

The River Zeta The River Zeta (Fig. 12) is the largest tributary of the Morača. The taxonomic composition of the benthic algae assemblages of the River Zeta were investigated in the 1970s [40]. In total, 66 species were registered. Diatoms predominated with 49 species, but Chlorophyta (22 spp.) and Cyanophyta (16 spp.) were also diverse [40].

The River Zeta is characterized by diverse invertebrate communities, of which only a few groups have been sporadically investigated, e.g., Oligochaeta [53], water mites [57–59], mayflies [46, 117, 118], stoneflies [46, 119], and caddisflies [116]. With regard to the dominant share of molluscs, chironomids, and oligochaetes, the River Zeta is a large lowland river. The Zeta is characterized by diverse Oligochaeta communities that include 40 known species [53]. The Zeta is home to abundant populations of *Bithynia radomani*, *Radomaniola curta*, and *Viviparus contectus* (syn. *V. mamillatus*). The mayfly and stonefly communities are represented by 13 and 8 species, respectively [46]. Ten species of caddisflies

Fig. 12 The River Zeta near the town of Spuž. A study carried in August 2019 at this site reveals that molluscs (*Theodoxus fluvitilis* and *Amphimelania holandrii*) in the biomass and chironomids in terms of abundance predominate in the lotic part of the water course with a gravel and sandy substrate and well-developed bushes of the willow moss *Fontinalis antipyretica*. Water mites, especially *Hygrobates fluviatilis* and torrenticolids are also found here abundant, reaching a density of 2,739 ind. m^{-2}. This locality, situated in the lower part of the Zeta, is characterized by high faunal diversity including rich interstitial water mite fauna [59]. In the deeper parts of the river, on the muddy substrate, the gastropods *Bithynia radomani* and *Viviparus contectus* predominate in the biomass. Chironomids, shrimps (*Atyaephyra vladoi*), and water mites (*Limnochares aquaticus* and the good swimmers *Mideopsis roztoczensis*, *Hygrobates longipalpis*, and *Neumania imitata*) were also abundant. Photograph by V. Pešić

have been reported from the mouth of the Zeta, being mainly the same that had been collected in the River Morača [116].

The finding of endemic species such as *Salmo zetensis* (the softmouth trout), which is endemic to the River Zeta [66, 120], and *S. marmoratus* (the marble trout), which is known also from the Morača and the Cijevna [66], is of particular interest. Both species are listed as critically endangered [66]. Other species of interest include *Austropotamobius pallipes* living in the lower part of the River Zeta, which represents its southernmost distribution limit [121, 122]. A second crayfish species, *Astacus astacus*, has been documented from the Liverovići reservoir and the upper course of the Zeta, and it is suspected to have undergone a decline of approximately 64% of the total biomass [122]. This is largely attributed to overfishing but possible also to the crayfish plague *Aphanomyces astaci*, still not reported in Montenegro, but widely distributed in Serbia and Bosnia and Herzegovina [123].

The autochthonous ichthyofauna of the River Zeta is represented by 20 species: *Alosa* sp., *Salmo farioides*, *Anguilla anguilla*, *Salmo marmoratus*, *Salmo zetensis*, *Alburnoides ohridanus*, *Alburnus scoranza*, *Gasterosteus gymnurus*, *Cobitis ohridana*, *Barbatula zetensis*, *Rhodeus amarus*, *Rutilus prespensis*, *Barbus rebeli*,

Chondrostoma ohridanum, *Squalius platyceps*, *Gobio skadarensis*, *Telestes montenigrinus*, *Pachychilon pictum*, *Phoxinus karsticus*, and *Gobio obtusirostris* [65]. Cyprinid fish that form half of the autochthonous fauna dominate in the lower part of the river, while the upper part is populated by *Salmo zetensis*, *S. marmoratus*, *S. farioides*, and some cyprinids [66].

The Mareza Canal The Mareza Canal is a small canalized tributary of the River Morača that originates from a number of perennial and intermittent springs located west of Podgorica, at the contact of the quaternary sediments and limestone of Velje Brdo. An extensive study on the macroinvertebrate communities of the Mareza Canal was conducted monthly from December 2008 to December 2009 [124]. The highest density was observed at the end of September immediately before the beginning of the rainy season, while the lowest density was recorded in December [124]. Gastropods (*Radomaniola curta*, *Bithynia radomani*, *Radix skutaris*, and *Valvata montenegrina*) predominated in all months, at approximately 56% of the total abundance, followed by gammarids (*Homeogammarus scutariensis* and *Gammarus balcanicus*), dipterans, and aquatic beetles. The share of other taxa did not exceed 5% of the total abundance [124].

The species of interest include the gastropod *Valvata montenegrina*, a species orginally described from this locality and listed by the UICN as an endangered species [125].

The River Cijevna/Cem The River Cijevna, the left tributary of the River Morača, originates in the Prokletije mountain range in Albania, and after it enters Montenegro, it forms an attractive canyon with high and steep sides. Mayflies and stoneflies are represented here by 11 and 7 species, respectively [46]. The mayfly fauna is mainly represented by a species restricted to cold rhitral habitats (*Baetis alpinus*, *B. lutheri*, and *Ephemera danica*) but also typical lowland species such as *Baetis fuscatus* [46]. The autochthonous ichthyofauna of the River Cijevna is represented by 17 species: *Salmo farioides*, *Anguilla anguilla*, *Salmo marmoratus*, *Alburnoides ohridanus*, *Alburnus scoranza*, *Gasterosteus gymnurus*, *Ninnigobius canestrinii*, *Cobitis ohridana*, *Barbatula zetensis*, *Rhodeus amarus*, *Rutilus prespensis*, *Barbus rebeli*, *Squalius platyceps*, *Telestes montenigrinus*, *Pachychilon pictum*, *Phoxinus karsticus*, and *Gobio obtusirostris* [65].

Rijeka Crnojevića Rijeka Crnojevića is the second largest tributary of Lake Skadar, originating from a spring of Obodska Pećina Cave. The cave host the richest fauna of aquatic troglobiotic species in Montenegro, including number of remarkable endemics (reviewed in [62]).

The upper part of Rijeka Crnojevica which has a stony and gravel substrate and a higher amount of coarse particulate organic matter hosts abundant populations of hydrobiid snails (*Radomaniola*), *Theodoxus fluviatilis*, and gammarid amphipods. In the lower and slower part of the river that is organically enriched, mollusks (*Radix skutaris*, *Viviparus contectus*, *Stagnicola montenegrinus*, and *Planorbarius corneus*), oligochaetes, and chironomids predominate.

The species of special interest here include the stone crayfish (*Austropotamobius torrentium*) recorded in the River Morača [126]. This species is a host for crayfish worms, *Brachiobdella hexadonta* and *B. pentadonta* [46]. The population of *Salmo farioides* belongs to one AdRc haplotype [127] requiring further conservation action.

5.2.3 The Trebišnjica Basin

The River Trebišnjica, the longest sinking river in Europe, has a rather small watershed in Montenegro (541 km^{-2} [84]). Most of the area is made up of dry-land calcareous mountainous landscapes with rather scarce surface running waters (with the exception of the River Nudoljska). All of the water disappears into complex underground structures that drain directly into the River Trebišnjica or Lake Bilećko or else sometimes appear as strong karstic springs (e.g., the springs at Zaslapnica and Rečevina). The group of strong karstic springs that includes the Trebišnjica spring-group below Bileća and the Čepelica river spring-group have been submerged under Lake Bilećko since 1968 when the Trebinje 1 Hydroelectric Power Plant was completed. Interestingly, some of these submerged springs had a fauna of subterranean snails more affiliated to the River Zeta. For example, the hydrobiid *Plagigeyeria zetaprotogona zetatridyma* is known from the Trebišnjica spring (probably "Nikšićko Vrelo" spring) and spring at Tunjevo in the valley of the Zeta River [128, 129]. Another example is the distribution of the genus *Anagastina* which includes several species (*A. gluhodolica*, *A. matjasici*, *A. scutarica*, *A. vidrovani*, and *A. zetaevallis*) that are restricted to the Lake Skadar basin [106]. Abundant populations of these species can be found in numerous springs in Nikšićko Polje and along the River Zeta. The only congeneric species found outside the Lake Skadar basin is *Anagastina hadouphylax*, known from a single site, a cave at Čepelica near Bileća [130].

Therefore, the biogeographic assignment of this region remains uncertain, though it seems to create a distinctive (sub)regional entity. According to Marić [65], the autochthonous ichthyofauna of the Montenegrin part of the Trebišnjica basin includes only five species (*Salmo farioides*, *Alburnus neretvae*, *Squalius svallize*, *Delminichthys ghetaldii*, and *Phoxinus karsticus*). The epigean and subterranean invertebrate fauna of the Montenegrin part of the Trebišnjica basin is poorly studied. It seems very likely that further work may discover many undescribed taxa associated with subterranean habitats and reveal interesting biogeographic patterns.

Acknowledgments We would like to express our sincere gratitude to all hydrobiologists who researched the Rivers of Montenegro. We thank Aleksandra Jabłońska for preparing Fig. 2 and Peter Glöer, Stamatis Zogaris, Ivana Pozojević and Saša Popović for permission to use their photographs.

References

1. Pešić V, Karaman G, Kostianoy A (eds) (2018) The Skadar/Shkodra Lake environment. The handbook of environmental chemistry, vol 80. Springer, Cham, p 508. https://doi.org/10.1007/978-3-319-99250-1
2. Paunović M, Tubić B, Kračun M, Marković V, Simić V, Zorić K, Atanacković A (2012) Ecoregions delineation for the territory of Serbia. Water Res Manag 2(1):65–74
3. Illies J (1978) Limnofauna Europaea. Gustav Fisher Verlag, Stuttgart
4. Zogaris S, Economou AN (2017) The biogeographic characteristics of the river basins in Greece. In: Skoulikidis NT, Dimitriou E, Karaouzas I (eds) The rivers of Greece: evolution, current status and perspectives, pp 93–95. https://doi.org/10.1007/698_2017_475
5. European Commission (2000) Directive 2000/60/EC, establishing a framework for community action in the field of water policy. Off J Eur Comm Brussels L 327:1–71
6. FEOW (2008) Freshwater ecoregions of the world-conservation science program. The Nature Conservancy and World Wildlife Fund, Washington, p 1
7. Abell R, Thieme ML, Revenga C, Bryer M, Kottelat M, Bogutskaya N, Coad B, Mandrak N, Balderas SC, Bussing W, Stiassny ML (2008) Freshwater ecoregions of the world: a new map of biogeographic units for freshwater biodiversity conservation. Bioscience 58(5):403–414
8. Schwan W (1997) Europe. In: Encyclopedia of European and Asian regional geology. Encyclopedia of earth science. Springer, Dordrecht
9. Neubauer F (2003) Geology of Europe. In: De Vivo B, Grasemann B, Stüwe K (eds) Volume geology. Encyclopedia of life supporting systems. UNESCO Publishing-Eolss Publishers, Oxford. (17 printed pages + 18 figures). www.eolss.net
10. Embleton C (1984) Geomorphology of Europe. Macmillan, London
11. Bognár A (2008) Basic geomorphological characteristics of the Dinarides Mountain System. Acta GGM Debrecina 3:121–125
12. Zupan Hajna N (2010) Geology. In: Michevc A, Prelovšek M, Zupan Hajna N (eds) Introduction to the Dinaric karst. Inštitut za raziskovanje krasa ZRC SAZU, Postojna
13. Frankl A, Lenaerts T, Radusinović S, Spalevic V, Nyssen J (2016) The regional geomorphology of Montenegro mapped using land surface parameters. Z Geomorphol 60(1):1–14
14. Vlahović I, Mandić O, Mrinjek E, Bergant S, Ćosović V, De Leeuw A, Enos P, Hrvatović H, Matičec D, Mikša G, Nemec W, Pavelić D, Pencinger V, Velić I, Vranjković A (2012) Marine to continental depositional systems of outer Dinarides foreland and intra-montane basins (Eocene-Miocene, Croatia and Bosnia and Herzegovina). Mitt Ges Geol Bergbaustud Österr 54:405–470
15. Radoman P (1985) Hydrobioidea a superfamily of Prosobranchia (Gastropoda), II. Origin, 1 zoogeography, evolution in the Balkans and Asia Minor. Monogr Inst Zool Belgrade 1:1–173
16. Buj I, Marčić Z, Caleta M, Sanda R, Geiger MF, Freyhof J et al (2017) Ancient connections among the European rivers and watersheds revealed from the evolutionary history of the genus Telestes (Actinopterygii; Cypriniformes). PLoS One 12(12):e0187366
17. Baldacci A (1929) L'Albania. Istituto per l'Europa Orientale, Roma
18. Biju-Duval B, Letouzey J, Montadert L, Courrir P, Mugniot JF, Sancho J (1974) Geology of the Mediterranean Sea basin. In: Burk CA et al (eds) The geology of continental margins. Springer, New York
19. Bianco PG (1990) Potential role of the palaeohistory of the Mediterranean and Paratethys basins on the early dispersal of Euro-Mediterranean freshwater fishes. Ichthyol Explor Freshw 1(3):167–184
20. Krijgsman W, Hilgen FJ, Raffi I, Sierro FJ, Wilson DS (1999) Chronology, causes and progression of the Messinian salinity crisis. Nature 400:652–655
21. Skoulikidis NT, Economou AN, Gritzalis KC, Zogaris S (2009) Rivers of the Balkans. In: Tockner K, Uehlinger U, Robinson CT (eds) Rivers of Europe. Elsevier/Academic Press, San Diego, pp 421–466

22. Tzedakis PC (2007) Seven ambiguities in the Mediterranean palaeoenvironmental narrative. Quat Sci Rev 26:2042–2066. https://doi.org/10.1016/j.quascirev.2007.03.014
23. Macklin MG, Lewin J, Woodward JC (2012) The fluvial record of climate change. Philos Trans R Soc A 370:2143–2172. https://doi.org/10.1098/rsta.2011.0608
24. Grabowski M, Mamos T, Bącela-Spychalska K, Rewicz T, Wattier RA (2017) Neogene paleogeography provides context for understanding the origin and spatial distribution of cryptic diversity in a widespread Balkan freshwater amphipod. PeerJ 5:e3016. https://doi.org/10.7717/peerj.3016
25. Grabowski M, Jabłońska A, Wysocka A, Pešić V (2018) The obscure history of the Lake Skadar and its biota: a perspective for future research. In: Pešić V, Karaman G, Kostianoy A (eds) The Skadar/Shkodra Lake environment. The handbook of environmental chemistry, vol 80. Springer, Cham, pp 47–61
26. Popov SV, Rögl F, Rozanov AY, Steininger FF, Shcherba IG, Kovac M (2004) Lithological-paleogeographic maps of Paratethys: 10 maps Late Eocene to Pliocene. Cour Forschungsinst Senck 250:1–46
27. Economou AN, Giakoumi S, Vardakas L, Barbieri R, Stoumboudi M, Zogaris S (2007) The freshwater ichthyofauna of Greece – an update based on a hydrographic basin survey. Mediterr Mar Sci 8:91–166. https://doi.org/10.12681/mms.164
28. Griffiths HI, Kryštufek B, Reed JM (2004) Balkan biodiversity: pattern and process in the European hotspot. Springer, Dordrecht
29. McInerney CE, Maurice L, Robertson AL, Knight LR, Arnscheidt J, Venditti C, Dooley JG, Mathers T, Matthijs S, Eriksson K, Proudlove GS, Hanfling B (2014) The ancient Britons: groundwater fauna survived extreme climate change over tens of millions of years across NW Europe. Mol Ecol 23:1153–1166. https://doi.org/10.1111/mec.12664
30. Mamos T, Wattier R, Burz«ski A, Grabowski M (2016) The legacy of a vanished sea: a high level of diversification within a European freshwater amphipod species complex driven by 15 My of Paratethys regression. Mol Ecol 3:795–810. https://doi.org/10.1111/mec.13499
31. Blondel J, Médail F (2009) Biodiversity and conservation. In: Woodward C (ed) The physical geography of the Mediterranean Basin. Oxford University Press, Oxford, pp 604–638
32. Blondel J, Aronson J, Bodiou J-Y, Boeuf G (2010) The Mediterranean region: biology diversity through space and time. Oxford University Press, Oxford
33. Tierno de Figueroa JM, López-Rodríguez MJ, Fenoglio S, Sánchez-Castillo P, Fochetti R (2013) Freshwater biodiversity in the rivers of the Mediterranean Basin. Hydrobiologia 719:137–186. https://doi.org/10.1007/s10750-012-1281-z
34. Poulakakis N, Kapli P, Lymberakis P, Trichas A, Vardinoyiannis K, Sfenthourakis S, Mylonas M (2015) A review of phylogeographic analyses of animal taxa from the Aegean and surrounding regions. J Zool Syst Evol Res 53:18–32. https://doi.org/10.1111/jzs.12071
35. Hadžiablahović S (2018) The diversity of the flora and vegetation of Lake Skadar/Shkodra. In: Pešić V, Karaman G, Kostianoy A (eds) The Skadar/Shkodra Lake environment. The handbook of environmental chemistry, vol 80. Springer, Cham, pp 203–238
36. Blaženčić J, Kashta L, Vesić A, Biberdžić V, Stevanović B (2018) Charophytes (Charales) of Lake Skadar/Shkodra: ecology and distribution. In: Pešić V, Karaman G, Kostianoy A (eds) The Skadar/Shkodra Lake environment. The handbook of environmental chemistry, vol 80. Springer, Cham, pp 169–202
37. Petrović D, Hadžiablahović S, Vuksanović S, Mačić V, Milanović Đ, Lakušić D (2019) Catalogue of habitat types of EU importance of Montenegro. Version 3. Podgorica-Banja Luka-Beograd. https://natura2000.me/downloads/docs-for-web/component-1/O1.5.1%20Catalogue%20of%20Habitat%20Types%20for%20Montenegro%20Ver%202019%2003%2019.pdf. Accessed 10 Oct 2019
38. Kalníková V, Palpurina S, Peterka T, Kubešová S, Plesková Z, Sabovljević M (2017) Bryophytes on river gravel bars in the Balkan mountains: new records and insights into ecology. Herzogia 30:370–386

39. Fanelli G, De Sanctis M, Gjeta E, Mullaj A, Attore F (2015) The vegetation of the Buna river protected landscape (Albania). Hacquetia 14(2):129–174
40. Petković S (1980) *Autumnal algofloristic aspect* of some Montenegrin rivers (Contribution to the algologic investigation of freshwaters in Montenegro). Polj Šum 6(1):71–89
41. Guiry MD, Guiry GM (2019) Algae Base. World-wide electronic publication, National University of Ireland, Galway. http://www.algaebase.org. Accessed 20 Feb 2019
42. Janauer G, Dokulil M (2006) Macrophytes and algae in running waters. In: Ziglio G, Siligardi M, Flaim G (eds) Biological monitoring of rivers. Wiley, Chichester, pp 89–109
43. Poulíčková A, Duchoslav M, Dokulil M (2004) Littoral diatom assemblages as bioindicators of lake trophic status: a case study from perialpine lakes in Austria. Eur J Phycol 39:143–152
44. Rakočević J (2018) Application of epilithic diatoms in the ecological assessment of mountain rivers: contribution to the development of biomonitoring tools for Montenegrin aquatic ecosystems using the case study of the Tara River. Nova Hedwigia 106(3–4):337–356
45. Haslam SM (1987) River plants of Western Europe. Cambridge University Press, Cambridge
46. Pešić V, Gadawski P, Gligorović B, Glöer P, Grabowski M, Kovács T, Murányi D, Płóciennik M (2018) The diversity of the Zoobenthos communities of the Lake Skadar/Shkodra basin. In: Pešić V, Karaman G, Kostianoy A (eds) The Skadar/Shkodra Lake environment. The handbook of environmental chemistry, vol 80. Springer, Cham, pp 255–293
47. Berlajolli V, Płóciennik M, Antczak-Orlewska O, Pešić V (2019) The optimal time for sampling macroinvertebrates and its implications for diversity indexing in rheocrenes case study from the Prokletije Mountains. Knowl Manag Aquat Ecosyst 420:6
48. Grabowski M, Bącela-Spychalska K, Pešić V (2014) Reproductive traits and conservation needs of the endemic gammarid *Laurogammarus scutarensis* (Schäferna, 1922) from the Skadar Lake system, Balkan Peninsula. Limnologica 47:44–51
49. Pešić V, Glöer P (2018) The diversity and conservation status of the molluscs of Lake Skadar/Shkodra. In: Pešić V, Karaman G, Kostianoy A (eds) The Skadar/Shkodra Lake environment. The handbook of environmental chemistry, vol 80. Springer, Cham, pp 295–310
50. Pešić V, Glöer P (2013) A new freshwater snail genus (Hydrobiidae, Gastropoda) from Montenegro, with a discussion on gastropod diversity and endemism in Skadar Lake. ZooKeys 281:69–90
51. Glöer P, Pešić V (2014) New subterranean freshwater gastropods of Montenegro (Mollusca: Gastropoda: Hydrobiidae). Ecol Monten 1(2):82–88
52. Grego J, Glöer P, Falniowski A, Hofman S, Osikowski A (2019) New subterranean freshwater gastropod species from Montenegro (Mollusca, Gastropoda, Moitessieriidae, and Hydrobiidae). Ecol Monten 20:71–90
53. Šundić D, Radujković B (2012) Study on freshwater oligochaeta of Montenegro and their use as indicators in water quality assessment. Nat Monten 11:117–383
54. Grosser C, Pešić V, Gligorović B (2015) A checklist of the leeches (Annelida: Hirudinea) of Montenegro. Ecol Monten 2(1):20–28
55. Grosser C, Pešić V, Berlajolli V, Gligorović B (2016) Glossiphonia balcanica n. sp. and Dina prokletijaca n. sp. (Hirudinida: Glossiphoniidae, Erpobdellidae) – two new leeches from Montenegro and Kosovo. Ecol Monten 8:17–26
56. Gelder SR (1999) Zoogeography of branchiobdellidans (Annelida) and temnocephalans (Platyhelminthes) ectosymbiotic on freshwater crustaceans, and their reactions to one another in vitro. Hydrobiologia 406:21–31
57. Pešić V, Smit H, Gerecke R, Di Sabatino A (2010) The water mites (Acari: Hydrachnidia) of the Balkan peninsula, a revised survey with new records and descriptions of five new taxa. Zootaxa 2586:1–100
58. Pešić V, Bańkowska A, Goldschmidt T, Grabowski M, Michoński G, Zawal A (2018) Supplement to the checklist of water mites (Acari: Hydrachnidia) from the Balkan peninsula. Zootaxa 4394(2):151–184

59. Zawal A, Pešić V (2018) The diversity of assemblages of water mites from Lake Skadar and its catchment. In: Pešić V, Karaman G, Kostianoy A (eds) The Skadar/Shkodra Lake environment. The handbook of environmental chemistry, vol 80. Springer, Cham, pp 311–323
60. Wrzesińska W, Hupało K, Jabłońska A, Mamos T, Zawal A, Pešić V, Grabowski M (2017) When DNA barcoding fails: a case of intergeneric mtDNA introgression in freshwater Balkan amphipods. In: 17th international colloquium on amphipoda, Trapani, Italy, 4–7 September 2017
61. Jabłońska A, Mamos T, Zawal A, Grabowski M (2018) Morphological and molecular evidence for a new shrimp species, *Atyaephyra vladoi* sp. nov. (Decapoda, Atyidae) in the ancient Skadar Lake system, Balkan Peninsula – its evolutionary relationships and demographic history. Zool Anz 275:66–79. https://doi.org/10.1007/698_2018_349
62. Pešić V, Karaman GS, Sket B (2018) The diversity and endemism of aquatic subterranean fauna of the Lake Skadar/Shkodra Basin. In: Pešić V, Karaman G, Kostianoy A (eds) The Skadar/Shkodra Lake environment. The handbook of environmental chemistry, vol 80. Springer, Cham, pp 339–361
63. Płóciennik M, Pešić V (2012) New records and list of non-biting midges (Chironomidae) from Montenegro. Biol Serb 34:36–50
64. Krusnik C (1987) Trichoptera (Insecta). In: Fauna of Mt. Durmitor, Book 2. Montenegrian Academy of Science and Arts, Special edition, Book 21, vol 13. Department of Natural Sciences, Podgorica, pp 201–224
65. Marić D (2019) Fauna of freshwater fish (Osteichthyes) of Montenegro. In: Montenegrian Academy of Science and Arts, Special edition, Book 149, vol 48. Department of Natural Sciences, Podgorica (in Serbian)
66. Marić D (2018) The ichthyofauna of Lake Skadar/Shkodra: diversity, economic significance, condition, and conservation status. In: Pešić V, Karaman G, Kostianoy A (eds) The Skadar/Shkodra Lake environment. The handbook of environmental chemistry, vol 80. Springer, Cham, pp 363–381
67. Crnobrnja-Isailović J, Polović L, Ljubisavljević K, Čađenović N, Čubrić T, Haxhiu I (2018) Diversity and conservation status of batrachofauna and herpetofauna in the Lake Skadar Region. In: Pešić V, Karaman G, Kostianoy A (eds) The Skadar/Shkodra Lake environment. The handbook of environmental chemistry, vol 80. Springer, Cham, pp 383–414
68. CHA (2019) Main road reconstruction project, rehabilitation and upgrade of the Danilovgrad – Podgorica road section Critical Habitat Assessment (CHA). https://dzscg.com/dokumenta/ebrd/PD-DG_Annex%20E%20(ENG).pdf
69. Rubinić B, Sackl P, Gramatikov M (2019) Conserving of wild birds in Montenegro. The first inventory of potential Special Protection Areas in Montenegro. AAM Consulting, Budapest. xiii +328 pp
70. Vizi O (2018) Ornithological features of Skadar Lake. In: Pešić V, Karaman G, Kostianoy A (eds) The Skadar/Shkodra Lake environment. The handbook of environmental chemistry, vol 80. Springer, Cham, pp 415–445
71. UK DETR (2001) Sustaining the variety of life: 5 years of the UK biodiversity action plan. – Export of the UK Biodiversity Group to the UK Government, the Scottish Executive, the National Assembly for Wales and the Northern Ireland Executive. – Product Code: 00 WADC 1286. Department of the Environment, Transport and the Regions, London
72. Ćirović D, Savić I, Bjedov V (2001) First finding of the beaver (*Castor fiber* L., 1758) on the territory of FR Yugoslavia. The European beaver in a New Millennium. In: Proceedings of the second European beaver symposium, Białowieza, Poland, 27–30 September 2000, p 58
73. Halley D, Rosell F, Saveljev A (2012) Population and distribution of Eurasian beaver (Castor fiber). Balt For 18:168–175
74. Beqiraj S, Dhora D (2007) Regional importance of the fauna of the cross-border River Buna. In: Rivers and citizens – cross-border experiences in environmental protection and sustainable development. ECE, Lecce

75. Miranda R, Copp GH, Williams J, Beyer K, Gozlan RE (2008) Do Eurasian otters *Lutra lutra* (L.) in the Somerset levels prey preferentially on non-native fish species? Fundam Appl Limnol 172:339–347
76. Wikramanayake E, Dinerstein E, Loucks C, Olson D, Morrison J, Lamoreux J, McKnight M, Hedao P (2002) Ecoregions in ascendance: reply to Jepson and Whittaker. Conserv Biol 16:238–243
77. Sket B, Zagmajster M (2006) Subterranean fauna of Montenegro: its biodiversity and biogeographical character. In: Pešić V, Hadžiabalahović S (eds) The book of abstracts and programme, II international symposium of ecologists of Montenegro, Kotor, 20–25 September 2006, p 23
78. Glöer P, Pešić V (2010) The freshwater snails of the genus *Bythinella* Moquin-Tandon (Gastropoda: Rissooidea: Hydrobiidae) from Montenegro. Arch Biol Sci 62(2):441–447
79. Freyhof J, Kottelat M (2008) *Eudontomyzon stankokaramani*. The IUCN Red List of Threatened Species 2008: e.T135530A4139291. https://doi.org/10.2305/IUCN.UK.2008.RLTS.T135530A4139291.en. Accessed 18 Aug 2019
80. Sket B (1994) Distribution patterns of some Crustacea in the territory of former Yugoslavia. Hydrobiologia 287:65–75
81. Fišer C, Zakšek V, Zagmajster M, Trontelj P, Sket B (2007) Taxonomy and biogeography of Niphargus steueri (Crustacea: Amphipoda). Limnology 8:297–309
82. Jugović J, Jalžić B, Prevorčnik S, Sket B (2012) Cave shrimps *Troglocaris* s. str. (Dormitzer, 1853), taxonomic revision and description of new taxa after phylogenetic and morphometric studies. Zootaxa 3421:1–31
83. EEA (European Environment Agency) (2016) Biogeographic regions of Europe. https://www.eea.europa.eu/data-and-maps/figures/biogeographical-regions-in-europe-2
84. IJC (Institute Jaroslav Cerni) (2001) Water master plan of Montenegro. Government of Montenegro, Podgorica. (in Serbian)
85. Kostianoy A et al (2020) Drainage basins of Montenegro under climate change. In: Pešić V, Paunović M, Kostianoy A (eds) The rivers of Montenegro. The handbook of environmental chemistry. Springer, Cham
86. Hrvačević S (2004) Resources of surface water in Montenegro. EPCG AD, Podgorica. (in Montenegrin)
87. Petković S, Petković S (1991) Composition, structure and character of microphytobenthic and microzoobenthic communities or river Tara and its tributaries. Polj Šum 37(3–4):13–36. (in Serbian)
88. Pešić V, Paunović M, Kostianoy A (2020) Rivers of Montenegro – from conflicts to science-based management. In: Pešić V, Paunović M, Kostianoy A (eds) The rivers of Montenegro. The handbook of environmental chemistry. Springer, Cham
89. Oláh J (2017) Trichoptera endemic in the Carpathian Basin and the adjacent areas. Folia Entomol Hung 78:111–255
90. Pešić V (2010) *Paladilhiopsis tarae*. The IUCN Red List of Threatened Species 2010: e.T155586A4804034. https://doi.org/10.2305/IUCN.UK.2010-4.RLTS.T155586A4804034.en. Accessed 08 Aug 2019
91. Kovács T, Zwick P (2008) Contribution to the knowledge of genus *Besdolus* (Plecoptera: Perlodidae). Aquat Insect 30(3):179–186
92. Marić D, Milošević D (2010) First record and morphological characteristics of the Balkan golden loach *Sabanejewia balcanica* (Cobitidae) in Montenegro. Period Biol 112(2):149–152
93. Petković S, Petković S (1976) Studies on microphyto and microzoobenthos of the preimpoundment basin of Piva Lake, Reservoir in Crna Gora Jugoslavia. Polj Šum 22(3):65–84. (in Serbian)
94. Marković V, Vasiljević B, Atanacković A, Tomović J, Zorić K, Tubić B, Paunović M (2012) Status assessment of the Lim River based on macroinvertebrate communities. In: ALWOIS conference 2012, Ohrid, Republic of Macedonia. http://www.balwois.com
95. Marić D, Rakočević J (2009) Hidrobiologija. Univerzitet Crne Gore, Podgorica
96. Simonović P, Povž M, Piria M, Treer T, Adrović A, Skrijelj R, Nikolić V, Simić V (2015) Ichthyofauna of the River Sava system. In: Milačič R, Ščančar J, Paunović M (eds)

The Sava River. The handbook of environmental chemistry, vol 31. Springer, Berlin, Heidelberg, pp 361–400
97. Radojičić B (2005) The waters of Montenegro. Faculty of Philosophy, Institute of Geography, Nikšić
98. Petković S, Petković S (1971) Hydrobiological research of Piva and Komarnica Rivers. Polj Šum 17(3):61–71. (in Serbian)
99. Petković S, Petković S (1977) The first seasonal results in forming of plankton communities in 1976. Polj Šum 23(1):61–73. (in Serbian)
100. Petković S (1986) Contribution to the knowledge of development of zooplankton in Pivsko jezero (reservoir in Montenegro) during the period 1976-1980. Polj Šum 32(4):33–56
101. Pešić V, Paunović M, Kostianoy AG (2020) The rivers of Montenegro – introductory remarks. In: Pešić V, Paunović M, Kostianoy AG (eds) The rivers of Montenegro. The handbook of environmental chemistry. Springer, Cham
102. Tanasijevic M (1985) Nova *Rhithrogena* – vrsta iz Crne Gore (Ephemeroptera, Heptageniidae). Godišnjak Biol Inst Sarajevo 37:195–200
103. Skenderović I, Adrović A, Skenderović I (2018) Macroinvertebrates of the upper part of the Ibar River in the summer period. Rožajski Zbornik 20:41–57. (in Montenegrin)
104. Tubić B, Zorić K, Vasiljević B, Tomović J, Atanacković A, Marković V, Paunović M (2012) Saprobiological analyzes of the Ibar River based on aquatic macroinvertebrates. In: BALWOIS conference 2012, Ohrid, Republic of Macedonia. http://www.balwois.com
105. Barović G, Spalević V, Pešić V, Vujačić D (2018) The physical and geographical characteristics of the Lake Skadar Basin. In: Pešić V, Karaman G, Kostianoy A (eds) The Skadar/Shkodra Lake environment. The handbook of environmental chemistry, vol 80. Springer, Cham, pp 11–23
106. Radoman P (1983) Hydrobioidea a superfamily of Prosobranchia (Gastropoda), I. Sistematics, vol 547. Serbian Academy of Sciences and Arts, Beograd. (Dep. Sci. 57): 1–256, 12 pls
107. Regnier C, Fontaine B, Bouchet P (2009) Not knowing, not recording, not listing: numerous unrecognized mollusk extinctions. Conserv Biol 23:1214–1221
108. Pešić V (2011) *Antibaria notata*. The IUCN Red List of Threatened Species 2011: e.T155897A4862242. https://doi.org/10.2305/IUCN.UK.2011-1.RLTS.T155897A4862242.en. Accessed 09 Aug 2019
109. Pešić V, Karaman GS, Kostianoy AG (2018) Introduction. In: Pešić V, Karaman G, Kostianoy A (eds) The Skadar/Shkodra Lake environment. The handbook of environmental chemistry, vol 80. Springer, Cham, pp 1–10
110. Jacobi GZ (1981) Zoobenthos from sublacustrine springs in Lake Skadar. In: Beeton AM, Karaman GS (eds) The biota and limnology of Lake Skadar. University Veljko Vlahović, Institute of Biological and Medicine Research Titograd, Montenegro, pp 251–263
111. Filipović D (1981) Benthic macroinvertebrates of the lower Morača River. In: Beeton AM, Karaman GS (eds) The biota and limnology of Lake Skadar. University Veljko Vlahović, Institute of Biological and Medicine Research Titograd, Montenegro, pp 294–299
112. Filipović D (1981) Distribution and quantitative relationships of the fauna of Ephemeroptera in the Morača River. In: Beeton AM, Karaman GS (eds) The biota and limnology of Lake Skadar. University Veljko Vlahović, Institute of Biological and Medicine Research Titograd, Montenegro, pp 300–304
113. Pešić V (2002) Two interesting species of the genus *Atractides* Koch 1837 (Acari, Actinedida) from Crna Gora (Balkan Peninsula). Lauterbornia 44:65–71
114. Kovács T, Murányi D (2013) New country data of some mayflies (Ephemeroptera) from Europe. Folia Hist Nat Mus Matraensis 37:15–19
115. Jacob U (1986) *Siphlonurus abraxas* n. sp. ein neues Taxon der lacustris Gruppe aus Montenegro (Jugoslawien) (Ephemeroptera, Siphlonuridae). Reichenbachia 23(25):139–143
116. Marinković-Gospodnetić M (1981) Trichoptera of the Morača and Plavnica river drainages. In: Beeton AM, Karaman GS (eds) The biota and limnology of Lake Skadar. University

Veljko Vlahović, Institute of Biological and Medicine Research Titograd, Montenegro, pp 307–309
117. Kimmins DE (1937) Some new Ephemeroptera. Ann Mag Nat Hist 10(19):430–440
118. Putz V (1980) Ergebnisse der Albanien-Expedition 1961 des Deutschen Entomologischen Institutes. 94. Beitrag: Ephemeroptera. Beitr Entomol 30(2):343–355
119. Kaćanski D, Baumann RW (1981) Notes on the plecoptera fauna of the Morača River drainage. In: Beeton AM, Karaman GS (eds) The biota and limnology of Lake Skadar. University Veljko Vlahović, Institute of Biological and Medicine Research Titograd, Montenegro, pp 304–307
120. Mrdak D, Nikolić V, Tošić A, Simonović P (2012) Molecular and ecological features of the soft-muzzled trout *Salmo obtusirostris* (Heckel, 1852) in the Zeta River, Montenegro. Biologia 67:222–233
121. Rajković M, Petrović A, Maguire I, Simić V, Simić S, Paunović M (2012) Discovery of a new population of the species complex of the white-clawed crayfish, *Austropotamobius pallipes/italicus* (Decapoda, Astacidae) in Montenegro, range extension, endangerment and conservation. Crustaceana 85:333–347
122. Petrovic A, Rajkovic M, Simic S, Maguire I, Simic V (2013) Importance of genetic characteristics in the conservation and management of crayfish in Serbia and Montenegro. Bulg J Agric Sci 19:1093–1104
123. Vujčić SB, Lolić SB, Bojić MM, Ilić PB, Novaković MS, Karaman M, Matavuly MN (2013) Invasive zoopathogenic mastigomycotina in Republika Srpska, Serbia and neighboring countries with special reference to *Aphanomyces astaci*. Zb Matice Srp Prir Nauke 124:341–354
124. Pavićević A (2011) Seasonal dynamics of macroinvertebrates of Mareze and Rimanić with special reference to aquatic Coleoptera. PhD Dissertation, University of Montenegro (in Montenegrin)
125. Pešić V (2010) *Valvata montenegrina*. The IUCN Red List of Threatened Species 2010: e.T164824A5929000. https://doi.org/10.2305/IUCN.UK.2010-4.RLTS.T164824A5929000.en. Accessed 18 Aug 2019
126. Simić V, Petrović A, Rajković M, Paunović M (2008) Crayfish of Serbia and Montenegro the population status and the level of endangerment. Crustaceana 81:1153–1176
127. Sušnik S, Snoj A, Wilson I, Mrdak D, Weiss S (2007) Historical demography of brown trout (*Salmo trutta*) in the Adriatic drainage including the putative *S. letnica* endemic to Lake Ohrid. Mol Phylogenet Evol 44:63–76. https://doi.org/10.1016/j.ympev.2006.08.021
128. Schütt H (1960) Neue Höhlenschnecken aus Montenegro. Arch Molluskenkd 89:145–152
129. Schütt H (1972) Ikonographische Darstellung der unterirdisch lebenden Molluskengattung *Plagigeyeria* Tomlin (Prosobranchia: Hydrobiidae). Arch Molluskenkd 102:113–123
130. Seddon MB (2010) *Vinodolia hadouphylax*. The IUCN Red List of Threatened Species 2010: e.T155478A4784600. https://doi.org/10.2305/IUCN.UK.2010-4.RLTS.T155478A4784600.en. Accessed 26 Aug 2019

Vegetation in Ravine Habitats of Montenegro

Branko Kraradžić, Zlatko Bulić, Snežana Jarić, Miroslava Mitrović, and Pavle Pavlović

Contents

Abstract Due to topographic, climatic, geological and soil complexity, vegetation of Montenegro is remarkably diverse. We reviewed plant communities of Montenegro by focusing attention on vegetation in ravine habitats. Some of the main rivers of Montenegro (the Morača, Komarnica, Piva and Tara Rivers) flow through spectacular canyons. Due to relatively low climate-induced and human-induced disturbances of habitats, the ravine vegetation in Montenegro represents an important biodiversity hotspot and significant pool of rare, paleo-endemic species.

B. Kraradžić (✉), S. Jarić, M. Mitrović, and P. Pavlović
Institute for Biological Research, "Siniša Stanković" – National Institute of Republic of Serbia, Belgrade, Serbia
e-mail: branko@ibiss.bg.ac.rs; nena2000@ibiss.bg.ac.rs; mmit@ibiss.bg.ac.rs; ppavle@ibiss.bg.ac.rs

Z. Bulić
Environmental Protection Agency of Montenegro, Podgorica, Montenegro
e-mail: zlatkobulic@t-com.me

Vladimir Pešić, Momir Paunović, and Andrey G. Kostianoy (eds.), *The Rivers of Montenegro*, Hdb Env Chem (2020) 93: 201–230, DOI 10.1007/698_2020_479,

We performed comparative analyses of biodiversity in mesic forests that occur in non-ravine habitats (alliances *Aremonio-Fagion* (Horvat 1950), Borhidi in Török et al. 1989 and *Galio odorati-Fagion sylvaticae* Knapp ex Tx. et Oberd. 1958) and in numerous canyons of Montenegro [alliances *Ostryo-Fagion* Borhidi 1963 and *Ostryo carpinifoliae-Tilion platyphylli* (Košir et al. 2008) Čarni in Willner et al. 2016]. Both alpha diversity (within-community diversity) and beta diversity (diversity among communities) are significantly greater in canyons than in non-ravine habitats.

Keywords Alpha diversity, Beta diversity, Mesic forests, Montenegro, Paleo-endemic species, Ravine habitats

1 Introduction

A great heterogeneity of environmental conditions and specific history of biota resulted in complex and diverse vegetation in numerous canyons of the Balkan Peninsula. Ravine vegetation in Montenegro, and more generally in the Balkan Peninsula, represents an important biodiversity hotspot and significant pool of rare, paleo-endemic species.

An apparent latitudinal gradient of species richness in Europe is a consequence of repeated Pleistocene glaciations that significantly impoverished flora and fauna from the southern Europe northwards. Numerous studies indicate that the southern European peninsulas have acted as major refugia of plant and animal taxa during repeated Pleistocene glaciations [1–5]. Periodic glacial and postglacial migration of biota in Europe was affected by physical barriers (the Alps and Iberian mountain ranges). Therefore, the postglacial recolonization of the northern Europe occurred primarily by an expansion of biota from the Balkan refugium. Efficient colonizators from the Balkan refugium additionally blocked the expansion of biota from the Apennines and Pyrenees.

Postglacial recolonization resulted in significant loss of alleles and homozygotic structure of northern populations. Reduced genetic polymorphism is a consequence of both, the founder and bottleneck effects of colonizing populations [6, 7]. These effects are associated with the genetic drift (stochastic changes in allelic frequencies) of small populations that leads to either fixation or loss of any given allele. Besides genetic drift effects, an apparent 'northern purity, southern richness' pattern of genetic polymorphism may be explained by the 'refugia within refugia' hypothesis [8, 9]. Spatially isolated canyons of the Balkan Peninsula formed a system of Pleistocene microrefugia for Tertiary biota even during the strongest glaciations. Allopatric speciation within the isolated canyons increased genetic variability of the southern European biota. Complex interactions of Tertiary plants within these microrefugia with plants that periodically migrated during

Pleistocene glaciations and interglacial periods resulted in diverse vegetation in ravine habitats. A high diversity of phanerophytes, a presence of numerous Tertiary relics and a high percentage of paleo-endemic species are the main characteristics of ravine vegetation in the Balkan Peninsula [10–18].

We reviewed environmental conditions and vegetation in both the ravine and non-ravine habitats of Montenegro. In order to detect effects of canyon habitats on vegetation, we performed comparative analyses of alpha and beta diversity in mesophylous beech forests. We selected the beech-dominated forests because they occur both in non-ravine habitats (alliances *Aremonio-Fagion* (Horvat 1950) Borhidi in Török et al. 1989 and *Galio odorati-Fagion sylvaticae* Knapp ex Tx. et Oberd. 1958) and also in numerous canyons of Montenegro [alliances *Ostryo-Fagion* Borhidi 1963 and *Ostryo carpinifoliae-Tilion platyphylli* (Košir et al. 2008) Čarni in Willner et al. 2016].

2 Climate

Climate variability is the main driving force of vegetation and soil diversification. Reversely, both the soil and vegetation may considerably modify climate. Due to strong interdependence, geographic patterns of climate, vegetation and soil are similar [19–22].

Classification of climate may be performed using different criteria [19]. Köppen [23] used limits of various climate indices (e.g. average temperature of the warmest month, average temperature of the coldest month, the number of days below or above specified temperature threshold) to separate the climatic regions that fit boundaries of corresponding vegetation regions as accurate as possible. Due to a great simplicity, the Köppen's classification of climate is widely accepted [24].

The territory of Montenegro, spanning relatively narrow latitude and longitude ranges (41°39′ to 43°33′ N and 18°26′ to 20°21′ E), belongs to a mild temperate (mesothermal) climatic zone (Köppen's 'C climate zone'). This climate zone covers regions where the average temperature is above 10°C during the warmest months and the coldest month average is above −3°C. Dominant 'C climate zone' in Montenegro can be divided into several variants [25–27]. The 'Cs' temperate climate with dry summers dominates in (sub-)Mediterranean lowland region. It may be divided into two subtypes. The 'Csa' climate covers regions where the average temperature of the warmest month exceeds 22°C (coastline region and Zeta-Bjelopavlići plane). The 'Csb' climate occurs at higher altitudes (above 500 m), where the average temperature of the warmest month is lower than 22°C. The 'Cf' climate, where the letter 'f' denotes a climate without a dry season, dominates in hilly continental regions (Pljevlja, Rožaje, Berane, Bijelo Polje, Plav municipalities).

Besides temperate climate, a large area of Montenegro (regions above 1,000 m) belongs to the boreal climate (Köppen's 'D climate'). The air temperature of the coldest month in the mountainous regions of Montenegro is below −3°C, and the temperature of the warmest month is above 10°C. The boreal climate of

Montenegro may be divided into 'Dfb', 'Dfc' and 'Dsb' variants, where the letter 'c' indicates a climate with 3 or fewer months with mean temperatures above 10°C.

The tundra climate (Köppen's 'ET climate') is sporadically represented only in those regions above 2,000 m (peaks of Durmitor, Komovi and Prokletije Mountains).

Information on the seasonal variability of temperature and precipitations is missing in the Köppen's climate classification. This information is extremely valuable, since a season with adverse climate conditions (e.g. extreme drought, perhumid period, extreme frost, etc.) may restrict the distribution of particular species or particular vegetation type. Walter [28, 29] proposed a simple graph for description of seasonal variability of climate. Months are plotted on abscissa of the graph, mean air temperatures on the left ordinate and precipitation averages on the second (right) ordinate. The monthly averages of temperature and precipitations are scaled such that $1°C = 2\ l/m^2$. Specific scaling of the monthly temperature and precipitation averages is used to detect arid and moist periods. A negative water balance occurs when the temperature exceeds half of precipitations. In such situations the climate is arid, because estimated evapotranspiration exceeds precipitation. Semi-arid (dry) periods occur when the temperature exceeds the precipitation that is scaled such that $1°C = 3\ l/m^2$. Humid periods occur when precipitation exceeds 100 mm. In such situations, precipitations are scaled such that $1°C = 10\ l/m^2$.

The Walter's climatic diagrams of Montenegro are presented in Fig. 1. We created the diagrams using a statistical average of weather conditions over 30 years (the reference climate period from 1961 to 1990).

Mediterranean climate prevails in maritime regions (Bar). Due to specific dynamics of precipitation, the Mediterranean climate is characterized by humid winters and arid summers. During summer, continental air masses tend to warm faster than those above Adriatic Sea. As hot air rises, a low pressure is formed and it is replaced by moist air from coastlines. Reversely, lifting of warm air above Adriatic Sea during autumn and winter months forms a low-pressure zone, with strong cyclonic activity. Such pattern of air circulation increases autumn and winter precipitation in maritime regions.

Sub-Mediterranean climate (Podgorica) is similar to eu-Mediterranean climate. These two climates differ with respect to annual temperature amplitudes that are greater in continental than in maritime regions.

Precipitation increases with increasing altitude [30, 31]. Mediterranean mountains (Orjen, Lovćen and Rumija) are characterized by extremely high precipitation during winter and autumn months. According to Ducić et al. [32], the village of Crkvice on the Orjen Mountain is the wettest place in Europe, with the highest annual precipitation total over 5,000 mm. The oro-Mediterranean climate is cold during winter and dry during summer. Paleo-endemic Balkan pine (*Pinus heldreichii* H.Christ.) that dominates in alti-montane zone of sub-Mediterranean region is adapted to both physiological drought during winter months (i.e. drought caused by extremely low temperatures that prevent plant activity) and physical drought during summer.

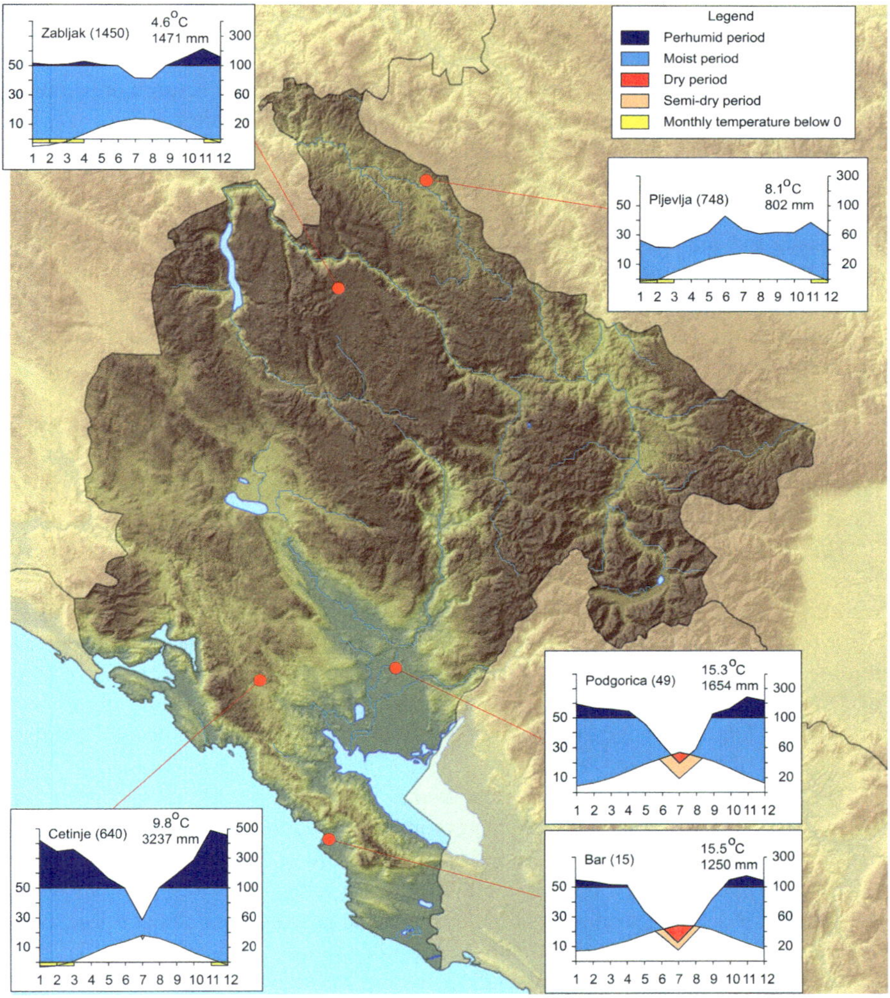

Fig. 1 Walter's climatic diagrams of Montenegro

Continentality increases from the coastline eastwards. Seasonal distribution of precipitations in Pljevlja is typical for continental regions. Contrary to maritime regions, where pluvial distribution has winter and autumn peaks, the continental regions are characterized by the summer peak of precipitations.

3 Geology

Diverse topography of Montenegro is a result of both complex tectonic movements, since the end of the Devonian period onwards, and erosion processes. The entire territory of Montenegro belongs to geotectonic karst units of Dinarides and peri-Adriatic (South Adriatic-Ionian) margin [33–37].

Karst of the South Adriatic-Ionian zone is composed of shallow marine limestone and anhydrite of Cretaceous age, dolomites, Middle Eocene limestone and Eocene-Oligocene Flysch deposits. This geotectonic zone covers the region of Ulcinj and Bar Municipalities.

Dinarides in Montenegro are divided into Budva-Cukali-Pindus zone (Budva and its surrounding), High Karst zone (Dalmatian-Herzegovinan zone), Pre-Karst and Bosnian Flysch zone and East Bosnian-Durmitor block.

Budva zone is mostly composed of the Oligocene Flysch. This zone also involves the Anisian Flysch, the Ladinian volcanic strata and thin Jurassic,and Cretaceous layers. Karstified limestones and dolomites are mixed with sedimentary and volcanic rocks. The South Adriatic-Ionian zone is thrust under the Budva zone [34].

The Dalmatian-Herzegovinan zone (frequently called the 'High Karst nappe') is composed of carbonate platform, developed during the Mesozoic (from the Middle Triassic to the end of Cretaceous). This zone is mainly built up of Mesozoic (Triassic, Jurassic and Cretaceous) limestones and dolomites. The middle belt of the High Karst Zone in the territory of Montenegro is built up of Upper Cretaceous-Paleogene Flysch beds. The High Karst geotectonic zone involves a large zeta tectonic depression with the Skadar (Shkodra) Lake, the largest lake of the Balkan Peninsula.

The Sarajevo sigmoid (or Pre-Karst and Bosnian Flysch) zone corresponds to the 'Kuči overthrust' which is composed of thin Permian and Triassic clastics and neritic carbonates, limestones from Upper Triassic to the Turonian (Upper Cretaceous) and transgressive Durmitor Flysch.

The East Bosnian-Durmitor unit includes a basement of Late Paleozoic-Mesozoic sediments consisting of Triassic siliciclastic rocks and carbonates, Late Triassic-Middle Jurassic mainly carbonate sediments and Middle-Late Jurassic radiolarites. The Lower Triassic in this unit is composed of sandstone. The Middle Triassic strata (Ladinian and Anisian) form complex layers. The Anisian stratum is formed of dolomite and bioclastic limestone, with red Bulog limestone and volcanic. The Ladinian mostly begins with chert and tuffs, followed by thin cherty limestone, dolomite and reef limestone. Jurassic layer is composed of limestone. The Bjelasica unit within this geotectonic block is composed mostly of Triassic volcanogenic-sedimentary formations. Specific vegetation of silicate oligotrophic pastures with *Festuca spadicea* Geners. ex Roem. & Schult (communities of the alliance *Achilleo-Arnicion* Horvat et Pawlowski in Horvat 1960) is widely distributed on Bjelasica Mountain [38].

Fluvial and glacial erosion processes of the karstified limestones and dolomites resulted in deep canyons (canyons of the Piva, Komarnica, Sušica, Tara Morača, Cijevna Rivers) and formation of numerous glacial lakes.

4 Soils

The heterogeneous soil in Montenegro is a result of the interaction between factors of pedogenetic processes (parent material, climate, topography, vegetation and other biotic components, including humans). Škorić et al. [39] described the soils of Yugoslavia, including the soils of the Montenegro Republic. More recently, Fuštić and Djuretić [40] described the soil heterogeneity in Montenegro.

The initial phases in soil development (Leptosol, Regosol and colluvial soil) have profile (A)-C or (A)-R, where (A) denotes initial, poorly developed horizon of humus accumulation; C is the parent substrate, virtually unaffected by the pedogenetic processes; and R refers to bare rock. These initial phases of soil development occur in chasmophytic habitats in canyons and altimontane regions of Montenegro.

More developed humus-accumulative soils have a typical humus-accumulative horizon (horizon A), in which well-humified colloidal organic matter is mixed with mineral soil as an organo-mineral complex. The humus-accumulative soils (A-C soils) in Montenegro involve kalkomelanosol (limestone-dolomite black soil), calcareous rendzina and rarely ranker soils on silicates.

Cambic soils with profile A-(B)-C, where the symbol '(B)' denotes the cambic horizon of argilogenesis (metamorphosis of parent material with increased clay content), are widely distributed in Montenegro. The cambic soils in Montenegro involve eutric cambisol, dystric cambisol, kalkocambisol (brown soil on limestone and dolomite) and *Terra rossa* (red soil). Eutric cambisol and kalkokambisol, together with rendzina and kalkomelanosol, dominate in western part of Montenegro (the High Karst Zone). *Terra rossa* occurs mainly in (sub-)Mediterranean area. Dystric cambisol is distributed mainly in continental part of Montenegro with the moist temperate or moist cold (boreal) climates.

Besides the automorphic soils that are wetted only by precipitation, various hydromorphic and halomorphic soils are distributed in Montenegro.

Hygromorphic soils, with the gley horizon (G), which exhibits signs of reduction and secondary oxidation in permanent or temporary anaerobic conditions, are distributed along the rivers and on the shores of Skadar Lake, Plav Lake and Šasko Lake. This group of temporary flooded soils involve fluvisols with (A)-G profile and semi-gleyed soils or humo-fluvisol (fluvial meadow soil) with A-C-G profile, transient peats and planohistol (low moor peat).

Halomorphic soils (sandy and loamy halomorphic soils) are exposed to periodic floods by saline and alkaline groundwater. This group of soils involves sand dunes and swamp soils along the estuaries with brackish water.

5 Zonal and Azonal Vegetation of Montenegro

Vegetation of Montenegro is diverse [38, 41–51]. It comprises both the zonal vegetations that reflect regional climatic conditions and different variants of azonal vegetation.

Evergreen oak forests and scrub (class *Quercetea ilicis* Br.-Bl. ex A. Bolòs et O. de Bolòs in A. Bolòs y Vayreda 1950) dominate in a narrow coastline region of Montenegro. However, due to enormous anthropogenic influence, this vegetation is reduced and significantly degraded. According to Rivas-Martínez, Penas and Díaz [52], the coastal zone in Montenegro belongs to the Adriatic province of the Eastern-Mediterranean biogeographical subregion.

Thermophylous oak forests (the class *Quercetea pubescentis* Doing-Kraft ex Scamoni et Passarge 1959 with xeric alliance *Carpinion orientalis* Horvat 1958 and more mesic alliances *Fraxino orni-Ostryion* Tomažić 1940 and *Quercion petraeae-cerris* Čarni et al. 2009) cover regions with sub-Mediterranean climate (Köppen's Csa and Csb climates).

Mesophylous beech and mixed broad-leaved European hornbeam, ash-leaved maple, beech and fir forests (class *Carpino betuli-Fagetea sylvaticae* Jakucs ex Passarge 1968) cover the continental mountain regions (regions between 600 and 1,100 m) that are characterized by moderate and moist climate (Köppen's Cfb climate). These forests belong to orders *Fagetalia sylvaticae* Pawlowski 1928 [with alliances *Aremonio-Fagion* (Horvat 1950) Borhidi in Török et al. 1989 and *Fagion sylvaticae* Luquet 1926 = *Galio odorati Fagion sylvaticae* Knapp ex Tx. et Oberd. 1958] and *Aceretalia pseudoplatani* Moor 1976 [with alliances *Fraxino excelsioris-Acerion pseudoplatani* P. Fukarek 1969 and *Ostryo carpinifoliae-Tilion platyphylli* (Košir et al. 2008) Čarni in Willner et al. 2016]. Forests of the alliance *Aremonio-Fagion* are more diverse than the beech-dominated forests of central and northern Europe [18, 53].

Acidophilous conifer forests on oligotrophic and leached soils (class *Vaccinio-Piceetea* Br.-Bl. in Br.-Bl. et al. 1939) dominate in continental mountains (Durmitor, Bjelasica, Ljubišnja Mountains) within the alti-montane zone of boreal microthermal climate (Köppen's Dfb climate).

Basophilous conifer forests (class *Erico-Pinetea* Horvat 1959) involve relic alliances *Pinion heldreichii* Horvat 1946 and *Erico carneae-Fraxinion orni* Horvat 1959. Contrary to continental boreal zone, where acidophilous spruce forests dominate, the alti-montane zone of the sub-Mediterranean region is covered with relic coniferous forests of the alliance *Pinion heldreichii* Horvat 1946. Alliance *Erico carneae-Fraxinion orni* Horvat 1959 comprises relict *Pinus nigra* forests on dolomite and ultramafic substrates. This alliance is distributed mostly in ravine habitats of the Dinarides, where it forms complex vegetation with other ravine alliances.

The tundra climate (Köppen's ET climate) of alpine belt is characterized by a short growing season, severe vinters, long-lasting snow cover and cold and strong winds. Long-lasting physiological drought that is caused by low temperatures and severe winds that cause physical drought are unfavourable for most trees. Specific life-form (i.e. cushion-like scrubs with short internodia and intense lateral branching and xeromorphic ericoid leaves) is an excellent adaptation to severe tundra climate. Čarni and Mucina [54] proposed numerous revisions of cryophilic syntaxa in the Balkan Peninsula. Cryophilic vegetation in Montenegro involves Arctic-Alpine vegetation of snow-rich habitats (class *Salicetea herbaceae* Br.-Bl. 1948), pine scrub (krummholz) in the subalpine belts (*Roso pendulinae-Pinetea mugo* Theurillat in Theurillat et al. 1995), graminoid tundra (calcareous grasslands of the class *Carici rupestris-Kobresietea bellardii* Ohba 1974 and acidophilous grasslands of the class *Juncetea trifidi* Hadač in Klika et Hadač 1944) and dwarfheath mountain tundra (*Loiseleurio procumbentis-Vaccinietea* Eggler ex Schubert 1960, which is mostly represented with the alliance *Juniperion sibiricae* Br.-Bl. in Br.-Bl. et al. 1939).

The most important species of the cryophilic vegetation in Montenegro are presented in Fig. 2.

Plant communities with a similar floristic composition may be distributed within different climate zones. Such 'azonal' vegetation is determined primarily by specific soil and hydrological conditions (waterlogged habitats, rock and scree habitats, halomorphic soils, etc.).

Pan-European chasmophytic vegetation on rock crevices (class *Asplenietea trichomanis* (Br.-Bl. in Meier et Br.-Bl. 1934) Oberd. 1977) and screes (classes *Thlaspietea rotundifolii* Br.-Bl. 1948 and *Drypidetea spinosae* Quézel 1964) is azonal, since chasmophytic communities appear in all vegetation zones, from colline (usually canyon) habitats to nival regions [55].

Halophytic vegetation in Montenegro is heterogeneous [56–61]. It involves the halophytic plant communities on cliffs and sand dunes along the Adriatic coast (*Cakiletea maritimae* Tx. et Preising in Tx. ex Br.-Bl. et Tx. 1952 and *Ammophiletea* Br.-Bl. et Tx. ex Westhoff et al. 1946) and vegetation of saline and brackish swamps and marshes (*Spartinetea maritimae* Beeftink 1962, *Thero-Salicornietea* Tx. in Tx. et Oberd. 1958 164, *Juncetea maritimi* Br.-Bl. in Br.-Bl. et al. 1952 166).

Azonal alluvial forests (class *Alno glutinosae-Populetea albae* P. Fukarek et Fabijanić 1968) and alluvial scrub communities (*Salicetea purpureae* Moor 1958) occupy riparian habitats along the main rivers of Montenegro [14, 45, 62]. Vegetation of freshwater springs and swamps (classes *Montio-Cardaminetea* Br.-Bl. et Tx. ex Klika et Hadač 1944 and *Phragmito-Magnocaricetea* Klika in Klika et Novák 1941) and vegetation of mires, fens and bogs (classes *Scheuchzerio palustris-Caricetea fuscae* Tx. 1937 and *Oxycocco-Sphagnetea* Br.-Bl. et Tx. ex Westhoff et al. 1946) cover numerous wetlands and lake shores [62–64].

Freshwater aquatic vegetation in numerous lakes and rivers of Montenegro involves free-floating vegetation of relatively nutrient-rich freshwater bodies (class Lemnetea O. de Bolòs et Masclans 1955) and vegetation of rooted floating or submerged macrophytes of stagnant mesotrophic, eutrophic freshwater bodies and slowly flowing shallow streams (class *Potamogetonetea* Klika in Klika et Novák 1941 with alliances *Callitricho hamulatae-Ranunculetalia aquatilis* Passarge ex Theurillat in Theurillat et al. 2015, *Batrachion fluitantis* Neuhäusl

Fig. 2 Cryophilic species that are adapted to microthermal tundra climate: *Pinus mugo* Turra (**a**), *Juniperus sibirica* Burgsd. (**b**), *Salix retusa* L. (**c**), *Soldanella alpina* L. (**d**), *Dryas octopetala* L. (**e**), *Parnassia palustris* L. (**f**), *Saxifraga aizoides* L. (**g**), *Vaccinium vitis idaea* L. (**h**), *Oxyria digyna* (L.) Hill (**i**). Photos made by B. Karadžić & S. Jarić

1959 and *Ranunculion aquatilis* Passarge ex Theurillat in Theurillat et al. 2015). Due to strong anthropogenic pressure, the native vegetation is degraded significantly in (sub-)urban and agricultural areas [65]. Adverse human impacts increase threat of rare plant species in Montenegro [66, 67].

6 Environmental Conditions and Vegetation in Ravine Habitats

According to Baede et al. [68], climate is an interactive system consisting of five major components: the atmosphere, the hydrosphere, the cryosphere, the land surface and the biosphere. Solar radiation is the primary energy source of the climate system. Energy flux of absorbed solar radiation (i.e. energy absorbed per unit area) varies with latitude. According to Kiehl and Trenberth [69], the average energy flux of absorbed radiation is 168 W/m^2. Besides latitude, specific topography of canyons also affects the energy flux.

Suppose that a canyon is oriented in east-west direction. In that case, the energy flux (i.e. the energy absorbed per unit area) is disproportionally greater on south-facing slope than on the opposite slope of the canyon (Fig. 3). Significant difference of the energy fluxes creates a strong thermal gradient.

Intensity of climate modifications in ravine habitats depends on the solar incidence angle and also on numerous topographic properties of a canyon (angle and aspect of canyon slopes; orientation, length and depth of the canyon). The solar incidence angle in Montenegro is high enough during entire vegetation season (approximately 40° during spring and autumn equinoxes and roughly 20° during summer solstice) to allow formation of strong thermal gradients in canyon habitats.

A strong thermal gradient in ravine habitats creates corresponding air moisture gradient, since air humidity and temperature are inversely correlated. Cool air above north-facing slopes and at the bottom of a canyon is more saturated with water

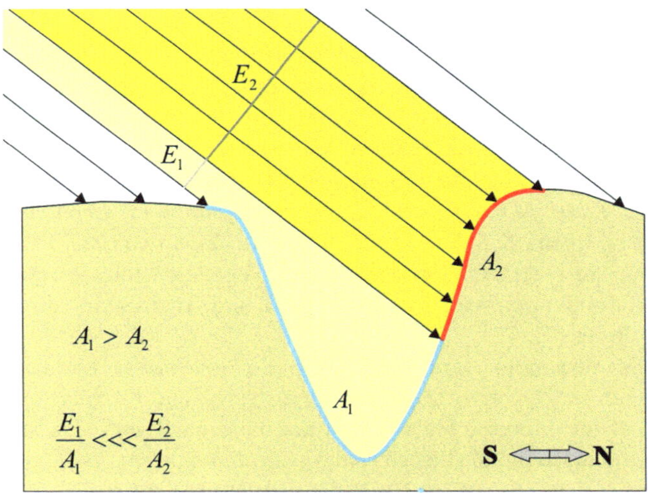

Fig. 3 Effects of ravine topography on microclimate. Absorbing surfaces that receive solar energy on northerly exposed (A_1) and south-facing (A_2) slopes of a canyon are denoted by blue and red lines. The energy flux (the energy absorbed per unit area) is disproportionally greater on south-facing slope than on the opposite slope of the canyon

vapour than a warmer air above the south-facing slopes. A high pressure of warm air at the canyon peaks blocks lifting of cool moist air from the bottom of the canyon. Therefore, the air humidity in canyon habitats is high, even during the extreme drought periods. Increased humidity in ravine habitats attenuates the temperature extremes and decreases continentality of microclimate.

Extreme fluvial erosion process in steep slopes of canyons prevents formation of deep soils. Consequently, soils in ravine habitats of Montenegro are represented mainly by initial phases of pedogenetic process (i.e. shallow skeletal Leptosol, Regosol and colluvial deposits, kalkomelanosol and calcareous rendzina). More developed cambic soils (eutric cambisol and kalkocambisol) occur sporadically.

Vegetation in numerous canyons of Montenegro is remarkably diverse. For example, almost all vegetation types of Montenegro (except halophytic, swamp and cryophilic vegetation types) are distributed in amazing canyon of the Tara River. Similar vegetation heterogeneity was observed in canyons of the Komarnica (Fig. 4), Piva, Sušica and Morača Rivers.

There are two, ecologically clearly distinguishable, types of canyons in Montenegro. The sub-Mediterranean canyons involve canyons of the Morača, Mala Rijeka and Cijevna Rivers. The continental group of canyons comprises canyons of the rivers that belong to the Danube River Basin. The Tara and Piva Rivers are joined at the confluence location 'Šćepan Polje', forming the Drina River. These three rivers, with their tributaries (Komarnica, Sušica, Janja, Drinjača, Žepa, Prača, Bistrica, Sutjeska, Jadar, Lim, Rzav, Ćehotina), form an impressive system of the West Balkan canyons (in the territories of Montenegro, Serbia and Bosnia and Herzegovina). This system of ravine habitats is one of the most important biodiversity hotspots in Europe [10, 14, 15, 17, 18].

Sub-Mediterranean canyons are exposed to typical Mediterranean climate with excessive summer drought, mild winters and precipitation peaks during autumn and winter months. Such climate conditions support the development of xeric vegetation with numerous Mediterranean species [13, 70–72]. The most important Mediterranean species in these canyons are *Phillyrea latifolia* L., *Petteria ramentacea* (Sieber) C. Presl., *Paliurus spina-christi* Mill., *Pistacia terebinthus* L., *Spartium junceum* L., *Punica granatum* L., *Euphorbia characias* L., *Salvia officinalis* L., *Satureja montana* L., *Euphorbia spinosa* L., *Cephalaria leucantha* (L.) Schrad. ex Roem. & Schult., *Helichrysum italicum* (Roth) G. Don, etc.

Mediterranean evergreen forest and scrub vegetation (classes *Quercetea ilicis* Br.-Bl. 1936. and *Paliuretea* Trinajstić 1978) are fragmentary distributed along the Cijevna River [71, 72].

The xeric deciduous forests of *Carpinus orientalis*, *Ostrya carpinifolia*, *Quercus pubescens*, *Quercus trojana*, *Quercus cerris* and *Fraxinus ornus* dominate in canyons of the Morača, Mala Rijeka and Cijevna Rivers. Traditionally, these forests were included in the alliance *Ostryo-Carpinion orientalis* Horvat 1954 of the class *Querco-Fagetea sylvaticae* Br.-Bl. et Vlieger in Vlieger 1937 [47, 48, 51, 73]. Horvat [48] divided the alliance into Adriatic (Dinaric) and Aegean (Vardar) suballiances. According to more recent classification [74], these forests belong to the alliance *Carpinion orientalis* Horvat 1954 of the class *Quercetea pubescentis*

Fig. 4 Spectacular canyon of the Komarnica River. Photo made by B. Karadžić

Doing-Kraft ex Scamoni et Passarge 1959. The alliance includes amphi-Adriatic xeric deciduous forests and scrub that form a belt in coastal regions above the evergreen vegetation of *Quercetea ilicis*. In comparison with oak-dominated forests, these forests thrive in areas with a mild climate during winter [75]. These forests are distributed along the coast of the Adriatic Sea and in regions where the maritime influence extends deep inland, along the Una, Neretva, Morača, Cijevna, Drim and Vardar Rivers. The alliance *Ostryo-Carpinion orientalis* Horvat 1959 is a synonym of widely accepted alliance *Fraxino orni-Ostryion* Tomažić 1940. Čarni et al. [75] demonstrated that the alliances *Fraxino orni-Ostryion* Tomažić 1940 and *Carpinion orientalis* Horvat 1954 differ with respect to both ecology and their floristic composition. The *Carpinion* alliance is distributed in Adriatic coastal areas and also, extrazonally, in inland areas (on steep southern slopes, around lakes where the climate is mitigated by the water body, etc.). Communities of the alliance *Fraxino orni-Ostryion* occupy more humid habitats, within the inner part of the mountain chains along the Adriatic coast at higher altitudes and extrazonally within areas with a maritime sub-Mediterranean climate.

Coniferous forests of endemic pine *Pinus heldreichii* Christ (alliance *Pinion heldreichii* Horvat 1946 of the class *Erico-Pinetea* Horvat 1959) are distributed at the upper course of the Cijevna River.

The chasmophytic vegetation in canyons of the Morača, Mala Rijeka and Cijevna Rivers is represented by communities of the classes *Asplenietea trichomanis* (Br.-Bl. in Meier et Br.-Bl. 1934) Oberd. 1977, *Adiantetea* Br.-Bl. et al. 1952 and *Thlaspietea rotundifolii* Br.-Bl. 1948.

Compared to sub-Mediterranean canyons, continental canyons of Montenegro (canyons of the Piva, Tara, Komarnica and Sušica Rivers) are by far more heterogeneous. The main types of forest vegetation in these canyons involve riparian forests (*Salicetea purpureae* Moor 1958), broad-leaved forests on xeric (dry) habitats (*Quercetea pubescentis* Doing-Kraft ex Scamoni & Passarge 1959), broad-leaved forests on mesic (moist) habitats (*Carpino betuli-Fagetea sylvaticae* Passarge in Passarge and Hofmann 1968) and coniferous forests on basic soils (*Erico-Pinetea* Horvat 1959). Contrary to sub-Mediterranean canyons, where xeric forests and scrub dominate, the percentage of mesic forests [alliances *Ostryo carpinifoliae-Tilion platyphylli* (Košir et al. 2008) Čarni in Willner et al. 2016, *Fraxino orni-Ostryion* Tomažić 1940, *Ostryo-Fagion* Borhidi 1963] is significantly greater in the continental canyons.

Alluvial forests of the class *Salicetea purpureae* Moor 1958 are represented by riparian forests and scrub of willow species, including *Salix elaeagnos* Scop., along gravelly banks of torrential rivers [alliances *Salicion elaeagno-daphnoidis* (Moor 1958) Grass 1993]. These forests and scrub are gradually replaced by the tall-herb fringe vegetation (alliance *Cicerbition pancicii* Lakušić 1978 with paleo-endemic species *Lactuca pancicii* (Vis.) N. Kilian & Greuter = *Mulgedium pancicii* Vis. = *Cicerbita pancicii* (Vis.) Beauverd and alliance *Filipendulo-Petasition* Br.-Bl. ex Duvigneaud 1949).

Xeric forests (forests and scrub of alliances *Carpinion orientalis* Horvat 1954 and *Fraxino orni-Ostryion* Tomažić 1940) occupy warm, south-exposed, slopes of numerous canyons. They are covered by patchily distributed individuals of *Ostrya carpinifolia* Scop., *Fraxinus ornus* L., *Frangula rupestris* (Scop.) Schur., *Cotinus coggygria* Scop., *Rhamnus saxatilis* Jacq., *Rhamnus cathartica* L., *Rhamnus alpina* subsp. f*allax* (Boiss.) Maire & Petitm., *Tilia platyphyllos* Scop., *Acer campestre* L., *Viburnum lantana* L., *Quercus cerris* L., *Quercus pubescens* Willd., etc.

Mesic forests in continental canyons of Montenegro belong to beech-dominated alliance *Ostryo-Fagion* Borhidi 1963, mesic alliances of hop-hornbeam (*Ostrya carpinifolia* Scop.) forests (*Ostryo carpinifoliae-Tilion platyphylli* (Košir et al. 2008) Čarni in Willner et al. 2016) and ash-maple scree and ravine forests

(*Fraxino excelsioris-Acerion pseudoplatani* P. Fukarek 1969). Due to floristic affinities to different types of forests (vegetation classes *Quercetea pubescentis* Doing-Kraft ex Scamoni & Passarge 1959, *Carpino betuli-Fagetea sylvaticae* Passarge in Passarge & Hofmann 1968, *Erico-Pinetea* Horvat 1959, *Asplenietea trichomanis* (Br.-Bl. in Meier et Br.-Bl. 1934) Oberd. 1977, *Thlaspietea rotundifolii* Br.-Bl. 1948, *Drypidetea spinosae* Quézel 1964), syntaxonomy of these forests is a controversial and yet unresolved issue [15–18, 74, 76]. These forests occur on screes, rocky cliffs and colluvial gravel and also on developed soils, mainly shallow rankers, rendzinas and brown calcareous soils. Beech (*Fagus sylvatica* L.) almost exclusively dominates in the mesic forests that occur outside ravine habitats (alliance *Aremonio-Fagion* (Horvat 1950) Borhidi in Török et al. 1989). Mesic forests in numerous canyons of Montenegro have a more diverse and more complex structure. They are covered by sparsely distributed individuals of *Fagus sylvatica* L., *Ostrya carpinifolia* Scop., *Juglans regia* L., *Carpinus betulus* L., *Acer platanoides* L., *Acer pseudoplatanus* L., *Fraxinus excelsior* L., *Sorbus aucuparia* L., *Acer obtusatum*, *Carpinus betulus* L., *Tilia platyphyllos* Scop., *Acer campestre* L., *Quercus cerris* L., *Quercus petraea* (Matt.) Liebl., etc. Heterogeneous environmental conditions prevent domination of one species and promote a polydominant structure of communities. Broad-leaved ravine *Tilio-Acerion* forests specify an EU priority type of ecosystems [77]. Compared with temporal *Tilio-Acerion* forests, the mesic forests in the southern European ravine habitats are by far more diverse [17].

Due to a wide distribution (Europe, North Africa, Middle East, the Arctic archipelagos and Greenland), the chasmophytic vegetation of crevices, rocky ledges and rocky cliffs [class *Asplenietea trichomanis* (Br.-Bl. in Meier et Br.-Bl. 1934) Oberd. 1977] comprises floristically and ecologically diverse communities. The chasmophytic communities in continental canyons of Montenegro belong to the Dinaric alliances *Moltkeetalia petraeae* Lakušić 1968, *Micromerion croaticae* Horvat in Blečić 1959, *Edraianthion* Lakušić 1968, *Amphoricarpion* Lakušić 1968 and *Edraiantho graminifolii-Erysimion comati* Mucina et al. 1990. The Alpic-Carpathian alliance *Potentillion caulescentis* Br.-Bl. in Br.-Bl. et Jenny 1926 is fragmentary distributed at higher altitudes in canyon slopes.

Scree vegetation is distributed across Europe, from the submontane to the alpine belts. The majority of calcareous and serpentine scree plant communities occur in southern Europe, in particular in the peri-Mediterranean mountain ranges, while in northern Europe (Scandinavia), acidophilus communities dominate. Scree vegetation in Montenegro is mainly represented by the xeric alliances *Achnatherion calamagrostis* Jenny-Lips 1930 and *Corydalo ochraleucae-Geranion macrorhyzi* Lakušic 1975 and *Teucrion montani* Csürös et Pop 1965 of the class *Thlaspietea rotundifolii* Br.-Bl. 1948. Communities of the class *Drypidetea spinosae* Quézel 1964 are distributed at higher altitudes in canyon habitats.

7 Biodiversity Components in Canyons and Adjacent Non-ravine Habitats

Despite its frequent and wide use, the term biodiversity is not defined unambiguously. Considering a complexity of different levels of biological organization (populations, species assemblages, biotic communities, ecosystems, landscapes), there are numerous formal definitions of biodiversity [78–80].

Biological variability is a multicomponent, complex parameter that involves intraspecies (population) and interspecies (cenotic) variety [22, 81–83]. Intraspecies or population variability is a result of genetic variety, diversity induced by the influence of environmental factors and variability which arises from the interaction of genetic background and environmental conditions [6]. Interspecies or coenotic variability can be divided into within-community variability (alpha diversity), variability between communities (beta diversity) and the total variability in a given region (gamma diversity). Alpha diversity is a function of the number of species present within a habitat (*species richness*) and the evenness with which the individuals are distributed among these species (*species equitability*) [84–86].

Beta diversity can be partitioned into two components: species turnover and nestedness [87–89]. *Species turnover* refers to the replacement of a set of species from one community by corresponding number of new species in another community. However, two communities may differ with respect to their species composition, not because of species replacement but due to missing species in the species poorer community. Since the species poorer community may be considered as a nested subsample of a richer community, the difference in species composition that is caused by missing species is denoted as *nestedness* (the nested component of beta diversity).

7.1 Sampling Procedures and Methodology

In order to detect effects of canyon topography on biodiversity, we selected beech-dominated forests that occur in both canyon and non-ravine habitats. We sampled 20 relevés (vegetation samples) of beech forests from non-ravine habitats of Durmitor and Komovi Mountains and 20 relevés from canyons of the Komarnica and Tara Rivers. The forests in non-ravine habitats belong to the alliance *Aremonio-Fagion* (Horvat 1950) Borhidi in Török et al. (1989) (formerly the alliance *Fagion Illyricum* Horvat 1950). These forests are distributed in Slovenia, Croatia, Bosnia and Herzegovina, Montenegro, western parts of Serbia, Albany and Greece [18, 49, 90]. Degraded variants of these forests may be included into a species poorer alliance *Fagion sylvaticae* Luquet 1926 = *Galio (Asperulo) odorati Fagion* Tx. 1955 that dominates in temperate Europe. The forests in canyon habitats belong to the mesic alliance *Ostryo-Fagion* Borhidi 1963.

We collected samples according to Braun-Blanquet methodology. Braun-Blanquet [91] proposed a combined cover-abundance scale to quantify the presence of a species in a site. The scale is semi-descriptive and therefore impractical for numerical analyses. To overcome this problem, we applied a simple transformation of Braun-Blanquet's scale to numeric scale of combined cover-abundance values [92].

Alpha diversity was assessed using two parameters: the species richness and Shannon's entropy

$$H_{(p)} = -\sum_{i=1}^{s} p_i \log p_i,$$

where p_i is the proportion of species i within particular site and s denotes number of species within the site. Equitability component of alpha diversity may be calculated using the equation

$$E = H/H_{\max} = -\sum_{i=1}^{s} p_i \log p_i / \log s,$$

where $H_{\max}$ is the greatest possible entropy. We detected components of beta diversity using the two alternative procedures, described by Baselga [87] and Podani et al. [88]. All statistical analyses were performed using the newest version of the 'FLORA' software package [93].

7.2 Results

Alpha diversity of analysed forests, expressed as Shannon's entropy, was slightly, but insignificantly, greater in canyons than in non-ravine habitats (Table 1). The difference in Shannon's entropy is a result of both a greater species richness

Table 1 Alpha diversity components in analysed forests

		Canyons	Non-ravine habitats
Complete community	Species richness	27.90 ± 5.04	25.15 ± 4.61
	Shannon's entropy	3.174 ± 0.42	3.11 ± 0.19
	Equitability	0.971 ± 0.01	0.942 ± 0.01
Phanerophytes	Species richness[a]	8.15 ± 2.15	5.61 ± 2.50
	Shannon's entropy[a]	1.92 ± 0.285	1.41 ± 0.55
	Equitability[a]	0.933 ± 0.05	0.836 ± 0.02
Herbaceous plants	Species richness	21.3 ± 7.46	20.00 ± 4.71
	Shannon's entropy	20.3 ± 7.46	19.9 ± 4.71
	Equitability	0.98 ± 0.08	0.957 ± 0.01

[a]Statistically significant difference between forests in canyons and non-ravine habitats

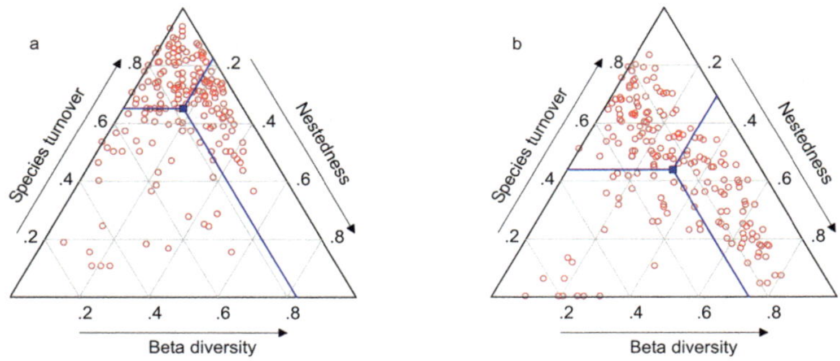

Fig. 5 Beta diversity components of mesic forests in canyons (**a**) and non-ravine habitats (**b**). Each point (red circle) represents a pair of communities. Its position is determined by values of beta diversity, species turnover and nestedness between the pair of communities. Blue rectangle corresponds to the centroid (average value) of the points

and a greater equitability of forests in canyon habitats. The greatest (and statistically significant) differences among these two groups of beech-dominated forests were obtained with respect to the alpha diversity of phanerophytes (trees and scrub). Both species richness and equitability of phanerophytes were substantially greater in canyons than in non-ravine habitats. *Fagus sylvatica* is the strongest competitor that almost exclusively dominates in mesic habitats outside canyons. Dominance of one species decreases equitability component of alpha diversity.

Similar, but statistically insignificant, trends of alpha diversity were observed for herbaceous plants. Forests in flat (up to 35°) habitats, outside canyons, are developed on relatively deep soil. Such edaphic conditions enable presence of typical "*Fagetalia*" species that belong to the group of vernal ephemeroides or shade-evading geophytes [94–97]. Some of these species [*Scilla bifolia* L., *Corydalis cava* (L.) Koerte, *Corydalis solida* (L.) Clairv., *Allium ursinum* L.] are rare, or they are absent in canyon habitats. Absence of some of vernal ephemeroides in canyons may be explained by adverse soil conditions (shallow skeletal soils). Inversely, a huge number of chasmophytic, shade-adapted species (*Asplenium scolopendrium* L., *Arabis turrita* L., *Ceterach officinarum* Willd., *Digitalis laevigata* Waldst. & Kit., etc.) that occur in canyons are absent in non-ravine beech forests.

A high alpha diversity in canyon habitats may be attributable to patchy soil patterns (bare rocks; screes; shallow and skeletal Leptosol, Regosol and colluvial deposits; kalkomelanosol relatively deep cambic soils) within each of analysed communities. Due to a range of opportunities, numerous species with different ecological requirements may coexist in patchy environment. Strong competitive interactions in such conditions increase asymmetry and reduce amplitude of response distributions of competing species [98].

Analyses of beta diversity confirmed that the forests in canyons are more diverse than forests in non-ravine habitats. Ternary graphs (Fig. 5) clearly indicate that both species turnover and overall beta diversity of analysed forests are significantly greater in canyons than in adjacent non-canyon habitats. A high rate of species replacement in canyons is a consequence of heterogeneous microclimatic conditions along strong thermal and moisture gradients. The gradients differentiated beech-dominated forests in canyons into three variants: Firstly, the xeric variant involves beech forests with evident affinity to oak-dominated communities [presence of *Quercus petraea* (Matt.) Liebl., *Quercus cerris* L, *Tilia platyphyllos* Scop., *Fraxinus ornus* L.]. Secondly, mesic variant involves typical communities of the alliance *Ostryo-Fagion* Borhidi 1963. Finally, the hygrophilous variant is represented by communities of *Fagus sylvatica* that codominates with *Fraxinus excelsior* L., *Acer platanoides* L., *Acer pseudoplatanus*, L. *Ostrya carpinifolia* Scop. and a range of mesic and hygrophilous species. Besides strong climatic gradients, the patchy soil patterns additionally increase species turnover rate in canyon habitats.

Difference in species richness (nestedness) was much greater in non-ravine habitats than in canyons. This indicates a degradation of beech forests in non-ravine habitats. Floristically rich forests of the alliance *Aremonio-Fagion* are frequently degraded to the species poorer forests of the alliance *Fagion sylvaticae*, due to inadequate exploitation of biotic resources. Contrary to non-ravine habitats, canyons are less exposed to human disturbances, since the land use in steep habitats is economically inefficient.

8 Endemism in Canyon Habitats

The Balkan Peninsula is one of the most important biodiversity hotspots in Europe. More than 8,000 species of vascular plants inhabit the Balkan Peninsula [99–101]. Among them, a huge number (approximately 2,600 taxa) belong to the group of the Balkan endemics. Balkan Peninsula is a biodiversity centre for 17 mainly monotypic genera, represented by 1 or only few phylogenetically isolated species. Endemic genera indicate the age of flora in the specified area and designate this area as a biodiversity centre and centre of flora diversification. Such monotypic and oligotypic genera are *Halacsya sendtneri* (Boiss.) Dörfler, *Paramoltkia doerfleri* (Wettst.) Greuter & Burdet, *Paraskevia cesatiana* (Fenzl & Friedr.) W. & G. Sauer, *Petromarula pinnata* (L.) A. DC. *Degenia velebitica* Deg., *Leptoplax emarginata* (Boiss.) O.E. Schultz, *Haberlea rhodopensis* Friv., *Jankaea heldreichii* (Boiss.) Boiss., *Wagenitzia lancifolia* Sieber ex Sprengel (Dostal), *Hymenonema laconicum* Boiss. & Heldr., *Hymenonema graecum* (L.) DC., *Thamnosciadium junceum* (Sm.) Hartvig, *Pancicia serbica* Vis., *Horstrissea dolinicola* Greuter, Gertsberger & Egli, *Petteria ramentacea* (Sieber) C. Presl, *Festucopsis sancta* (Janka) Melderis, *Festucopsis serpentini* (C.E. Hubbard) Melderis, *Lutzia cretica* (L.) Greuter & Burdet and *Phitosia crocifolia* Kamari & Greuter.

According to Velčev, Kožuharov and Ančev [102], the Balkan endemics may be divided into two main categories: paleo-endemics and neo-endemics. The first group involves Tertiary relics, with low genetic variability, diploid or low-ploid chromosome number and apparent disjunct distribution with its closest relatives. Due to long timescale isolation and absence of genetic contacts with their relatives, these species formed stable genotypes under particular conditions. Neo-endemics are species that appeared during Quaternary climatic stresses. These species are characterized by polyploidy (tetraploid or high-ploid chromosome numbers).

High mountains of Montenegro were exposed to a very strong influence of Pleistocene glaciations. The traces of former glaciers (moraines, glacial lakes and cirques) clearly indicate that the ice not only covered the peaks but extended all the way down to the valleys [37, 103–108]. Due to periodic Pleistocene glaciations, numerous species from boreo-Arctic regions migrated southwards and merged with cryophytes of southern Europe that migrated altitudinally towards lowland regions. Backward migrations during interglacial periods resulted in disjunct distribution of European cryophytes (glacial relics). Populations of glacial relics became separated ca. 10,000 years ago, i.e. at the end of the Würm glaciation. The most important Arctic-Alpine relics in Montenegro and adjacent regions are *Salix reticulata* L., *Salix herbacea* L., *Juniperus sibirica* Lodd., *Dryas octopetala* L., *Oxyria digyna* (L.) Hill., *Trollius europaeus* L., *Potentilla crantzii* Beck, *Gnaphalium supinum* L., *Saxifraga aizoides* L., *Parnassia palustris* L., *Pinguicula alpina* L., *Veronica alpina* L., *Huperzia selago* (L.), Trev. and numerous species within the genera *Carex* L., *Juncus* L. and *Eriophorum* L. Arctic-Alpine species repeatedly merged and separated during glacial-interglacial periods. Relatively wide hybridization zone of Alpine and Arctic (boreal) species during glacial periods prevented speciation. However, some glacial relics in southern European mountains were isolated even during glacial periods. More or less continuous forest zone persisted in the southern European regions even during the coldest period of the Würm glaciation. The forest zones were barriers for migration of high-mountain cryophytes. Therefore, the southern European cryophilic flora responded to climate changes with altitudinal, instead of latitudinal, migrations. Due to isolated mountain peaks and migration barriers, the gene flow of glacial relics in southern Europe was reduced. This caused a rapid speciation of high-mountain regions of southern Europe [100, 109–111]. The most important endemic cryophytes in Montenegro and adjacent regions are *Geum bulgaricum* Pančić, *Sempervivum kosaninii* Praeger, *Sempervivum heuffelii* Schott, *Achillea clavennae* L., *Wulfenia blecicii* Lakušić, *Viola orphanidis* Boiss., etc. In comparison with the mountains of Greece, the number of local endemics in the mountains of Montenegro, and generally in the central part of the Balkan Peninsula, is markedly lower. This may be attributed to the absence of glaciation impact in Mediterranean centres of orophytic endemism, in which glaciations did not play any significant role [100, 111].

Tertiary relics comprise species that dominated in Europe at the beginning of the so-called Quaternary period. According to the International Commission on Stratigraphy [112], this period (early Pleistocene age) started 2.5 million years ago. Tertiary flora almost disappeared from Europe during Quaternary period that is characterized by marked temperature oscillations. Repeated swings of glacial and interglacial periods significantly reduced floristic diversity in temperate Europe.

Amazing canyons of the Tara, Piva, Komarnica, Cijevna and Morača Rivers formed a system of Pleistocene microrefugia for Tertiary biota even during the strongest glaciations.

Species-rich forests covered these canyons continuously throughout Pleistocene. A high diversity of phanerophytes indicates that these forests are remnants of Tertiary subtropical vegetation. The most important phanerophytes in ravine habitats of Montenegro are *Ostrya carpinifolia* Scop., *Acer heldreichii* Orph., *Fagus sylvatica* L., *Juglans regia* L., *Corylus colurna* L., *Celtis australis* L., *Pinus heldreichii* Christ, *Fraxinus ornus* L., *Carpinus orientalis* Miller, *Quercus pubescens* Willd., *Quercus trojana* Webb., *Acer intermedium* Panč., *Cotinus coggygria* Scop., *Rhamnus saxatilis* Jacq., *Cotoneaster integerrima* Medik., *Cotoneaster tomentosus* (Ait.) Lindl., *Frangula rupestris* (Scop.) Schur., *Chamaecytisus leiocarpus* (Kern.) Rothm., *Rhamnus fallax* Boiss.[=*Rhamnus alpinus* L. subsp. *fallax* (Boiss.) Maire & Petitmengin], *Euonymus verrucosus* Scop., *Staphylea pinnata* L., *Viburnum lantana* L., *Ilex aquifolium* L., *Taxus baccata* L., etc.

Tertiary relics that occur in canyons of the Balkan Peninsula, including numerous canyons of Montenegro, usually belong to a group of endemic species with relatively narrow distribution. A high percent of endemic species in canyons may be explained by the 'chasmophytic divergence' hypothesis [17, 113]. The hypothesis is based on two main assumptions: (1) strong selective pressure in adverse saxatile habitats that are spatially isolated and (2) long-lasting evolutionary process in relatively stable canyon habitats that were not exposed to climate-induced and human-induced disturbances.

Shallow and skeletal soil, bare rock and screes dominate in canyons, due to intense fluvial erosion. The vegetation of chasmophytes (i.e. plants that are adapted to adverse rocky habitats) is widely distributed in Paleo-Temperate region [77]. Despite the wide distribution, chasmophytic vegetation is fragmented and isolated. Gene flow between population in fragmented and geographically isolated habitats is obstructed. Surina [9] emphasized that thermophilic Tertiary relics responded to periodic Pleistocene climate oscillations by latitudinal range shifts. However, expansion of chasmophytes in non-ravine habitats was prevented by more efficient competitors. Conversely, stress conditions in saxatile habitats prevent colonization of canyons by stronger competitors from adjacent non-ravine habitats. Such ecological situation contributed to isolation and reduced gene flow between chasmophytic populations. Even within the same saxatile region, the effective gene flow between chasmophytic populations may be considerably inhibited through the frequent failure of seed to establish new individuals. The first step in colonization of bare rock must be the establishment of seedlings. This occurs in microsites with crevices, in which root systems can find anchorage. Most of the holes suitable for ecesis are already occupied in saxatile habitats. Such situation obstructs individuals to produce progeny. Due to a low effective population size, the effects of genetic drift and inbreeding depression may lead to the accident elimination of alleles and increased proportion of homozygous (often non-adaptive) alleles. Ecological plasticity of rare chasmophytes is narrowed because of reduced genetic variability. Therefore, most of rare chasmophytes belong to the group of critically endangered taxa.

Strong selective pressures and reduced gene flow due to both geographical isolation of saxatile areas and ecesis problems resulted in fast speciation of chasmophytic taxa with relatively narrow distribution. Long-lasting isolation of chasmophytes resulted in diversification of chasmophytic genera (e.g. *Amphoricarpos* Visiani and *Edraianthus* DC.) within a narrow geographical range [9, 114–116].

The most important paleo-endemic species in ravine habitats of Montenegro are *Amphoricarpus autariatus* Blečić & Mayer, *Daphne malyana* Blečić, *Ramonda serbica* Pančić, *Centaurea derventana* Visiani et Pančić, *Lactuca pancicii* (Vis.) N. Kilian & Greuter (b), *Degenia velebitica* (Degen) Hayek, *Moltkia petraea* (Tratt.) Griseb., *Reichardia macrophylla* Visiani et Pančić, *Scabiosa fumarioides* Visiani et Pančić, *Lathyrus binnatus* Pančić, *Edraianthus tenuifolius* (Waldst. et Kit.) DC, *Athamanta haynaldii* Borb. et Uchtr. = *Athamanta turbith* (L.) Brot. subsp. *haynaldii* (Borbás & Uechtr.) Tutin, *Pseudofumaria alba* (Mill.) Lidén = *Corydalis ochroleuca* W.D.J.Koch, *Campanula lingulata* Waldst. et Kit., *Dianthus petraeus* Waldst. et Kit., *Doronicum columnae* Ten., *Iris bosniaca* (Beck) Dörfl. = *Iris reichenbachii* Heuff., *Melampyrum hoermenianum* Maly, *Arabis procurrens* Wald. et Kit., *Minuartia bosniaca* (Beck) Maly, *Onosma stellulatum* Wald. et Kit. *Polygala murbeckii* Deg., *Teucrium arduini* L., *Valeriana bertiscea* Pančić, *Valeriana pancicii* Halacsy et Bald., etc. A part of numerous paleo-endemic plants in canyons of Montenegro is presented in Fig. 6.

A huge number of species in canyon habitats belong to melliferous plants. Pollinators (a group of species that are extremely important in ecosystem functioning) are in decline globally [117–119]. Conservation of floral resources (melliferous plants), especially in native and undisturbed canyon habitats, may attenuate such unfavourable trend.

9 Conclusions

Main rivers of Montenegro (the Morača, Komarnica, Piva and Tara Rivers) flow through amazing canyons. Due to low climate-induced and human-induced disturbances of habitats, the ravine vegetation in Montenegro represents an important biodiversity hotspot and significant pool of rare, paleo-endemic species.

Comparative analyses of biodiversity in mesic forests revealed that both components of alpha diversity (species richness and equitability) were much greater in canyons than in non-ravine habitats. Beech (*Fagus sylvatica* L.) almost exclusively dominates in non-ravine, relatively flat, habitats with deep soils. Patchy soil patterns in canyons increase competitive abilities of other species and increase alpha diversity by enabling coexistence of numerous trees.

Analyses of beta diversity also confirmed that forests in canyons are more diverse than forests in non-ravine habitats. A high rate of species turnover in canyons is attributable to heterogeneous microclimatic conditions along strong thermal and moisture gradients.

Fig. 6 Endemic plants in canyon habitats of Montenegro. *Amphoricarpos autariatus* Blečić et Mayer (**a**), *Lactuca pancicii* (Vis.) N. Kilian & Greuter (**b**), *Degenia velebitica* (Degen) Hayek (**c**), *Moltkia petraea* (Tratt.) Griseb. (**d**), *Daphne malyana* Blečić (**e**), *Reichardia macrophylla* Visiani et Pančić (**f**), *Scabiosa fumarioides* Visiani et Pančić (**g**), *Edraianthus tenuifolius* (Waldst. et Kit.) DC (**h**) and *Melampyrum hoermenianum* Maly (**i**). Photos made by B. Karadžić S. Jarić

Difference in species richness (nestedness) was much greater in non-ravine habitats than in canyons. This indicates a human-induced degradation of beech forests in non-ravine habitats.

Acknowledgements This study was supported by the Ministry of Education, Science and Technological Development of the Republic of Serbia, Grant no. 173018. We thank the Environmental Protection Agency of Montenegro for their valuable cooperation.

References

1. Hewitt GM (1996) Some genetic consequences of ice ages, and their role in divergence and speciation. Biol J Linn Soc 58:247–276
2. Hewitt G (1999) Postglacial recolonization of European Biota. Biol J Linn Soc 68:87–112
3. Hewitt GM (2000) The genetic legacy of the quaternary ice ages. Nature 405:907–913
4. Hewitt GM (2011) Mediterranean peninsulas: the evolution of hotspots. In: Zachos FE, Habel JC (eds) Biodiversity hotspots. Springer, Berlin, pp 123–147
5. Schmitt T (2007) Molecular biogeography of Europe: pleistocene cycles and postglacial trends. Front Zool 4:11
6. Tamarin RH, Leavitt RW (1991) Principles of genetics. Wm. C. Brown Publishers, Dubuque
7. Whitlock MC, Bürger R (2004) Fixation of new mutations in small populations. In: Ferrière R, Dieckmann U, Couvet D (eds) Evolutionary conservation biology. Cambridge University Press, Cambridge, pp 155–170
8. Gómez A, Lunt DH (2007) Refugia within refugia: patterns of phylogeographic concordance in the Iberian Peninsula. In: Weiss S, Ferrand N (eds) Phylogeography of southern European refugia. Springer, Berlin, pp 155–188
9. Surina B, Schönswetter P, Schneeweiss GM (2011) Quaternary range dynamics of ecologically divergent species (*Edraianthus serpyllifolius* and *E. tenuifolius, Campanulaceae*) within the Balkan refugium. J Biogeogr 38:1381–1393
10. Lakušić R (1971) Specifičnosti flore i vegetacije crnogorskih kanjona. Glasnik Republičkog zavoda za zaštitu prirode i Prirodnjačkog muzeja u Titogradu 4:157–169
11. Lakušić R (1974) Prirodni sistem populacija vrsta roda *Edraianthus* DC. Godišnjak Biološkog Instituta u Sarajevu 26:1–129
12. Mišić V (1981) Forest vegetation in gorges and canyons of the Eastern Serbia. IBISS, Beograd. (in Serbian)
13. Pulević V, Lakušić R (1983) Floristićke zabilješke iz kanjona rijeke Cijevne. Glasnik Republičkog zavoda zaštitu prirode (Podgorica) 16:15–26
14. Lakušić R, Redžić S (1989) Flora i vegetacija vaskularnih biljka u refugijalno-reliktnim ekosistemima kanjona rijeke Drine i njenih pritoka. Glasnik odeljenja prirodnih nauka, Crnogorska akademija nauka i umjetnosti 7:107–205
15. Karadžić B, Mijović A, Popović R, Perišić S, Marinković S (2001) Forest vegetation in West Serbian canyons: biodiversity hotspots. In: Ragodlou K (ed) Forest research: a challenge for an integrated European approach. NAGREF-Forest Research Institute, Thessaloniki, pp 513–518
16. Košir P, Čarni A, di Pietro R (2008) Classification and phytogeographical differentiation of broad-leaved ravine forests in southeastern Europe. J Veg Sci 19:331–342
17. Karadžić B (2017) Chasmophytic forests of *Ostrya carpinifolia* in west-Serbian canyons. Biol Nyssana 8:73–81
18. Karadžić B (2018) Beech forests (order Fagetalia sylvaticae Pawlowski 1928) in Serbia. Bot Serbica 42:91–107
19. Tuhkanen S (1980) Climatic parameters and indices in plant geography. Acta Phytogeographica Suec 67:1–105
20. Peel MC, Finlayson BL, McMahon TA (2007) Updated world map of the Köppen-Geiger climate classification. Hydrol Earth Syst Sci 11:1633–1644
21. Pavlović P, Kostić N, Karadžić B, Mitrović M (2017) The Soil of Serbia. Springer, Heidelberg, p 225

22. Pianka ER (1988) Evolutionary ecology. Harper and Row, New York
23. Köppen W (1936) Das geographisches System der Klimate. In: Köppen W, Geiger R (eds) Handbuch der Klimatologie I(C). Gebruider Borntraeger, Berlin
24. Kottek M, Grieser J, Beck C, Rudolf B, Rubel F (2006) World map of the Köppen-Geiger climate classification updated. Meteorol Z 15:259–263
25. Burić M, Micev B, Mitrović L (2012) Atlas klime Crne Gore. Leksikografski centar, knj. 2, CANU, Podgorica
26. Burić D, Ducić V, Mihajlović J (2013) The climate of Montenegro: Modificators and types-Part one. Bull Serb Geogr Soc 93:83–102
27. Burić D, Ducić V, Mihajlović J (2014) The climate of Montenegro: Modificators and types-Part two. Bull Serb Geogr Soc 94:73–90
28. Walter H (1955) Die Klimagramme als Mittel zur Beurteilung der Klimaverhältnisse für ökologische, vegetationskundliche und landwirtschaftliche Zwecke. Ber Dtsch Bot Ges 68:331–344
29. Walter H (1962) Die Vegetation der Erde in oekophysiologischer Betrachtung. Gustav Fischer Verlag, Jena
30. Smith RB (1979) The influence of mountains on the atmosphere. Adv Geophys 21:87–229
31. Hill FF (1983) The use of average annual rainfall to derive estimates of orographic enhancement of frontal rain over England and Wales for different wind directions. J Climatol 3:113–129
32. Ducić V, Luković J, Burić D, Stanojević G, Mustafić S (2012) Precipitation extremes in the wettest Mediterranean region (Krivošije) and associated atmospheric circulation types. Nat Hazard Earth Syst Sci 12:687–697
33. Radulović V, Radulović M (1997) Karst of Montenegro. In: Stevanović Z (ed) 100 years of hydrogeology in Yugoslavia, Belgrade, pp 147–185
34. Dimitrijević E (2002) Geological atlas of Serbia with its surrounding. Ministry of Environment and Spatial Planning, Republic of Serbia
35. Schmid SM, Bernoulli D, Fügenschuh B, Matenco L, Schuster R, Schefer S, Tischler M, Ustaszewski K (2008) The Alpine-Carpathian-Dinaridic orogenic system: correlation and evolution of tectonic units. Swiss J Geosci 101:139–183
36. Ustaszewski K, Schmid SM, Fügenschuh B, Tischler M, Kissling E, Spakman W (2008) A map-view restoration of the Alpine-Carpathian-Dinaridic system for the Early Miocene. Swiss J Geosci 101:273–294
37. Djurović M, Djurović P (2015) The impact of glacial processes in the evolution of the poljes of Montenegro. Acta Geobalcanica 1:25–31
38. Blečić V (1960) Beitrag zur Kenntniss der Weidenvegetation des gebirges Bjelasica. Bot Serbica 2:109–118
39. Škorić A, Filipovski G, Ćirić M (1985) Classification of Yugoslav soils, vol 13. Academy of Sciences and Arts of Bosnia and Herzegovina, Interacademic Board for Soil Science, Sarajevo, pp 71–72
40. Fuštić B, Djuretić G (2000) Soils of Montenegro. Biotechnical Institute, University of Montenegro, Montenegro, pp 1–626
41. Blečić V (1956) Crna Gora – Biljni pokrov. In: Krleža M (ed) Enciklopedija Jugoslavije, vol 2. Leksikografski zavod FNRJ, Zagreb, pp 407–410
42. Blečić V (1957) Prilog poznavanju šumske vegetacije planine Ljubišnje. Bull Nat Hist Mus Belgrade 10:25–42
43. Blečić V, Tatić B (1957) Šuma molike u Crnoj Gori (*Pinetum peucis montenegrinum*). Bull Nat Hist Mus Belgrade 10:43–53
44. Blečić V (1958) Šumska vegetacija i vegetacija stena i točila reke Pive. Bull Nat Hist Mus Belgrade 11:5–101
45. Blečić V (1960) Der Weisserlenwald und der Sauerklee (*Oxali-Alnetum incanae*) im Quellgebiet der Flüsse Tara und Lim. Bot Serbica 2:101–108

46. Blečić V (1961) Beitrag zur kenntnis der Fichtenwälder aus montenegrinischen Prokletija. Bot Serbica 1:227–236
47. Blečić V, Lakušić R (1976) Prodromus biljnih zajednica Crne Gore. Glasnik Republićkog zavoda za zaštitu prirode i Prirodnjačkog muzeja u Titogradu 9:57–98
48. Horvat I (1963) Šumske zajednice Jugoslavije. In: Potočić Z (ed) Šumarska enciklopedija, vol 2. Jugoslavenski leksikografski zavod, Zagreb, pp 560–590
49. Horvat I, Glavač V, Ellenberg H (1974) Vegetation Südosteuropas. Gustav Fischer, Jena/Stuttgart
50. Jovanović B, Lakušić R, Rizovski R, Trinajstić I, Zupančić M (1986) Prodromus Phytocenosum Jugoslaviae ad mappam vegetationis Bribir-Ilok. Naučno veće vegetacijske karte Jugoslavije
51. Lakušić R (1987) Šumske zajednice Jugoslavije: SR Crna Gora. In: Potočić Z (ed) Šumarska enciklopedija, vol 3. Jugoslavenski leksikografski zavod, Zagreb, pp 388–395
52. Rivas-Martínez S, Penas A, Díaz TE (2004) Biogeographic map of Europe. Scale 1:16.000.000. Cartographic Service, University of León, León, ES
53. Török K, Podani J, Borhidi A (1989) Numerical revision of the *Fagion illyricum* alliance. Vegetatio 81:169–180
54. Čarni A, Mucina L (2015) Validations and typifications of some South European syntaxa. Hacquetia 14:289–299
55. Lakušić D, Karadžić B (2010) New associations of serpentine chasmophitic vegetation (*Asplenietea trichomanis* Br.-Bl. 1934 corr. Oberd. 1977) on Kopaonik Mt in Serbia. Bot Serbica 34:67–79
56. Mijović A, Popović Z, Karadžić B, Mijatović M, Perišić S (2006) Distribution of xerohalophytic vegetation along the sea ward and landward zone in south – Adriatic sandy beach (Montenegro). Biotechnol Biotechnol Equip 20:30–35
57. Mijović A, Popović Z, Vukov T, Smiljanić M, Matić R, Bojović S (2012) Sand topography influences the distribution of xerohalophytic vegetation on a southern Adriatic beach in Montenegro. Plant Biosyst 146:664–673
58. Caković D, Milošević D (2013) Studija biodiverziteta i zaštite prirode obalnog područja Crne Gore. Program integralnog upravljanja obalnog područja Crne Gore (CAMP CG). Ministarstvo održivog razvoja i turizma Crne Gore, 348 p
59. Bubanja N, Šinžar-Sekulić J, Stevanović V (2016) Assessing the influence of environmental parameters on aquatic plants of ponds in the hinterland of Long Beach in Montenegro. Limnetica 35:385–396
60. Šilc U, Mullaj A, Alegro A, Ibraliu A, Dajić Stevanović Z, Luković M, Stešević D (2016) Sand dune vegetation along the eastern Adriatic coast. Phytocoenologia 46:339–355
61. Stešević D, Bubanja N, Caković D, Jogan N, Luković M, Šilc U (2017) Synecology of *Cutandia maritima* (L.) Barbey, a rare psammophytic species along the Montenegrin coast (East Adriatic coast). Hacquetia 16:181–187
62. Lakušić R, Pavlović D (1976) Vegetacija Skadarskog jezera. Glasnik Republićkog zavoda za zaštitu prirode i Prirodnjačkog muzeja u Titogradu 9:45–50
63. Martinčič A (2006) Moss flora of the Prokletije mountains (Serbia, Montenegro). Hacquetia 5:113–130
64. Hadžiablahović S (2018) The diversity of the flora and vegetation of Lake Skadar/Shkodra. In: Pešić V, Karaman G, Kostianoy A (eds) The Skadar/Shkodra Lake environment. The handbook of environmental chemistry, vol 80. Springer, Cham, pp 203–238
65. Caković D, Stešević D, Iković V, Knežević M, Latinović N (2012) Contribution to the knowledge of weed flora in Bjelopavlići plain. Agr Forest 58(4):25–41
66. Petrović D, Stešević D (2010) Materials for the red book of vascular flora of Montenegro (second contribution). Biol Nyssana 1:27–34
67. Petrović D, Stešević D, Vuksanović S (2008) Material for the Red Book of the flora of Montenegro. Proceedings of III international symposium of ecologist of the Republic of Montenegro, Natura Montenegrina, vol 7, pp 605–631

68. Baede APM, Ahlonsou E, Ding Y, Schimel D (2001) The climate system: an overview. In: Houghton JT, Ding Y, Griggs DJ, Noguer M, van der Linden PJ, Dai X, Maskell K, Johnson CA (eds) Climate change 2001: the scientific basis. Contribution of Working Group I to the Third Assessment Report of the Intergovernmental Panel on Climate Change. Cambridge University Press, Cambridge
69. Kiehl JT, Trenberth KE (1997) Earth's annual global mean energy budget. Bull Am Met Soc 78:197–208
70. Stevanović V, Bulić Z (1992) New data on chorology and phytocenology of species *Ramonda serbica* Panč. (*Gesneriaceae*) in Montenegro. Glasn Republ zav za zaš prir-Prir muz Podgorica 25:7–16
71. Bulić Z (1998) Flora kanjona Cijevne u Crnoj Gori. Glasn Republ zav za zaš prir-Prir muz Podgorica 26:5–31
72. Bulić Z, Lakušić D, Stevanović V (2008) Comparative analysis of the vascular floras of the Morača and Cijevna Canyons (Montenegro). Arch Biol Sci 60:485–492
73. Lakušić R, Pavlović D, Redžić S (1982) Horološko-ekološka i florističa diferencijacija šuma i šikara sa bjelograbićem (*Carpinus orientalis* Mill.) i crnim grabom (*Ostrya carpinifolia* Scop.) na prostoru Jugoslavije. Glasnik Republičkog zavoda za zastitu prirode i Prirodnjačkog muzeja u Titogradu 15:103–116
74. Mucina L, Bültmann H, Dierssen K, Theurillat JP, Raus T, Čarni A, Šumberová K, Willner W, Dengler J, García RG, Chytrý M, Hájek M, Di Pietro R, Lakushenko D, Pallas J, Daniëls FJA, Bergmeier E, Guerra AS, Ermakov N, Valachović M, Schaminée JHJ, Lysenko T, Didukh YP, Pignatti S, Rodwell JS, Capelo J, Weber HE, Solomeshch A, Dimopoulos P, Aguiar C, Hennekens SM, Tichý L (2016) Vegetation of Europe: hierarchical floristic classification system of vascular plant, bryophyte, lichen, and algal communities. Appl Veg Sci 19:3–264
75. Čarni A, Košir P, Karadžić B, Matevski V, Redžić S, Škvorc Ž (2009) Thermophilous deciduous forests in Southeastern Europe. Plant Biosyst 143:1–13
76. Karadžić B, Jarić S, Pavlović P, Mitrović M (2015) Aquatic and wetland vegetation along the Sava River. In: Sčančar J, Paunović M, Milačić R (eds) The Sava River. Springer, Heidelberg, pp 249–316
77. Rodwell JS, Schaminée JHJ, Mucina L, Pignatti S, Dring J, Moss D (1998) The scientific basis of the EUNIS habitat classification. Report to the European Topic Centre on Nature Conservation, Unit of Vegetation Science, Lancaster
78. Schulze ED, Mooney HA (1993) Biodiversity and ecosystem function. Springer, Berlin
79. DeLong Jr DC (1996) Defining biodiversity. Wildl Soc Bull 24:738–749
80. Gaston KJ, Spicer JI (2004) Biodiversity: an introduction. Blackwell Publishing, Oxford
81. Whittaker RH (1972) Communities and ecosystems. Macmillan, New York
82. Begon M, Harper JL, Townsend CR (1996) Ecology: individuals, populations and communities. Blackwell Science, Oxford
83. Karadžić B, Marinković S (2009) Kvantitativna ekologija. IBISS, Beograd, 489 p
84. Margalef DR (1958) Information theory in ecology. Int J Gen Syst 3:36–71
85. Lloyd M, Ghelardi RJ (1964) A table for calculating the "equitability" component of species diver-sity. J Anim Ecol 33:217–225
86. Pielou EC (1966) Species-diversity and pattern-diversity in the study of ecological succession. J Theor Biol 10:370–383
87. Baselga A (2010) Partitioning the turnover and nestedness components of beta diversity. Global Ecol Biogeography 19:134–143
88. Podani J, Ricotta C, Schmera D (2013) A general framework for analyzing beta diversity, nestedness and related community-level phenomena based on abundance data. Ecol Complex 15:52–61
89. Marinković N, Karadžić B, Pešić V, Gligorović B, Grosser C, Paunović M, Nikolić V, Raković M (2019) Faunistic patterns and diversity components of leech assemblages in karst springs of Montenegro. Knowl Manage Aquat Ecosyst 420:26

90. Bergmeier E, Dimopoulos P (2001) *Fagus sylvatica* forest vegetation in Greece: syntaxonomy and gradient analysis. J Veg Sci 12:109–126
91. Braun-Blanquet J (1965) Phlanzensoziologie. Grundzuge der Vegetatinskunde, 3rd edn. Springer, Wien, NY
92. van der Maarel E (1979) Transformation of cover-abundance values in phytosociology and its effects on community similarity. Plant Ecol (Vegetatio) 39:97–114
93. Karadžić B (2013) FLORA: a software package for statistical analysis of ecological data. Water Res Manage 3:45–54
94. Walter H, Breckle S-W (1985) Ecological systems of the geobiosphere 1. Ecological principles in global perspective. Springer, Berlin
95. Popović Z, Mijović A, Karadžić B (2005) Dry matter distribution patterns and photosynthetic traits of two vernal ephemeroids in a deciduous forest. Period Biol 107:287–292
96. Popović Z, Mijović A, Karadžić B, Mijatović M (2006) Response of growth dynamics of two spring geophytes to light regime in a lime-beech forest. J Integr Plant Biol 48:527–535
97. Popović Z, Bojović S, Matić R, Stevanović B, Karadžić B (2016) Comparative ecophysiology of seven spring geophytes from an oak-hornbeam forest. Braz J Bot 39:29–40
98. Karadžić B, Marinković S, Kataranovski D (2003) Use of the β-function to estimate the skewness of species responses. J Veg Sci 14:799–805
99. Stevanović V (1996) Analysis of the Central European and Mediterranean orophytic element on the mountains of the West and Central Balkan Peninsula, with special reference to endemics. Bocconea 5:77–97
100. Stevanović V (1999) The Red data book of Flora of Serbia, 1. Extinct and critically endangered taxa. Ministry of Environmental Protection, Biological faculty of the Belgrade University, Institute for Nature Protection, Belgrade
101. Stevanović V, Tan K, Petrova A (2007) Mapping the endemic flora of the Balkans-a progress report. Bocconea 21:131–137
102. Velčev VI, Kožuharov SI, Ančev ME (1992) Atlas of endemic plants in Bulgaria. Bulgarian Academy of Sciences, Sofia
103. Milivojević M, Menković L, Čalić J (2008) Pleistocene glacial relief of the central part of Mt. Prokletije (Albanian Alps). Quat Int 190:112–122
104. Djurović P (2009) Reconstruction of the pleistocene glaciers of Mount Durmitor in Montenegro. Acta Geogr Slov Ljubljana 49:263–279
105. Stepišnik U, Žebre M (2011) Glaciokras Lovčena. E-GeograFF 2, Univerza v Ljubljani, Filozofska fakulteta, Ljubljana
106. Hughes PD, Woodward JC, van Calsteren PC, Thomas LE (2011) The glacial history of the Dinaric Alps, Montenegro. Quat Sci Rev 30:3393–3412
107. Djurović P (2012) Comment on: "The glacial history of the Dinaric Alps, Montenegro" by P.D. Hughes, J.C.Woodward, P.C. van Calsteren, L.E. Thomas [Quaternary Science Reviews 30 (2011), 3393–3412]. Quaternary Sci Rev 49:109–110
108. Hughes PD, Woodward JC, van Calsteren PC, Thomas LE (2012) Reply: the glacial history of the Dinaric Alps, Montenegro. Quat Sci Rev 49:111–112
109. Médail F, Diadema K (2009) Glacial refugia influence plant diversity patterns in the Mediterranean Basin. J Biogeography 36:1333–1345
110. Rakaj M (2009) Floristic and chorological news from north Albania. Bot Serbica 33:177–183
111. Georghious K, Delipetrou P (2010) Patterns and traits of the endemic plants of Greece. Bot J Linn Soc 162:130–422
112. International Commission on Stratigraphy (2013) International chronostratigraphic chart. www.stratigraphy.org
113. Davis HP (1951) Cliff vegetation in the Eastern Mediterranean. J Ecol 39:63–93
114. Blečić V, Mayer E (1967) Die europäischen Sippen der Gattung *Amphoricarpos* Visiani. Phyton 12:150–158

115. Surina B, Rakić T, Stefanović S, Stevanović V, Lakušić D (2009) One new species of the genus *Edraianthus*, and a change in taxonomic status for *Edraianthus serpyllifolius f. pilosulus (Campanulaceae)* from the Balkan Peninsula. Syst Bot 34:602–608
116. Caković D, Stešević D, Schönswetter P, Frajman B (2015) How many taxa? Spatiotemporal evolution and taxonomy of *Amphoricarpus* (*Asteraceae, Carduoideae*) on the Balkan Peninsula. Org Divers Evol. https://doi.org/10.1007/s13127-015-0218-6
117. Jarić S, Mačukanović-Jocić M, Mitrović M, Pavlović P (2013) The melliferous potential of forest and meadow communities on Mount Tara (Serbia). Environ Entomol 42:724–732
118. Jarić S, Mitrović M, Pavlović P (2014) An ethnobotanical and ethnomedicinal study on the use of wild medicinal plants in rural areas of Serbia. In: Pieroni A, Quave C (eds) Ethnobotany and biocultural diversities in the Balkans. Springer, New York, pp 87–112
119. Winfree R, Bartomeus I, Cariveau DP (2011) Native pollinators in anthropogenic habitats. Ann Rev Ecol Evol Syst 42:1–22

The Intermittent Rivers of South Montenegro: Ecology and Biomonitoring

Vladimir Pešić, Ana Pavićević, Ana Savić, and Sead Hadžiablahović

Contents

Abstract Intermittent rivers (IRs), rivers that periodically cease to flow, are common in the southern part of Montenegro and form a high proportion of the river networks in the Mediterranean and sub-Mediterranean parts of the country. IRs comprise of lentic, lotic, and dry patches that support rich aquatic, semiaquatic, and terrestrial communities that are largely neglected by ongoing biomonitoring programs. In this chapter, we present the recent research on the communities that inhabit the intermittent rivers of South Montenegro, with a focus on aquatic and

V. Pešić (✉)
Department of Biology, University of Montenegro, Podgorica, Montenegro
e-mail: vladopesic@gmail.com

A. Pavićević
Ministry of Sustainable Development and Tourism, Podgorica, Montenegro
e-mail: anpavicevic@gmail.com

A. Savić
Department of Biology and Ecology, Faculty of Sciences and Mathematics, University of Niš, Niš, Serbia
e-mail: anka@pmf.ni.ac.rs

S. Hadžiablahović
Nature and Environment Protection Agency of Montenegro, Podgorica, Montenegro

Faculty of Philosophy, University of Montenegro, Podgorica, Montenegro
e-mail: seadh@t-com.me

Vladimir Pešić, Momir Paunović, and Andrey G. Kostianoy (eds.), *The Rivers of Montenegro*, Hdb Env Chem (2020) 93: 231–252, DOI 10.1007/698_2019_415,

terrestrial invertebrates and plants. For each group, we outline the characteristic communities found in the IRs. Future research should focus on including IRs in ongoing monitoring programs ensuring that their status assessment encompasses both their flowing and dry phases and their aquatic and terrestrial biotas.

Keywords Aquatic and riparian plants, Drought, Intermittent streams, Macroinvertebrates, Meiofauna, Montenegro, Terrestrial invertebrates

1 Introduction

Intermittent rivers (IRs) are defined as the rivers that periodically cease to flow. In this chapter, the term intermittent rivers has been used following Datry et al. [1] and refers to all temporary, ephemeral, seasonal, and episodic streams and rivers in defined channels. As has been emphasized by many researchers, the extent of intermittent rivers in the global river network is increasing, presumably due to climate change [1–4]. Intermittent rivers represent over half of the global river network [5, 6]. In arid and semiarid regions, including the Mediterranean region, the proportion of IRs can exceed 70% of the whole hydrographic network [1].

Regardless of the fact that intermittent rivers make up a high proportion of river networks at both the regional and global level, their better recognition, especially for the purpose of ongoing biomonitoring programs, has been hampered extensively, notably by the following: (1) the characterizing of their ecological status, managed to meet the requirements of the European Union Water Framework Directive 2000/60/EC (WFD) [7], is based on methods developed for perennial waters, and the accuracy with which such methods characterize intermittent rivers is largely unknown [8], and 2) despite increasing research on IRs in the last decade, there is still a lack of knowledge about the species and communities inhabiting intermittent rivers in many parts of the world [8].

The first step in studying intermittent rivers is mapping their occurrence, followed by their adequate typology classification and recognizing the main types of hydrological alterations caused by anthropogenic factor, so as to discriminate between natural and artificial IRs [8]. It is still unclear exactly how much of the entire hydrographic network in Montenegro is made up of intermittent rivers (Fig. 1). Moreover, detailed information on IRs is hampered by the lack of quantitative hydrological metrics, as none of the hydrological monitoring points in Montenegro are located on intermittent reaches of the river network. A more promising qualitative approach to monitoring, including remote sensing by the use of aerial or satellite photography, has been shown in this volume [9].

The number of studies [1–3] has shown that both the flowing and non-flowing phases of IRs are biogeochemically and ecologically active and provide essential ecosystem services, for example, in contributing to nutrient and carbon cycling

Fig. 1 Photographs of some intermittent rivers in South Montenegro. (**a**) The lower part of Cijevna/Cemi River, (**b**) the Sušica River (between Podgorica and Danilovgrad), (**c**) the lower part of the Sitnica River in Podgorica, (**d**) the Miđanska (Vladimirska) River near Vladimir town in Ulcinj, (**e**) the lower part of the Ribnica River near its confluence with the River Morača. Photos by V. Pešić (**a**, **d**, **e**), S. Hadžiablahović (**b**, **c**)

[6]. During a non-flowing phase, the dry riverbed, often for extended period, accumulates riparian leaf litter (LL) and sediments [1, 6]. When the flow resumes after the drying events, the organic matter (and nutrients) are transported downstream. As showed by Datry et al. [6], these rewetting events are associated with intense biological activity. The oxygen consumption rate of rewetted LL taken from the Rimanić stream (discussed below; Fig. 2) was 0.389 mg O_2 g^{-1} dry mass h^{-1} [6]. This value is almost half of that reported for the stream of Ribnica (0.6398 mg O_2 g^{-1} dry mass h^{-1}). The estimate of CO_2 release from the Rimanić stream upon LL rewetting was 8.13 g CO_2 m^{-2} day^{-1} [6] and was lower than the rate reported for the Ribnica stream (16.5674 gCO_2 m^{-2} day^{-1}). The latter stream is located in the center of the city of Podgorica and has a higher quantity of accumulated leaf mass, 18.8333 g m^{-2} than the Rimanić stream which had 15.2000 g m^{-2} [6].

Intermittent rivers are common in the Mediterranean and sub-Mediterranean parts of Montenegro. Most of the small rivers in the southern part of Montenegro dry up in the summer. One such stream that has been studied and emphasized in this chapter is the Rimanić stream, which is located in the central part of Montenegro and which flows into the River Zeta near Spuž (Fig. 2). This region is one of the driest in Montenegro. The Rimanić stream has two sections, the upper part being an intermittent reach where the surface flow is lost for a part of the year (its non-flowing phase – from the end of May to the end of October in the studied years). After accumulating groundwater, the stream becomes perennial again ca. 1.5 km upstream of its confluence with the River Zeta [10]. It is known that in extremely dry years, the bed of the River Morača, the biggest tributary of Lake Skadar and the longest river in South Montenegro, also dries up in some places (presumably as results of sinking), notably from Duga Monastery to the confluence with the River Zeta and, further downstream, between Botun and the discharge into Lake Skadar/Shkoder.

Many of these rivers provide habitats for diverse and unique plant and animal communities. However, most of these habitats, both in the Balkans and in Montenegro, have been neglected in biodiversity and ecology studies, meaning that the quantification of biodiversity and ecosystem processes of IRs has been restricted to a few studies and a limited number of streams [10–13].

In this chapter, we present recent research on the species and communities that inhabit intermittent rivers in the southern part of Montenegro, with a focus on aquatic and terrestrial invertebrates and plants.

2 The Macroinvertebrate and Meiofauna Communities of Intermittent Streams

The invertebrate communities of intermittent and perennial rivers differ in terms of both taxa richness and abundance [14, 15]. During their flowing phase, intermittent sites usually host smaller taxa numbers than the equivalent perennial sites

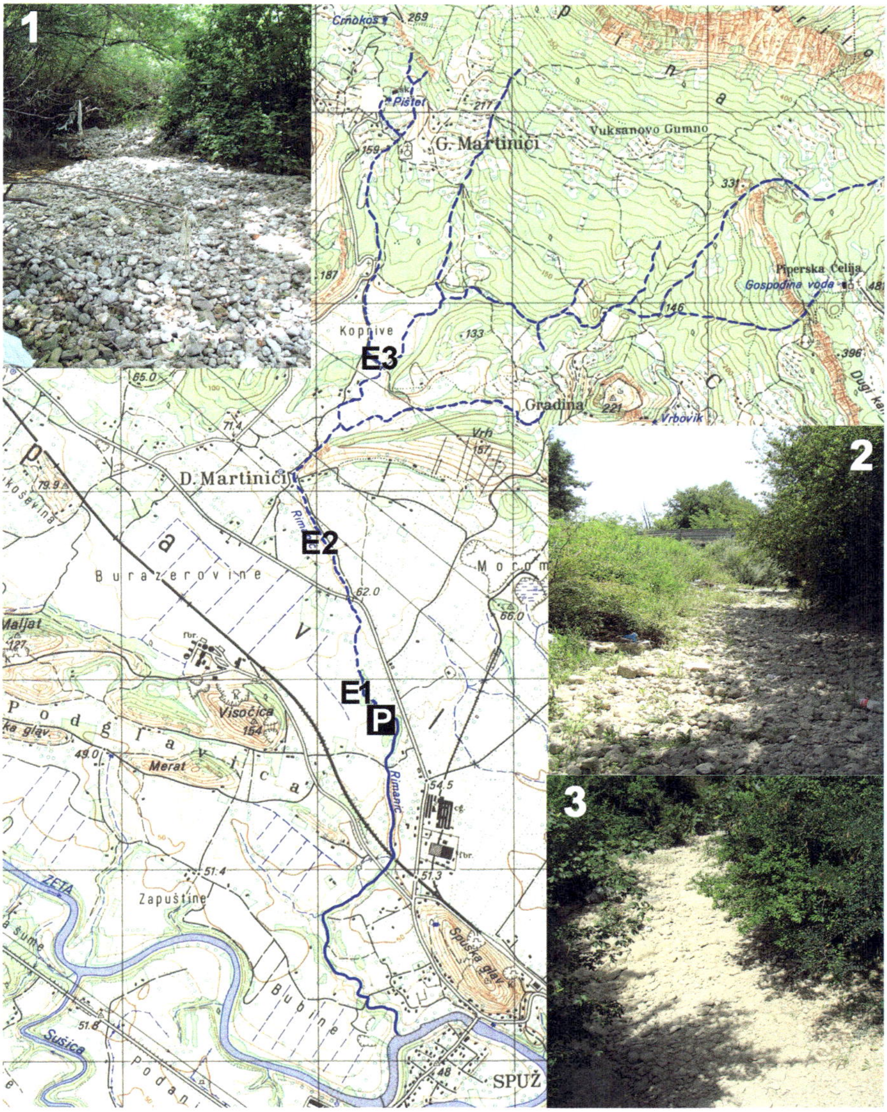

Fig. 2 Topographic map (at a scale of 1:25,000) showing the locations and photographs of the investigated sites, E1, E2, and E3, along the intermittent reach of the Rimanić stream. The interrupted lines represent the intermittent reach of the Rimanić stream. *P* – show the location where the stream becomes perennial. Modified after [10]

[1, 8]. However, this does not necessarily lead to a decrease in community metrics at the level of the entire fauna of the intermittent rivers. Sometimes, intermittent sites can harbor greater biodiversity than perennial sites when the annual cycle is considered [3, 16]. This is because different taxa colonize the same locations over a

temporal gradient, as and when appropriate habitats become available [8, 17]. Lotic taxa colonize stream riverbed when the flow resumes, lentic taxa dominate in the pool phase when flow interruption occurs, while terrestrial taxa colonize dry sediments. Therefore, when the lotic, lentic, and terrestrial invertebrates are considered together, the diversity of intermittent rivers can in fact exceed the richness of perennial ones [8].

Intermittent rivers dry up periodically, and the communities living within them become disturbed by the alternating flowing and dry phases. In such streams drying and/or flow cessation can be considered "predictable" events, and thus the species and communities that inhabit these streams are likely to be adapted to such conditions. In terms of ongoing but also future climate change, it may mean that intermittency can enhance the adaptability of IR assemblages and also encourage the appearance of specialists [8, 18].

There a number of studies that have described the process of community recovery from drying in intermittent streams [19–22]. All these studies have shown that when the flow resumes, aquatic assemblage richness increases rapidly as a result of recolonization by lotic taxa. The process of recolonization includes horizontal migration from nearby permanent water bodies (most commonly for species which are weak fliers like mayflies and stoneflies) but also a vertical migration from refuges within the saturated hyporheic sediment [8, 23]. Recolonization mechanisms are usually species-specific [19] and can include active (such as in flying adult insects) or passive (such as water mites) pathways of migration. In many insect orders, like stoneflies, mayflies, caddisflies, and midges, colonization occurs via laying eggs [24].

The time required for the full recovery and stabilization of the communities can vary significantly, ranging from some months to several years [14, 21, 22, 25, 26]. The speed and success of the recolonization depend on a number of factors including the stream size, the distance from perennial recolonization source, and the degree of impact and duration of the drought period [20, 21, 23]. Several studies have shown that smaller streams like those that occur in the Mediterranean part of Montenegro require a shorter recovery times from drought than larger ones do [25, 27, 28].

Pavićević [11] studied the mechanism of recovery after a drought of the invertebrate communities in the intermittent reach of the Rimanić stream. The study was conducted in 2009 and 2010 by using a quantitative approach by Surber net at three sites: E1, located at the border with the perennial reach of the stream, and two further sites, E2 and E3 located upstream along the intermittent reach (see Fig. 2 for the exact positions of the studied sites).

In total 14,527 specimens were collected from the studied sites. Plecoptera and Simuliidae dominated in the macroinvertebrate assemblages, while ostracods and copepods were predominant in the meiofauna samples [11]. The mean site abundance of the macroinvertebrate assemblage at the near-perennial site E1, being 1,323.3 ind. m^{-2}, was higher than those at either E2 or E3, which recorded 589.4 and 1,073.3 ind. m^{-2}, respectively. The abundance of the macroinvertebrate assemblage was lowest in November at site E2, totaling only 75.6 ind. m^{-2}, and highest in

April at E3, where a total of 2,912 ind. m^{-2} was recorded. A one-way ANOVA analysis did not reveal significant differences between the studied sites ($p = 0.375$).

The mean site abundance of the meiofauna assemblage at the near-perennial E1, which was 1,095.4 ind. m^{-2}, was higher in comparison with sites E2 and E3 located upstream, which showed averages of 509.1 and 636.8 ind. m^{-2}, respectively. The abundance of meiofauna assemblages was highest in December at site E1, where a total of 3,409 ind. m^{-2} was recorded, while it was lowest in November, when no specimens were found at either E1 or E2. A one-way ANOVA revealed no significant differences between the studied sites ($p = 0.415$).

One of the central questions in understanding the process of the recolonization of invertebrate assemblages in intermittent streams is whether there is a spatial biodiversity pattern controlled by flow intermittency in space and time. The spatial arrangement of the intermittent reach(es) in relation to the perennial reach(es) can influence IRs community composition [8]. In the study on the recovery of invertebrate communities in the intermittent reach of the Rimanić stream, we predicted that as distance from a downstream perennial recolonization source increased, differences between the studied assemblages along the intermittent gradient of Rimanić stream would become more pronounced [11]. However, ANOSIM results revealed no significant differences between the macroinvertebrate communities ($R = 0.114$ $p = 0.06$) nor between the meiofauna communities ($R = -0.054$ $p = 0.81$) found at the studied sites, but did reveal differences in terms of the sampling date, both for the macroinvertebrate ($R = 0.241$ $p = 0.02$) and the meiofauna ($R = 0.544$ $p = 0.001$) communities, respectively.

The study revealed that the invertebrate communities of the intermittent reach of the Rimanić stream recovered slowly when the flow resumed after a drying event [11]. The lowest abundance of the macroinvertebrate assemblages at all three sites were detected in the first month after the flow resumes (mean, range; 162, 75.6–316.8 ind. m^{-2}). At the site of E1, located at the border with the perennial reach, the highest abundance of the macroinvertebrate assemblage appeared in December, totaling 2,008.8 ind. m^{-2}, and in January, with a value of 3,013.2 ind. m^{-2}, 2 months after the flow resumed. At the two other sites, located upstream along the intermittent reach, the abundance of macroinvertebrate assemblages increased with the duration of the flowing phase, reaching their maximum values, 813.6–1,324 ind. m^{-2} at E2 and 2,783–2,912 ind. m^{-2} at E3, in April and May, before the surface flow ceased (Fig. 3b).

The abundance of the meiofauna assemblage at the near-perennial site of E1 was highest in December, being 3,409.2 ind. m^{-2}, 2 months after the flow resumed, indicating the time needed to recolonize a dry riverbed. The two sites of E2 and E3, located further upstream, exhibit their highest abundance in January, totaling 1,209.6 and 1,396.8 ind. m^{-2}, respectively, thus showing a lag of 1 month in comparison to E1. However, in contrast to the macroinvertebrate communities, it is evident that the abundance of meiofauna assemblages decreased as the flowing phase continued, reaching their lowest densities, in April and May, just before the surface flow ceased (Fig. 3a).

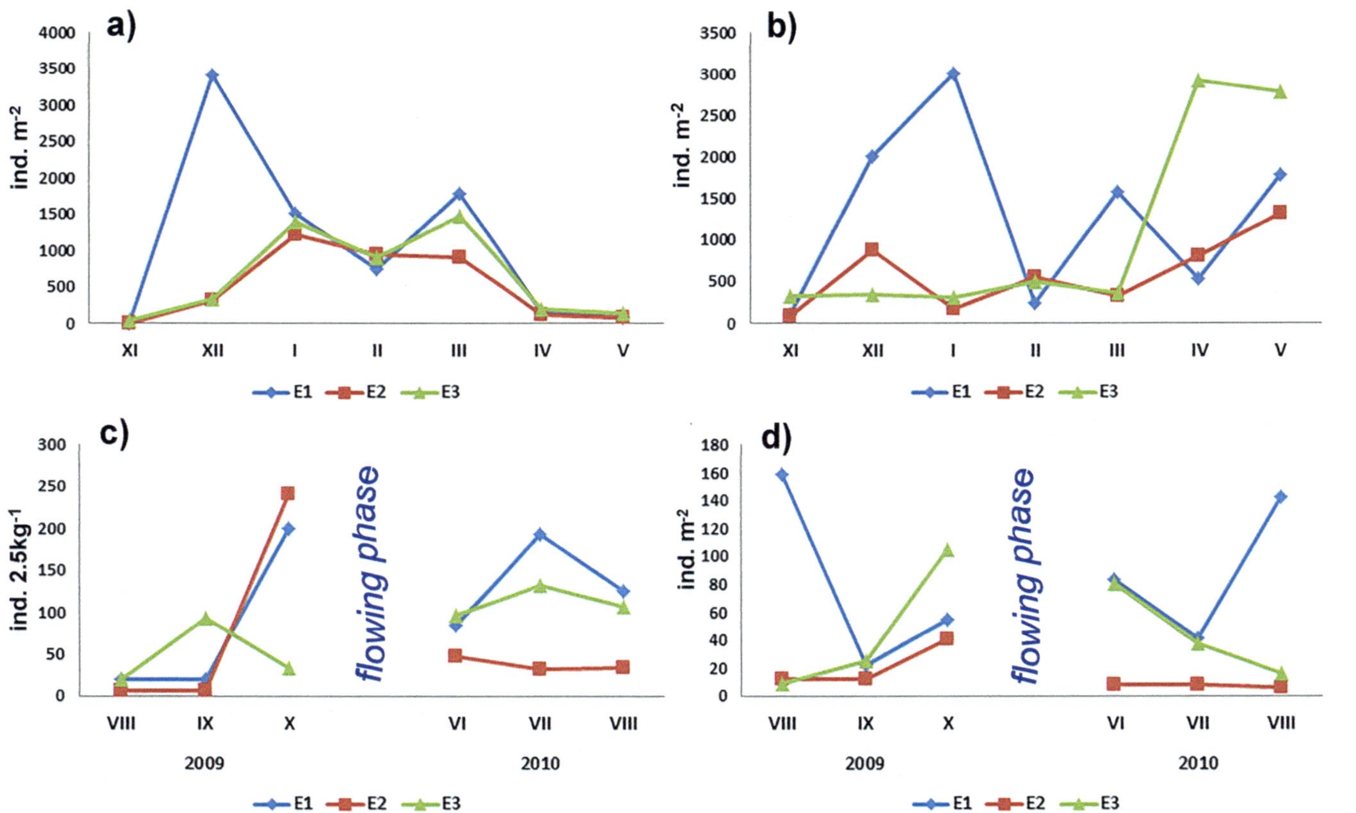

Fig. 3 Monthly fluctuations in the abundance of assemblages at the investigated sites, E1, E2, and E3, along the intermittent reach of the Rimanić stream. (**a**) Meiofauna, (**b**) macroinvertebrates, (**c**) "seedbank," (**d**) semiaquatic and terrestrial invertebrates

The lotic taxa that have the ability to drift or migrate from the perennial part of the stream dominated in the macroinvertebrate communities of intermittent reach of the Rimanić stream [11]. Some taxa that are known to migrate vertically such as Oligochaeta, chironomid midges, and water mites can also migrate upstream [19]. The dominance of a small number of taxa (that exhibit a high degree of resilience) in the initial stages of the recovery is a common feature of the macroinvertebrate communities in the early stages of recolonization. Early colonizers that were found after 15 days in the Rimanić stream include limnephilid caddisflies, a common shredder *Asellus aquaticus*, and hydrobiid snails. Moreover, the initial communities often include inundation-tolerant terrestrial taxa that remain within the channel when the flow resumes such as springtails (Collembola) and oribatid mites. The latter group dominated at E3, the site furthest upstream in the macroinvertebrate communities present after 15 days and 75 days, comprising from 40% in November to almost 53% of the total abundance in January.

Simuliidae and Plecoptera dominated at site E1 in the macroinvertebrate communities present after 45 and 75 days, respectively, comprising 78% of the total abundance in December and 86% of the total abundance in January. The high abundance of these two groups is favored by the seasonal disturbance that affects their life history strategies [26, 29]. The presence of stoneflies nymphs in rehydrated samples from the dry riverbed of the Rimanić stream indicates that some species are semivoltine and probably underwent nymphal quiescence when the stream was dry. As the flowing phase duration increases, the abundance of stoneflies and simulid midges decreased, disappearing completely from the assemblage (as was the case with the simulid midges) or reaching a minimum abundance (as was the case with stoneflies) immediately before the surface flow completely ceased.

Figure 3b shows the abundance of invertebrate communities at the upriver sites of E2 and E3 starting to increase after 145 days, toward the end of the wet flowing phase. Shredders were represented mainly by *Asellus aquaticus* prevailed in most samples collected at site E3, where they formed approximately 52.6% of the total abundance in April and 33.7% of the total abundance in May. At site E2 the proportion of shredders decreased, to the benefit of collectors/gatherers and scrapers (*Hydraena nigrita*).

The macroinvertebrate community that was present after 145 days at the near-perennial site of E1 showed two peaks of abundance, the first one in March (when it was dominated by stoneflies and simulid midges) and the second just before complete drying, when aquatic beetles, mayflies of the subfamily Leptophlebiidae, the amphipod *Gammarus balcanicus*, and a snail species *Radomaniola curta* were the dominant species found.

Some groups, such as adults of the aquatic Heteroptera, appear just before the complete drying of the river at the end of May. Ephemeropterans are often regarded as slow colonizers [30], and in the Rimanić stream, they reached their highest abundance at the end of the flowing phase. As the duration of the flowing phase increased, the abundance of chironomid midges also increased, reaching their highest abundance just before the complete drying up of the stream. Furthermore, the result of the study showed that the proportion of predators, such as water mites and leeches, also increases as the flowing phase increases, reaching a maximum just before the stream completely dries up.

3 The Structure of a Water Beetle Assemblage in an Intermittent Stream: A Case Study of the Rimanić Stream

Water beetles are common in various types of freshwater ecosystems, including also intermittent rivers [31]. The high affinity of water beetles for intermittent water bodies has been highlighted in several studies (see [32–34]). This is especially common for predaceous water beetles that possess traits such as a short larval and longer adult phase that enhance their adaptability to the "predictable" environment of intermittent rivers [35]. However, only a few species is known to survive in a dry sediment (e.g., [36]), and most species use a winged phase to disperse to favorable wet habitats [37].

A comprehensive study on the water beetle was conducted in 2009 and 2010 at four sites along the course of the Rimanić stream [11]. The first site was located in the perennial reach and three others along the intermittent reach of the stream (see Fig. 2 for the exact positions of the investigated sites). A total of 1,028 adults and 130 larvae of 22 species were collected from the perennial reach [11], while 582 adults and 47 larvae belonging to 14 species were collected from intermittent reach [10]. Among them, four species, *Hydroporus pubescens*, *Helophorus montenegrinus*, *H. obscurus*, and *Laccobius obscuratus*, appeared only at intermittent sites. On the other hand, 12 species were exclusive to the perennial site. The study reveals that all crawling water beetles (*Haliplus lineatocollis*), whirligig beetles (*Gyrinus substriatus*), dryopid beetles (*Pomatinus substriatus*), and elmid beetles (*Esolus angustatus*, *Limnius intermedius*, *Oulimnius fuscipes*, and *Elmis bosnica*) appeared exclusively at the perennial site [11].

The results of the study on the water beetle community structure of the Rimanić stream reveal the highest site-specific number of taxa (α-diversity) at the perennial site (mean $\pm$ SD $= 6.6 \pm 4.29$ taxa), while the lowest number of taxa was found at the intermittent site E2 (2.29 ± 1.38) [11]. The number of taxa at the uppermost intermittent site E3 (3.00 ± 2.83) was higher than at E1 (2.43 ± 2.94) which was situated at the border with the perennial reach. A Kruskal–Wallis test revealed the presence of statistically significant differences in the number of taxa between the studied sites ($p = 0.027$).

The average Shannon diversity was highest at the perennial site (0.91 ± 0.62). On the other hand, with increasing distance from the perennial reach, the site-specific Shannon diversity of the communities from the intermittent reach increases, from 0.34 at site E1 to 0.63 at site E3. The Kruskal–Wallis test did not reveal that this metric significantly differed between the studied sites ($p = 0.083$).

At the level of the whole assemblage of the intermittent reach, the mean number of taxa was 2.57 ± 2.38, and the average Shannon diversity was 0.49 ± 0.53; both values were lower than those of the perennial reach. A Mann–Whitney U-test showed significant differences in the number of taxa ($p = 0.005$) and the average Shannon diversity ($p = 0.046$) between the assemblage of the perennial site and the total assemblage of the intermittent reach.

SIMPER analysis revealed the dominance of *Hydraena nigrita* at all sites, but the subdominant taxa of the perennial reach were different, i.e., *Scarodytes halensis* and *Esolus angustatus*. The study showed that the average dissimilarity of the water beetle assemblage between the perennial site and the consequent intermittent sites increased as the distance from the perennial reach increased (the values for the sites were E1 75.24%, E2 83.01%, and E3 87.17%). On the other hand, the internal similarity of the water beetle assemblages along the intermittent reach decreased as the distance from the perennial recolonization source increased. The community from the near-perennial site of E1 had a higher internal similarity, 35.46%, while the communities from the sites of E2 and E3, which were located further upstream, showed internal similarity of 16.84% and 9.91%, respectively.

Moreover, the results of the study revealed a different pattern for the water beetle adults and larvae of the Rimanić stream. While the ANOSIM analysis of the entire water beetle assemblages did not show a clear differentiation between the assemblages of the studied sites ($p = 0.282$ $R = 0.03$), in the case of larvae, the latter procedure revealed significant differences between the investigated sites ($p = 0.001$ $R = 0.515$). SIMPER analysis of the larvae assemblages revealed that elmid beetles dominated in the perennial reach, while dytiscid beetles dominated in the intermittent reach.

Pavićević and Pešić [10] in their study on the aquatic beetle assemblages in the intermittent reach of the Rimanić stream showed that the density and richness of this macrobenthic group increased from the late autumn to the spring, reaching a maximum abundance in May, just before the complete drying up of the intermittent reach. ANOSIM analysis showed no significant changes in community structure between the sites situated at different distances from the perennial recolonization source but did reveal changes in terms of date [10]. This suggests that the spatial arrangement of the perennial reach(es) is unlikely to affect the aquatic beetle assemblages of an intermittent reach. This might be explained that the main mechanism for recolonization in water beetles is dispersion by winged adults [37], allowing them to fly long distances and move between perennial and intermittent reaches.

4 The Invertebrate "Seedbank" Composition in Intermittent Rivers

During a dry phase, the richness of the aquatic assemblage rapidly declines as most desiccation-sensitive aquatic invertebrates disappear in the unwater sediment. However, the "seedbank," which comprises the life stages of desiccation-tolerant aquatic organisms, still persists within the dry sediment [38]. Several studies have shown that this "seedbank" can maintain moderate richness in instream communities [38, 39] emphasizing its importance in the face of ongoing climate change [40]. Regardless of its significance as a mechanism that allows the persistence of IRs communities during a dry phase, the "seedbank" assemblages of intermittent rivers in the Balkans, especially in its Mediterranean and sub-Mediterranean parts, have not been a focus of research among aquatic or terrestrial ecologists.

One rare study was conducted in 2009 and 2010 by Pavićević [11] who studied the "seedbank" assemblages from rehydrated sediments (RS) collected from three sites along the intermittent reach of the Rimanić stream (see Fig. 2 for the exact site positions). At each of the studied sites, 2.5 kg of sediment (at a depth of 0–20 cm) was collected. The samples were transported to the laboratory within 1 h and placed in 8 L containers. Each container was inundated with 5 L of filtered, dechlorinated tap water and fitted with an airstone for continuous aeration. After 17 days of inundation, the sediment in each container was poured through a 63 mm mesh sieve.

In total, 5,561 individuals were recorded from the rehydrated (RH) samples of the Rimanić stream [11]. Oribatid mites, springtails (Collembola), and ants (Formicidae) were the only terrestrial taxa that survived as adults or hatched during the 17-day inundation. After 17 days of inundation, the average invertebrate densities were 75.8 (in the range of 5–217.4) aquatic invertebrates and 7.05 (in the range of 0.3–30.6) terrestrial invertebrates per 2.5 kg sediment sample.

Ostracoda was the most abundant group in the RH samples, accounting for 71.6% of the total taxa abundance, followed by oribatid mites (7.03%) and gastropod snails (6.44%) [11]. The remaining fauna groups (Oligochaeta, leeches, thrombidid mites, copepods, amphipod gammarids, mayflies, dipterans, and stoneflies) individually accounted for less than 5% of the total taxa abundance [11]. PCA conducted on the faunistic data revealed that PC axes one and two explained 72.4% and 10.1% of the variance, respectively. PC1 was strongly determined by Ostracoda (0.943), whereas PC2 was mostly determined by oribatid mites (−0.565).

The study revealed the highest mean site abundance of the RS community at the near-perennial site of E1, with 107.4 individuals per 2.5^{-1} kg, while the lowest mean abundance was found at E2, showing only 60.95 individuals per 2.5^{-1} kg. A two-way ANOVA revealed no significant differences in the mean site abundance of the RS community, either between the studied sites ($p = 0.551$) or the two consecutive years of the study, 2009/2010 ($p = 0.510$).

The highest abundance of RS communities was found at E1 and E2, showing 200.6 and 241.1 ind. 2.5^{-1} kg, respectively, at the end of a dry phase. At these two sites, the lowest number on individuals was recovered from the RH samples collected in August and September 2009 (Fig. 3c). On the other hand, at the furthest upriver of the sites, E3, the lowest number of individuals was recovered from the RH samples taken in August and October, numbering 5.7 ind. 2.5^{-1} kg, while the highest abundance of the RS community was observed in July 2010, showing 131.9 ind. 2.5^{-1} kg. ANOSIM analysis did not find a clear differentiation between the studied assemblages, either for the investigated sites ($R = 0.057$ $p = 0.23$) or for the sampling dates ($R = 0.105$ $p = 0.21$). A similar pattern was observed when terrestrial taxa were excluded from the analyses, in that no significant differences between the fauna data were found either for the studied sites ($R = 0.065$ $p = 0.20$) or for the sampling dates ($R = 0.078$ $p = 0.28$).

The study on rehydrated sediments from the dry riverbed of Rimanić stream revealed that the resumption of flow, simulated by the inundation experiment, led to the appearance of aquatic invertebrate taxa in all samples [11]. The fauna of the aquatic invertebrates that appeared after 17 days of inundation was formed of desiccation-resistant taxa (ostracods, copepods, and nematodes) and a smaller number of desiccation-sensitive taxa that survived short dry periods (such as leeches, gammarid amphipods, and dragonflies larvae) [11].

Most of the desiccation-sensitive taxa become eliminated shortly after the flow ceased [41]. These taxa react to the surface drying by burrowing toward the saturated subsurface of the sediment [41] and/or into the hyporheic zone [42, 43]. In the Rimanić stream, most of these desiccation-sensitive taxa were found in the rehydrated samples from site E1, located at the border with the perennial reach. The latter site is located in a forest, and here the stream dried out more gradually and retained regions with small pools (Fig. 4) that saturated the subsurface volume of the sediment during the dry season. The other two sites were more open and unshaded, and their channels tended to dry up quickly and completely (showing "environmental harshness"). Stubbington and Datry [40] reviewed all the "seedbank" studies and found that the proportion of flowing river taxa present in rehydrated sediments collected from dry riverbeds declined as environmental harshness increased.

In the above-described inundation experiment, it was predicted that as the distance from the downstream perennial recolonization source increased, differences between the assemblages recovered from the inundated samples would become more pronounced. However, NMDS analysis revealed no clear pattern in the studied assemblages across the spatial variation, indicating that distance from the downstream perennial source did not significantly affect the RS assemblages [11].

The lack of a spatial pattern of variation to the RH assemblages might be the result of the taxon-specific traits of the two main groups appearing in the rehydrated samples, i.e., ostracods and oribatid mites. In ostracods specific traits include forming desiccation-resistant stages that allow surviving in arrested physiological stages, for several months until the flow resumes, as well as egg bank formation and erratic hatching [44, 45]. The heterogeneity of ostracod density in the rehydrated samples from the Rimanić stream is likely to have been the results of variable hatching "strategies" in selecting the appropriate patches for resting stage settlements that can greatly vary over spatial and temporal scales. For example, the August 2010 density of ostracods in the RH samples was much greater, approximately 80 ind. 2.5^{-1} kg, than in the same month of the preceding year, where the value was 12.7 ind. 2.5^{-1} kg.

The dry riverbed sediment is a habitat colonized by oribatid mites that represents one of the major components of soil fauna in temperate regions. Most oribatid mites, usually over 80% of them, live in the upper horizons, between 0 and 5 cm, at least in temperate climates [46]. In the RH sediments from the dry riverbed of the Rimanić stream, oribatid mites appeared in almost all samples. The content and quality of

Fig. 4 Intermittent pool phase of the Rimanić stream supports diverse communities of lentic invertebrates but also vertebrates (newts) trapped in a disconnected groundwater-feed pools. Photo by V. Pešić

litter and microclimatic conditions in the uppermost horizons are likely the main drivers of the spatiotemporal variation of the oribatid assemblages at the studied sites.

Recently, Stubbington et al. [8] emphasized that the ratio of oribatid mites (Acari) to Ostracoda could be used as an indicator for recognizing and characterizing the responses of RS assemblages to intermittence. This interesting and promising approach, however, needed to be tested in a higher number of intermittent streams across a wider geographical area.

5 The Terrestrial Invertebrate Communities of Intermittent Rivers

Most of the studies on the invertebrate communities of intermittent rivers have dealt with the aquatic phase, while only a few of them have concentrated on the terrestrial fauna colonizing the dry riverbeds [47, 48]. Dry riverbeds are common in the Mediterranean and therefore can provide habitats for many terrestrial organisms. Based on the available studies, the terrestrial communities that inhabit the dry stream bed are typically dominated by ground-dwelling beetles of Carabidae and Staphylinidae, ants, and spiders, but other taxonomic groups such as crustaceans, centipedes, millipedes, slugs, springtails, true bugs, true flies, and worms are also represented [49].

The two most widely used methods for sampling terrestrial invertebrates in the dry riverbed of intermittent streams are pitfall trapping and quadrat sampling [47, 48]. Quadrat samples are considered to be more appropriate for slow-moving taxa (Diplopoda, Oligochaeta, and some Carabidae), while pitfall trap can overestimate the abundance of the highly mobile invertebrates such as carabid beetles and spiders (see [48]).

The study on the terrestrial community of the intermittent course of the Rimanić stream was conducted at three sites (see Fig. 2) in 2009 and 2010 [12]. At each site, three quadrates (1 m^2) were sampled using a hand vacuum cleaner (Fig. 5c) to collect terrestrial invertebrates. In total, 2,197 individuals were collected. Springtails (Collembola) with 1,590 individuals were the most abundant group, accounting for 72.4% of the total number of specimens, followed by spiders (15.1% of the total abundance), oribatid mites (8.5%), and Formicidae (6.6%). The remaining taxa individually accounted for less than 5% of the total taxa abundance.

The mean site abundance of the terrestrial community was highest at E1, with 83.8 ind. m^{-2}, followed by those at E3 and E2, with 45.6 and 14.27 ind. m^{-2}, respectively. The highest abundance of the terrestrial community at E1 was found in August, with 159 and 142.3 ind. m^{-2} in 2009 and 2010, respectively. In the other months, the abundance of the terrestrial community at E1 was lower, ranging from 22.3 ind. m^{-2} in September to 83.7 ind. m^{-2} in June, at the beginning of the dry phase. At E2, the highest abundance, 40.3 ind. m^{-2}, was detected in October, at the end of the non-flowing phase (Fig. 3d). In other months, the abundance at this site was lower, ranging from 5.3 to 12.3 ind. m^{-2}. At E3, which was furthest upstream, the higher abundance of terrestrial invertebrate assemblages was recorded in June at the beginning of the dry phase, with 80.3 ind. m^{-2}, and in October, at the end of the dry phase, which gave a result of 105 ind. m^{-2}. In the other months, the abundance was lower ranging from 9.3 to 37.7 ind. m^{-2}.

NMDS analysis of the entire terrestrial invertebrate assemblages showed a clear differentiation between the assemblages at the studied sites, but did not reveal a differentiation in terms of sampling dates (Fig. 5). Moreover, the ANOSIM procedure indicated significant differences between the fauna data of the investigated sites ($R = 0.258$ $p = 0.03$), but not between the sampling periods ($R = 0.008$ $p = 0.467$).

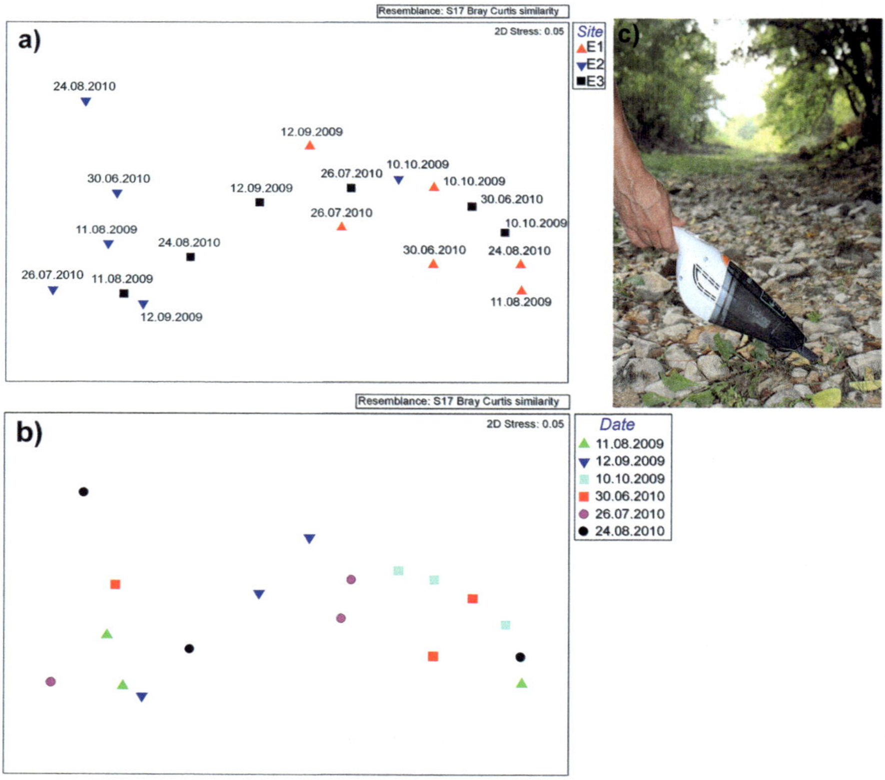

Fig. 5 Results of the non-metric multidimensional scaling of the terrestrial communities sampled using a hand vacuum cleaner (**c**) at the investigated sites along the intermittent reach of the Rimanić stream. (**a**) By sampling sites and (**b**) by date of sampling

It is likely that the observed pattern might suffer from a sampling bias, as estimates of taxa representation in quadrate samples might be less reliable. It is known that using quadratic sampling can be the subject of multiple sources of sampling bias (see [50–52] for discussion). For example, it is likely that the riparian shading that exhibits a cooling influence at the sampling sites may have contributed to the higher sampling efficiency of springtails (Collembola). Site E1, located at the border with the perennial reach, is highly shaded during the dry phase, while the two other sites located upstream along the intermittent gradient were unshaded. Secondly, the behavioral differences among taxa can affect their susceptibility to sampling, and, for example, nocturnal species might be missed in daytime quadrat samples [48]. In the case of some small organisms like oribatid mites, sampling efficiency, using forceps and aspirators, may have contributed to their lower presence in the quadrat samples. The use of vacuum cleaners in this case may have the advantage of nonselective collection, especially of small organisms such as oribatid mites and springtails, which are abundant and diverse in temperate areas.

6 The Aquatic and Riparian Plant Communities of Intermittent Rivers

Intermittent rivers host very diverse plant communities that can inhabit the same localities within different hydrological regimes (Fig. 1). In general, intermittent rivers host only a small number of strictly aquatic plants. Of the five surveyed intermittent rivers in the central part of Montenegro, the Sitnica (Matica), the Sušica, the Rimanić, the Cijevna, and the Rujela, only the intermittent reach of the River Sitnica (Fig. 1c) hosts well-developed aquatic vegetation represented by communities of *Ranunculus trichophyllus* (the threadleaf crowfoot). The community of this perennial plant belongs to the "3260 Water courses of plain to montane levels with the *Ranunculion fluitantis* and *Callitricho-Batrachion* vegetation" habitat type which is common along Montenegrin perennial rivers [53]. It should be noted that a well-developed community of the semiaquatic *Myosotis scorpioides* was found here, as well as fragments of *Mentha aquatica* and *Veronica beccabunga* communities, respectively, and fragments of *Cyperus longus* community. The upper part of the Sitnica that flows year-round is called the River Matica. This section is characterized by the well-developed aquatic vegetation of the class *Potamogetonetea*, represented by several communities, e.g., *Nymphaeo albae-Nupharetum luteae*, *Potamogetonetum denso-nodosi*, and *Potamogetonetum lucentis*, belonging to the "3150 Natural eutrophic lakes with *Magnopotamion* and *Hydrocharition* type vegetation" habitat type. *Potamogeton* (pondweeds) and *Nymphaeaceae* (waterlilies) are known to tolerate short-term drying [54]. Communities of these species do not occur downstream in the part of the watercourse that, when it enters Tološko polje field, is called the River Sitnica and whose lower part (below the Komanski Most bridge) regularly dries up in the summer.

The riparian vegetation of IRs is dominated by shrubs and trees with shorter canopies, but this is known to vary depending on the type of substrate, moisture, and dominant climatogene vegetation [55, 56]. Intermittent river riparian zones may support the presence of some rare species but also a high number of invasive species in particular. The lower part of the River Ribnica (Fig. 1e) that flows through the urbanized zone of Podgorica hosts a number of invasive species such as the alien tree species *Acer negundo*, *Ailanthus altissima*, *Melia azedarach*, *Ficus carica*, *Platanus orientalis*, and *Robinia pseudoacacia*. The percent of non-native species at this site is notably higher when compared to other sites located in more rural areas.

In some cases, riparian vegetation may experience succession along a perennial-intermittent gradient. The woody riparian vegetation of the upper part of the River Sitnica (here called the River Matica), which flows throughout the year, is dominated by communities of *Fraxinus angustifolia* and *Salix alba*, respectively, often accompanied by *Ulmus glabra* and *Quercus robur* (often cited as the endemic *Q. robur* subsp. *scutariensis*). These communities belong to a multilayered riverine forest of the "92A0 *Salix alba* and *Populus alba* galleries" habitat type which is common along the lowland River Zeta. Downstream, in the middle part which flows throughout the year (here the river is called the Sitnica), the riparian vegetation becomes more sparse, represented by fragments of *Fraxinus angustifolia* and *Salix alba* communities, respectively, and with a strong influence of shrub vegetation of the *Carpinus orientalis* community (e.g., *Paliurus spina-christi*, *Punica granatum*, *Clematis vitalba*, and so on). Alien woody species such as *Ailanthus altissima* and *Ficus carica* are also common. Finally, the riparian vegetation of the lower part of the River Sitnica that dries up during the summer is dominated by the shrub vegetation of *Vitex agnus-castus* (the chaste tree) (Fig. 1c). Fragments of other shrub communities such as those of *Rubus ulmifolius* and *Sambucus racemosa*, respectively, were also present.

The course of the River Sušica, which is the largest tributary of the River Zeta, is predominantly dry in the summer (Fig. 1b). The woody riparian vegetation of the River Sušica is dominated by communities of *Fraxinus angustifolia* and *Salix alba*, respectively. The community of *Fraxinus angustifolia* is often accompanied by up to 30-m-tall poplar trees, *Populus nigra*, that has a significant share in this community. *Ulmus glabra*, and *Quercus robur* are also common.

The woody and shrub riparian vegetation of the upper part of the intermittent reach of the Rimanić stream, which is environmentally harsher, is dominated by *Carpinus orientalis* and *Quercus cerris*. *Robinia pseudoacacia*, *Acer campestre*, *Punica granatum*, *Ficus carica*, *Cornus sanguinea*, *Salix viminalis*, *Fraxinus ornus*, and *Rubus ulmifolius* are also common. The riparian vegetation of the lower part, adjacent to the perennial reach, is dominated by *Carpinus orientalis* and *Ulmus glabra*, which is a common element of residual alluvial forests. A significant occurrence of *Cornus sanguinea* and *Acer campestre* is also present. Among other species *Rubus ulmifolius*, *Ficus carica*, *Salix alba*, *S. viminalis*, *Quercus cerris*, and *Paliurus spina-christi* have been found.

The riparian vegetation of the River Cijevna (Fig. 1a) along its perennial course is represented by individual trees of *Salix alba* and *Fraxinus angustifolia*. In the lower part, which dries in the summer, the riparian vegetation is dominated by a community of *Vitex agnus-castus*. Fragments with communities of *Tamarix dalmatica* and *Rubus ulmifolius*, respectively, can also be found here. Fragments of a community of *Salix alba* occur at the mouth of the River Cijevna where it joins the River Morača. There is a characteristic presence of fig trees (*Ficus carica*) on the rocks along the watercourse, from the middle part of the river toward the mouth into the River Morača.

The riparian vegetation along the entire watercourse of the River Rujela, which completely dries up during the summer, is dominated by the community of *Vitex agnus-castus* (*Viticetum agni-casti*), which in some parts is mixed with *Rubus ulmifolius* (*Rubo ulmifolii-Viticetum agni-casti*). The primary climatogenic vegetation of the Ćemovsko polje field through which the River Rujela flows is dominantly formed by a community of the Macedonian oak (*Quercus trojana*) whose elements (*Prunus webbii*, *Pistacia terebinthus*, and so on), as individual trees, are still present on the banks of this river.

In summary, riparian monitoring programs should also include those communities found along IRs. An indexing approach that allows the ranking of monitoring sites from good to poor such as using a Riparian Index of Biotic Integrity [57] or a

Riparian Quality Index [58] should include parameters related to specific aspects of riparian IR communities responses to intermittence and human impact [59].

7 Conclusions

Due to the increasing severity of drying events, it is likely that more perennial rivers, also in Montenegro, will become intermittent. Moreover, increased demand for water in the future resulting in increased river extraction and drainage may lead to the increasing prevalence of artificial intermittent rivers in the local and regional hydrological network. At the same time, many natural intermittent rivers will be altered by hydrological changes, resulting in a changing hydrological regime from near-perennial to episodic flow. Such example is the Rujela stream near Tuzi (but in last years also the Rimanić stream), which was previously intermittent and now ephemeral and has flowing water for a short period in response to rainfall. Other threats mentioned in the literature include water extraction, channel modification, and land-use change [1–3] to which intermittent rivers of South Montenegro are heavily exposed. All these threats and their unknown consequences for biota require the inclusion of IRs into ongoing biomonitoring programs, accompanied by the developing of adequate management and policy strategies for their long-term protection.

References

1. Datry T, Larned ST, Tockner K (2014) Intermittent rivers: a challenge for freshwater ecology. BioScience 64(3):229–235. https://doi.org/10.1093/biosci/bit027
2. Acuna V, Datry T, Marshall J, Barcelo D, Dahm CN, Ginebreda A, McGregor G, Sabater S, Tockner K, Palmer MA (2014) Why should we care about temporary waterways? Science 343 (6175):1080–1082. https://doi.org/10.1126/science.1246666
3. Larned ST, Datry T, Arscott DB, Tockner K (2010) Emerging concepts in temporary-river ecology. Freshw Biol 55:717–738. https://doi.org/10.1111/j.1365-2427.2009.02322.x
4. Shumilova O, Zak D, Datry T et al (2019) Simulating rewetting events in intermittent rivers and ephemeral streams: a global analysis of leached nutrients and organic matter. Glob Change Biol 25:1591–1611. https://doi.org/10.1111/gcb.14537
5. Datry T, Singer G, Sauquet E et al (2017) Science and management of intermittent rivers and ephemeral streams (SMIRES). Res Ideas Outcomes 3:e21774. https://doi.org/10.3897/rio.3.e21774
6. Datry T, Foulquier A, Corti R, von Schiller D, Tockner K, Mendoza-Lera C et al (2018) A global analysis of terrestrial plant litter dynamics in non-perennial waterways. Nat Geosci 11:497–503. https://doi.org/10.1038/s41561-018-0134-4
7. European Commission (2000) Directive 2000/60/EC, establishing a framework for community action in the field of water policy. Off J Eur Comm L 327:1–71
8. Stubbington R, Chadd R, Cid N, Csabai Z, Miliša M, Morais M, Munné A, Pařil P, Pešić V, Tziortzis I, Verdonschot RC (2018) Biomonitoring of intermittent rivers and ephemeral streams in Europe: current practice and priorities to enhance ecological status assessments. Sci Total Environ 618:1096–1113. https://doi.org/10.1016/j.scitotenv.2017.09.137

9. Kostianoy AG, Soloviev DM, Pešić V (2020) Remote sensing of intermittent rivers in Montenegro. In: Pešić V, Paunović M, Kostianoy A (eds) The rivers of Montenegro. The handbook of environmental chemistry. Springer, Cham
10. Pavićević A, Pešić V (2012) Water beetle distribution along a perennial distance gradient in an intermittent stream from the Mediterranean part of Montenegro. Arch Biol Sci 64:345–351. https://doi.org/10.2298/ABS1201345P
11. Pavićević A (2011) Seasonal dynamics of macroinvertebrates of Mareze and Rimanić with special reference to aquatic Coleoptera. PhD dissertation, University of Montenegro (in Montenegrin)
12. Pavićević I (2012) Colonization of dry riverbeds by invertebrates on the example of the Rimanic stream. Diploma thesis, University of Montenegro (in Montenegrin)
13. Zawal A, Pešić V (2018) The diversity of assemblages of water mites from Lake Skadar and its catchment. In: Pešić V, Karaman G, Kostianoy A (eds) The Skadar/Shkodra Lake environment. The handbook of environmental chemistry, vol 80. Springer, Cham, pp 311–323
14. Feminella JW (1996) Comparison of benthic macroinvertebrate assemblages in small streams along a gradient of flow permanence. J N Am Benthol Soc 15:651–669
15. Datry T, Arscott DB, Sabater S (2011) Recent perspectives on temporary river ecology. Aquat Sci 73:453–457
16. Korhonen JJ, Soininen J, Hillebrand H (2010) A quantitative analysis of temporal turnover in aquatic species assemblages across ecosystems. Ecology 91:508–517
17. Bogan MT, Lytle DA (2007) Seasonal flow variation allows 'time-sharing' by disparate aquatic insect communities in montane desert streams. Freshw Biol 52:290–304. https://doi.org/10.1111/j.1365-2427.2006.01691.x
18. Welborn GA, Skelly DK, Werner EE (1996) Mechanisms creating community structure across a freshwater habitat gradient. Annu Rev Ecol Syst 27:337–363. https://doi.org/10.1146/annurev.ecolsys.27.1.337
19. Williams DD, Hynes HBN (1976b) The recolonization mechanisms of stream benthos. Oikos 27:265–272
20. Minshall GW, Andrews DA, Manuel-Faler CY (1983) Application of island 19 biogeographic theory to streams: macroinvertebrate recolonization of the Teton River, Idaho. In: Barnes JR, Minshall GW (eds) Stream ecology: application and testing of general ecological theory. Plenum Press, New York, pp 279–297
21. Wood PJ, Petts GE (1999) The influence of drought on chalk stream macroinvertebrates. Hydrol Process 13:387–399
22. Churchel MA, Batzer DP (2006) Recovery of aquatic macroinvertebrate communities from drought in Georgia Piedmont headwater streams. Am Midl Nat 156:259–272
23. Perrow MR, Skeate ER, Leeming D, England J, Tomlinson ML (2007) Life after low flow – ecological recovery of the river Misbourne. Br Wildl 18:335–346
24. Williams DD, Feltmate BW (1992) Aquatic insects. CAB International, Wallingford
25. Gore JA (1979) Patterns of initial benthic recolonization of a reclaimed coal strip-mined river channel. Can J Zool 57:2429–2439
26. MacKay RJ (1992) Colonization by lotic macroinvertebrates: a review of processes and patterns. Can J Fish Aquat Sci 49:617–628
27. Williams DD, Hynes HBN (1977) Benthic community development in a new stream. Can J Zool 55:1071–1076
28. Malmqvist B, Rundle S, Bronmark C, Erlandsson A (1991) Invertebrate colonization of a new, man-made stream in southern Sweden. Freshw Biol 26:307–324
29. López-Rodríguez MJ, Tierno de Figueroa JM, Fenoglio S, Bo T, Alba-Tercedor J (2009) Life strategies of 3 Perlodidae species (Plecoptera) in a Mediterranean seasonal stream in southern Europe. J N Am Benthol Soc 28:611–625
30. McArthur JV, Barnes JR (1985) Patterns of macroinvertebrate colonization in an intermittent rocky mountain stream in Utah. Great Basin Nat 45:117–123

31. Larson DJ, Alarie Y, Roughley RE (2000) Predaceous diving beetles (Coleoptera: Dytiscidae) of the Nearctic region, with emphasis on the fauna of Canada and Alaska. NRC Research Press, Ottawa. 982 pp
32. Moreno JL, Millán A, Suarez ML, Vidal-Abarca MR, Velasco J (1997) Aquatic Coleoptera and Heteroptera assemblages in waterbodies from ephemeral coastal streams ('ramblas') of south-eastern Spain. Arch Hydrobiol 141:93–107
33. Beauchard O, Gagneur J, Brosse S (2003) Macroinvertebrate richness patterns in North African streams. J Biogeogr 30:1821–1833
34. Vidal-Abarca MR, Sánchez-Montoya MM, Guerrero C, Gómez R, Arce MI, García-García V, Suárez ML (2013) Effects of intermittent stream flow on macroinvertebrate community composition and biological traits in a naturally saline Mediterranean stream. J Arid Environ 99:28–40
35. Ribera I (2008) Habitat constraints and the generation of diversity in freshwater macro-invertebrates. In: Lancaster J, Briers RA (eds) Aquatic insects: challenges to populations. CAB International, Wallingford, pp 289–311
36. Davy-Bowker J (2002) A mark and recapture study of water beetles (Coleoptera: Dytiscidae) group of semi-permanent and temporary ponds. Aquat Ecol 36:435–446. https://doi.org/10.1023/A:1016550127986.
37. Bilton DT (2014) Dispersal in dytiscidae. In: Yee DA (ed) Ecology, systematics, and natural history of predaceous diving beetles (Coleoptera: Dytiscidae). Springer, New York, pp 387–407
38. Stubbington R, Gunn J, Little S, Worrall TP, Wood PJ (2016) Macroinvertebrate seedbank composition in relation to antecedent duration of drying and multiple wet-dry cycles in a temporary stream. Freshw Biol 61:1293–1307. https://doi.org/10.1111/fwb.12770
39. Datry T, Corti R, Philippe M (2012) Spatial and temporal aquatic–terrestrial transitions in the temporary Albarine River, France: responses of invertebrates to experimental rewetting. Freshw Biol 57:716–727. https://doi.org/10.1111/j.1365-2427.2012.02737.x
40. Stubbington R, Datry T (2013) The macroinvertebrate seedbank promotes community persistence in temporary rivers across climate zones. Freshw Biol 58:1202–1220. https://doi.org/10.1111/fwb.12121
41. Strachan SR, Chester ET, Robson BJ (2015) Freshwater invertebrate life history strategies for surviving desiccation. Springer Sci Rev 3:57–75
42. Stubbington R (2012) The hyporheic zone as an invertebrate refuge: a review of variability in space, time, taxa and behavior. Mar Freshwater Res 63:293–311
43. Vander Vorste R, Malard F, Datry T (2016) Is drift the primary process promoting the resilience of river invertebrate communities? A manipulative field experiment in an intermittent alluvial river. Freshw Biol 61:1276–1292
44. Williams DD (1998) The role of dormancy in the evolution and structure of temporary water invertebrate communities. Arch Hydrobiol 52:109–124
45. Bilton DT, Freeland JR, Okamura B (2001) Dispersal in freshwater invertebrates. Annu Rev Ecol Syst 32:159–181
46. Krivolutsky DA (ed) (1995) Oribatid mites: morphology, development, phylogeny, ecology, methods of study, the characteristic of modelling species *Nothrus palustris* C.L. Koch, 1839. Nauka, Moscow. 224 pp
47. Steward AL, Marshall JC, Sheldon F, Harch B, Choy S, Bunn SE, Tockner K (2011) Terrestrial invertebrates of dry river beds are not simply subsets of riparian assemblages. Aquat Sci 73:551–566. https://doi.org/10.1007/s00027-011-0217-4
48. Corti R, Datry T (2016) Terrestrial and aquatic invertebrates in the riverbed of an intermittent river: parallels and contrasts in community organisation. Freshw Biol 61:1308–1320. https://doi.org/10.1111/fwb.12692
49. Steward AL, Langhans SD, Corti R, Datry T (2017) The biota of intermittent rivers and ephemeral streams: terrestrial and semiaquatic invertebrates. In: Datry T, Bonada N, Boulton AJ (eds) Intermittent rivers and ephemeral streams: ecology and management. Elsevier, Amsterdam, pp 245–271

50. Desender K, Segers R (1985) A simple device and technique for quantitative sampling of riparian beetle populations with some Carabid and Staphylinid abundance estimates on different riparian habitats (Coleoptera). Rev Ecol Biol Sol 22:497–506
51. Melbourne BA (1999) Bias in the effect of habitat structure on pitfall traps: an experimental evaluation. Aust J Ecol 24:228–239
52. Missa O, Basset Y, Alonso A, Miller SE, Curletti G, De Meyer M et al (2009) Monitoring arthropods in a tropical landscape: relative effects of sampling methods and habitat types on trap catches. J Insect Conserv 13:103–118
53. Pešić V, Grabowski M, Hadžiablahović S, Marić S, Paunović M (2018) The biodiversity and biogeographical characteristics of the river basins of Montenegro. In: Pešić V, Paunović M, Kostianoy A (eds) The rivers of Montenegro. The handbook of environmental chemistry. Springer, Cham
54. Rodwell JS (ed) (1995) British plant communities. Aquatic communities, swamps and tall-herb fens, vol 4. Cambridge University Press, Cambridge. 283 pp
55. Dodds WK, Gido K, Whiles MR, Fritz KM, Matthews WJ (2004) Life on the edge: the ecology of Great Plains prairie streams. BioScience 54:205–216. https://doi.org/10.1641/0006-3568(2004)054%5b0205:LOTETE%5d2.0.CO;2
56. Stromberg JC, Merritt D (2016) Riparian plant guilds of ephemeral, intermittent and perennial rivers. Freshw Biol 61:1259–1275. https://doi.org/10.1111/fwb.12686
57. Miller SJ, Wardrop DH, Mahaney WM, Brooks RP (2006) A plant-based index of biological integrity (IBI) for headwater wetlands in central Pennsylvania. Ecol Indic 6:290–312
58. González del Tánago M, García de Jalón D (2011) Riparian Quality Index (RQI): a methodology for characterising and assessing the environmental conditions of riparian zones. Limnetica 30:235–254. https://doi.org/10.23818/limn.30.18
59. Bruno D, Gutiérrez-Cánovas C, Velasco J, Sánchez-Fernández D (2016) Functional redundancy as a tool for bioassessment: a test using riparian vegetation. Sci Total Environ 566:1268–1276. https://doi.org/10.1016/j.scitotenv.2016.05.186

Application of Google Earth in Mapping Intermittent Rivers of Montenegro

Andrey G. Kostianoy, Dmitry M. Soloviev, and Vladimir Pešić

Contents

Abstract Intermittent rivers that periodically cease to flow represent the main part of the hydrological network of South Montenegro. Their inclusion in ongoing biomonitoring programs are hampered, among other things, because of the shortcomings given about the spatial distribution of temporary and perennial river channels. Google Earth has been shown to be a useful tool for mapping intermittent and perennial rivers. To test the classification accuracy with Google Earth imagery, we have selected ten intermittent rivers located in the southern part of Montenegro. The results revealed the moderate accuracy of Google Earth imagery in the mapping spatial extent of intermittent and perennial river sections. The results demonstrated a different discriminatory capacity of GE imagery depending on the presence of forested area and river width. Nevertheless, this study emphasized that using satellite and aerial images in combination with field surveys could provide an accurate and effective classification mode for mapping

A. G. Kostianoy (✉)
P.P. Shirshov Institute of Oceanology, Russian Academy of Sciences, Moscow, Russia

S.Yu. Witte Moscow University, Moscow, Russia
e-mail: kostianoy@gmail.com

D. M. Soloviev
Marine Hydrophysical Institute, Russian Academy of Sciences, Sevastopol, Russia
e-mail: solmit@gmail.com

V. Pešić
Faculty of Sciences, University of Montenegro, Podgorica, Montenegro
e-mail: vladopesic@gmail.com

Vladimir Pešić, Momir Paunović, and Andrey G. Kostianoy (eds.), *The Rivers of Montenegro*, Hdb Env Chem (2020) 93: 253–264, DOI 10.1007/698_2020_488,

intermittent rivers in karstic regions such as South Montenegro, as a tool that would facilitate the protection of these threatened and unique ecosystems.

Keywords Google Earth imagery, High-resolution optical imagery, Intermittent rivers, Montenegro, Rivers, Satellite remote sensing

1 Introduction

Intermittent rivers (IRs) are defined as rivers that periodically cease to flow, often dominating in local and regional hydrologic catchment [1, 2]. In particular in arid and semiarid regions, the proportion of IRs can exceed 70% of the entire hydrographic network [1, 2]. It is still unclear what proportion is in Montenegro but intermittent rivers predominated in the river network of the Mediterranean and sub-Mediterranean parts of Montenegro.

Pešić et al. [2] in this book provide an overview of communities inhabiting intermittent rivers in the South Montenegro emphasizing recent progress and gaps in their inclusion in ongoing monitoring programs. They stressed that the first step in any monitoring should include mapping the spatial distribution of temporary and perennial river channels at the level of local or regional catchment [2]. As stressed by Fritz et al. [3] the accurate mapping of the spatial distribution of the temporary and perennial segments in a river network is basic information for any monitoring. This is important for proposing specific action plans that would lead to biodiversity conservation through an integrated basin management.

As demonstrated by many studies, the dynamic characteristics of temporary rivers that include both terrestrial and aquatic phases present serious challenges to the assessment of their ecological states [4, 5]. This could potentially affect the accuracy of monitoring results [6] especially if these challenges were not properly addressed by ongoing national surface water monitoring.

The data providing the accurate hydrography come mainly from the gauging stations (e.g., [4, 7]). New approaches include citizen involvement through citizen science programs that have been shown to be a useful tool for mapping intermittent rivers (e.g., [8]).

In both Montenegrin basins, there are only 24 gauging stations from which a long time series of monthly and annual flows could be obtained [9]. This number of stations is insufficient for the area such as Montenegro [9], especially given the fact that numerous rivers cannot be measured, most especially in the upper parts of the basins. All gauging stations in Montenegro are located in perennial river segments. The consequence of this limitation is that many streams in Montenegro cannot be analyzed from the point of view of hydrology, biology, pollution, and ecology. This was the case in our study area, in which there were several active official gauging stations (three in river channels and four in reservoirs) all of which were located in perennial river segments.

In the absence of hydrological monitoring points located on IRs in Montenegro, the alternative is a remote sensing approach that includes aerial photography and satellite imagery. There are a small number of publications related to satellite remote sensing approaches to the investigation of water resources and land cover in Montenegro. Recently, several papers have been published on the Boka Kotorska Bay [10] and the Skadar Lake [11].

Remotely sensed imagery has been used extensively in studying various aspects of riverine ecosystems including intermittent rivers too [12]. Spence and Mengistu [13] highlighted the usefulness of using unmanned aerial vehicles and systems (UAVSs) might be an appropriate approach to mapping intermittent rivers in the river network.

In the last decade, using Google Earth (GE) which allows users to view satellite images has gained a lot of popularity due to its high resolution and free cost [14]. Thus, this software was found applicable in various fields of river ecosystems studies such as river geomorphology [15, 16] and sedimentology [14], or mapping a river erosion (e.g., [17]).

In this chapter we show the capabilities of high-resolution satellite optical imagery in the mapping of IRs in the Montenegro and the factors that affect accuracy of GE imagery. Moreover, this article aims to show the potential of the GE database of historical remotely sensed data.

2 Data and Methods

High-resolution satellite optical, infrared, and radar imagery is widely used for the analysis of land cover. For the analysis of rivers in Montenegro, it is possible to use the high-resolution satellite sensors ETM+ Landsat-7, OLI and TIRS Landsat-8, and MSI Sentinel-2A like it was done for the Boka-Kotorska Bay [10] and Skadar/ Shkodra Lake [11]. Spatial resolution of the processed true color MSI Sentinel-2A maps is 10 m, while for other sensors it is 20 m. The other option is to use the synthetic aperture radar images acquired, for example, by SAR-C Sentinel-1A and Sentinel-1B with spatial resolution 20 m. SAR data are widely used for the analysis of floods on rivers [13–19]. Our first attempts to use the abovementioned satellite imagery for the analysis of IRs in Montenegro has failed because it became clear that spatial resolution of 10–20 m is not enough for rivers and streams whose width did not exceed 10 m. One of such images is presented in Fig. 1a for the Cijevna/Cemi River where area of our interest is marked by red rectangle, which is absolutely blurred due to low spatial resolution.

A single solution to solve this problem is to use much higher spatial resolution of optical sensors installed, for example, on QuickBird (0.6–2.6 m), WorldView-3, -4 (30 cm), GeoEye-1 (50 cm), Pleiades-1A, -1B (50 cm), and other satellites. Such kind of satellite imagery is very spectacular due to a fantastic quality of high-resolution imagery where every car is clearly detectable. Such kind of imagery is used for environmental monitoring, including natural and anthropogenic disasters,

Fig. 1 (**a**) Satellite view of the area westward of Podgorica Airport from MSI Sentinel-2A on 31 March 2019. Red rectangle shows the area of investigation presented on Fig. 3e, f. (**b**) Satellite view of the Ribnica River on 15 August 2017. (**c**) Satellite view of the Sitnica River on 14 August 2016 (**d**) Photograph of Sitnica River near Komanski Most taken on September 20, 2019 (Photo by V.Pešić) (Figs. 1b, c Source: Google Earth Pro)

in oil and gas industry, engineering and construction, mapping sport, tourism, and historical places, as well as for defense and intelligence. This kind of imagery is unique for scientific research, but they are used very rarely in science because of a very high cost. In case when a regular monitoring is required, the cost of a series of satellite imagery significantly exceeds the budgets of scientific projects.

We found an unexpected solution to this problem by using the Google Earth Pro option "Historical imagery", which allows to switch between images acquired on different dates for a certain location since 2003 till present.

3 Results

To test the classification accuracy with Google Earth (GE) imagery in mapping the spatial distribution of temporary rivers, we have randomly selected ten intermittent river sections located in the southern part of Montenegro. We checked satellite imagery acquired in 2015–2019. Moreover, the field survey campaign was conducted to inspect intermittent river segments in order to compare filed data with those obtained from using GE imagery.

Here we briefly describe the difference in GE imagery accuracy in recognizing the intermittent conditions of studied rivers and their main features.

Cijevna/Cemi River is a left tributary of the Morača River. Part of the investigated river section is indicated in Fig. 1a (viewed from MSI Sentinel-2A with a spatial resolution of 10 m) by a red rectangle showing the area before the mouth into the Morača River which regularly dry up. Here the river has features of single channel and with low curvature. The riparian vegetation is represented by the shrubs of the chaste tree (*Vitex agnus-castus*) community (see [2]).

The *Ribnica* is a small river that flows through the urban part of Podgorica. It is regularly dried up in the summer. In features of a single channel, with well-riparian vegetation with the presence of trees (e.g., *Acer negundo, Ailanthus altissima, Melia azedarach, Ficus carica, Platanus orientalis*, and *Robinia pseudoacacia* [1]), occasionally forming a forest canopy that covers parts of the investigated rivers segment. This affects GE imagery accuracy and not allow the use of aerial images to distinguish intermittent from perennial reaches along the whole studied river section. The best visibility was found at the confluence of the Ribnica and Morača River where the river bed is the widest (Fig. 1b).

The *Sitnica River* is a small tributary of the Morača River. The river has features of a single channel with a moderate curvature. The neighboring riparian vegetation is represented by the shrub vegetation of *Vitex agnus-castus* (the chaste tree) [2] which does not form a canopy over the intermittent part of the river (from Komanski Most – bridge toward its mouth into the Morača River, see Fig. 1d). The transient zone between the upper (perennial) and the lower (intermittent) part is represented by the series of disconnecting pools that are visible also in the satellite images taken in summer (see Fig. 1c).

Sušica River (*Orašja jama*) This is the largest tributary of the River Zeta and arises from the karst springs at the foot of Dubrava Mountain (including the Orašja jama cave). During January and February, if the flow volumes are low enough, the water can freeze. The riparian vegetation is well-developed and represented with communities of *Fraxinus angustifolia* and *Salix alba,* often accompanied by up to 30-m tall trees of poplar (*Populus nigra*) [2]. However, the relatively wide channel allows us to use satellite images to identify the presence and/or lack of a water flow as it is presented in Fig. 2a, b.

Sušica River (*Dobro Polje*). The river has a feature of a single channel. Woody riparian vegetation is well developed, covering partly the investigated river segment. The small width of the stream as well as the small amount of water during most of the dry season (Fig. 2d) does not allow to use the aerial images to identify with certainty the absence and/or presence of a water flow.

Rimanić Stream. The upper part of the Rimanić Stream in Bjelopavlići Valley is being an intermittent reach, while its lower part ca. 1.5 km upstream of its confluence with the River Zeta is perennial reach. The neighboring riparian vegetation along intermittent reach is well presented with domination of trees (e.g., *Carpinus orientalis, Quercus cerris, Robinia pseudoacacia, Acer campestre* [2]) that form a canopy that covers a large part of the investigated stream section. As a result GE imagery accuracy does not allow with certainty to distinguish intermittent from perennial reaches.

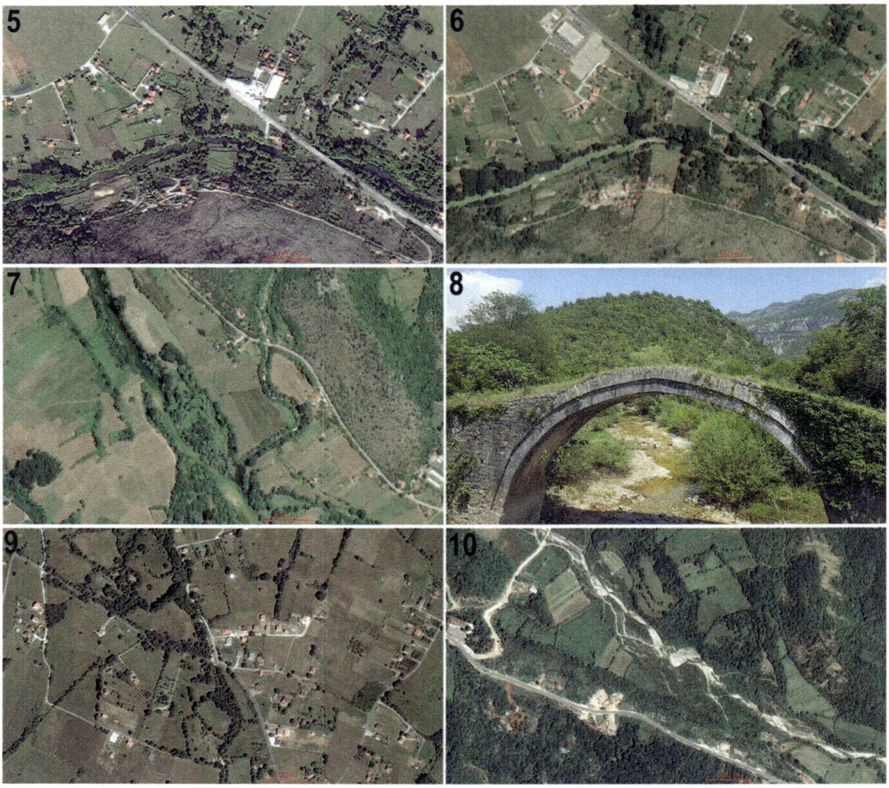

Fig. 2 (**a**) Satellite view of the Sušica River (Orašja jama) on 13 October 2015. (**b**) Satellite view of the Sušica River (Orašja jama) on 1 July 2019. (**c**) Satellite view of the Sušica River (Dobro Polje) on 1 July 2019. (**d**) Photograph of Sušica River near the bridge in Dobro Polje village (Photo by V. Pešić). (**e**) Satellite view of the Rimanić River on 14 August 2016. (**f**) Satellite view of the Miđanska (Vladimirska) River on 8 June 2019. (Fig. 2a–c, e, f Source: Google Earth Pro)

Miđanska (Vladimirska) Rijeka. It is only of studied rivers located in the Coastal region (the municipality of Ulcinj). The river has features of single channel and with vegetation consisted of shrub vegetation of *Vitex agnus-castus* (the chaste tree).

Orahovštica River. This is a small river that flows to Skadar/Shkodra Lake near Virpazar town. The stream features a single channel with a low curvature near its mouth into Lake Skadar/Shkodra. The riparian woody vegetation is well developed forming a canopy that covers river segments. This affects GE imagery accuracy and does not allow the use of aerial images to distinguish intermittent from perennial reaches along the whole studied river section (see Fig. 3a, b).

Gračanica River (Nikšić). The river, which in some segments has unstable riverbank due to the intensive erosive processes (as a result of sand and gravel extraction activities), has a relatively wide channel. The riparian vegetation of the selected river section is represented by poorly developed shrub vegetation.

Fig. 3 (**a**) Satellite view of the Oraovštica River on 2 October 2017. (**b**) Satellite view of the Oraovštica River on 1 July 2019. (**c**) Satellite view of the Gračanica River (Nikšić) on 25 September 2018. (**d**) Satellite view of the Gračanica River (Danilovgrad) on 1 July 2019. (**e**) Satellite view of the Cijevna River on 17 October 2016. (**f**) Satellite view of the Cijevna River on 20 October 2017. Source: Google Earth Pro

Gračanica River (Danilovgrad). This ephemeral stream that regularly flows for a short time following rainfalls has features of a single channel and moderate curvature. Riparian vegetation is well developed and in some places forms a canopy that cover investigated river segment. The small channel width and episodic water flow affect GE imagery accuracy in identifying perennial and/or intermittent hydrological regime of watercourse.

Characteristics of the studied intermittent rivers/streams are presented in Table 1 for years 2015–2019. The width of rivers (water area) was measured in the same places. Cases when there was a dry river bed were not taken into account when calculating average and minimum river width. These dates are indicated in the same column.

Table 1 Characteristics of studied intermittent rivers/streams obtained by using Google Earth

Intermittent river	Latitude/longitude	Channel stability	Forest canopy	Stream width (W), m Average (min–max)
Cijevna/Cemi River	42°22′01.72″N 19°13′24.35″E	Stable	Lacking	Number of images with water = 11 W 32.34 m (27.23–39.25) Dry on 06.08.2015 Dry on 18.08.2015 Dry on 14.08.2016 Dry on 15.08.2017 Dry on 20.10.2017 Dry on 20.07.2019
Ribnica River	42°26′21.20″N 19°15′34.63″E	Stable	Moderately developed	Number of images with water = 14 W 8.04 m (6.54–9.92) Dry river bed was not identified
Sitnica River	42°26′38.01″N 19°12′14.46″E	Stable	Lacking	Number of images with water = 13 W 13.26 m (6.19–16.89) Dry on 06.08.2015 Dry on 18.08.2015 Dry on 06.08.2016 Dry on 14.08.2016 Dry on 15.08.2017 Dry on 20.10.2017
Sušica River (Orašja jama)	42°30′25.25″N 19°07′52.32″E	Stable	Moderately developed	Number of images with water = 3 W 16.21 m (14.39–18.72) Dry on 15.06.2015 Dry on 24.08.2015 Dry on 16.09.2016 Dry on 06.07.2017 Dry on 15.08.2017 Dry on 02.10.2017 Dry on 01.07.2019 Dry on 20.07.2019 Dry on 29.10.2019
Sušica River (Dobro Polje)	42°37′23.34″N 19°02′34.04″E	Stable	Well developed	Number of images with water = 1 W 11.58 m Dry on 06.07.2017 Dry on 25.09.2018 Dry on 01.07.2019
Rimanić stream	42°31′54.86″N 19°11′24.07″E	Stable	Well developed	Number of images with water = 0 Dry on 13.10.2015 Dry on 20.04.2016

(continued)

Table 1 (continued)

Intermittent river	Latitude/longitude	Channel stability	Forest canopy	Stream width (W), m Average (min–max)
				Dry on 14.08.2016 Dry on 16.09.2016 Dry on 17.10.2016 Dry on 15.08.2017 Dry on 10.05.2018 Dry on 21.02.2019 Dry on 20.07.2019
Miđanska (Vladimirska) River	42°00′52.43″N 19°16′24.18″E	Stable	Lacking	Number of images with water = 3 W 4.90 m (2.45–7.71) Dry on 15.06.2015 Dry on 08.09.2015 Dry on 20.10.2017 Dry on 23.06.2018 Dry on 25.10.2018 Dry on 20.10.2019
Oraovštica River	42°14′45.64″N 19°04′15.65″E	Stable	Moderately developed	Number of images with water = 5 W 8.92 m (8.05–9.74) Dry on 02.10.2017 Dry on 27.03.2019
Gračanica River (Nikšić)	42°46′01.40″N 19°01′19.16″E	Unstable	Lacking	Number of images with water = 0 Dry on 23.09.2016 Dry on 06.07.2016 Dry on 25.09.2018 Dry on 28.03.2019 Dry on 01.07.2019
Gračanica River (Danilovgrad)	42°33′49.73″N 19°04′00.16″E	Stable	Moderately developed	Number of images with water = 0 Dry on 20.04.2016 Dry on 06.07.2017 Dry on 01.07.2019

As mentioned above, Google Earth software has the option "Historical imagery." This function allows the study of temporal changes of the perennial and intermittent channel extent as well as the degree of flow permanence over a time. It is worth mentioning that Google Earth allows you to check the earliest historical images in 2003. Additionally, Google Earth Pro allows users to view an image with a resolution of 0.5–1 m.

The benefits of using historical images are, for example, that it is possible for the investigated area to compare GE imagery of the same river section in different seasons and/or in different years.

For example, Fig. 3e taken on October 17, 2016, shows that Cijevna/Cemi River had water flow, while in the same period next (see Fig. 3f taken on October 20, 2017) year, the compared river section was dry.

4 Discussion

The results of our study show that the classification accuracy of Google Earth (GE) imagery in the mapping spatial extent of selected intermittent rivers of South Montenegro was moderate. Of the ten selected rivers, GE imagery allows with certainty to identify the channel extent and the degree of flow permanence for five rivers (the Cijevna, the Sitnica, the Sušica (Orašja Jama), the Miđanska rijeka, and the Gračanica (Nikšić) Rivers). For five rivers, i.e., the Ribnica, the Sušica (Dobro Polje), the Rimanić, the Gračanica (Danilovgrad), and the Oraovštica Rivers, using GE imagery show limited capacities in identifying channel extent and degree of flow permanence over the whole investigated river sections. These difficulties are related to existing forest cover that partially or completely cover certain parts of the selected river segment not allowing us to use the satellite images for distinguishing temporary and perennial river segments.

The results revealed different discriminating capacities of GE imagery depending on the presence of forested area and river width. On rivers where woody riparian vegetation does not cover the river segments, it is possible to detect water flows by using satellite images. In our study, the minimum water width that was detected was 2.5 m (see Table 1), which allows us to use the aerial images to identify temporary and perennial segments with certainty.

The best results were obtained for the Cijevna, the Sitnica, and the Miđanska Rivers where the lack of woody vegetation and the relatively large channel width of the studied segments allow for a higher hydrological accuracy. It should be noted that most intermittent rivers in the Mediterranean and sub-Mediterranean parts of Montenegro are small in size, which complicates the accuracy of using GE imagery. Further, these limitations are reflected in the limited possibilities in quantification of river features that can be done using GE imagery. For example, the width of a river channel and a slope are measurable, but the water depth of the selected river sections is not.

As demonstrated in this chapter, the use of the function "Historical imagery" has great application in the study of spatial and temporal changes in the extent of perennial and intermittent rivers over time. The timing of images is extremely useful for ecological assessment. However, the fact that most IRs are small in size and dry up during the growing season when riparian vegetation is well developed indicates limited capacities of using only GE imagery. As stressed by Benda et al. [18], using aerial images through a virtual watershed approach should be accompanied by field surveys taken in the low-flow season, in order to build a more accurate and effective mode for temporal and perennial mapping reaches.

Recent studies stressed that due to the climate change likely more and more perennial rivers in Montenegro will become intermittent or ephemeral. This will require the development of adequate management and policy strategies for their long-term protection [2]. Using a tool such as Google Earth imagery that is quick, easy, and free to access can also help social mobilization activities including volunteers and citizens in intermittent rivers monitoring projects.

Acknowledgments A.G. Kostianoy was partially supported in the framework of the P.P. Shirshov Institute of Oceanology RAS budgetary financing (Project N 149-2019-0004).

References

1. Datry T, Larned ST, Tockner K (2014) Intermittent rivers: a challenge for freshwater ecology. Bioscience 64(3):229–235. https://doi.org/10.1093/biosci/bit027
2. Pešić V, Pavićević A, Savić A, Hadžiablahović S (2020) The intermittent rivers of South Montenegro: ecology and biomonitoring. In: Pešić V, Paunović M, Kostianoy A (eds) The rivers of Montenegro. The handbook of environmental chemistry. Springer, Cham
3. Fritz KM, Hagenbuch E, D'Amico E, Reif M, Wigington PJ, Leibowitz SG, Comeleo RL, Ebersole JL, Nadeau TL (2013) Comparing the extent and permanence of headwater streams from two field surveys to values from hydrographic databases and maps. J Am Water Resour Assoc 49(4):867–882. https://doi.org/10.1111/jawr.12040
4. Stubbington R, Chadd R, Cid N, Csabai Z, Miliša M, Morais M, Munné A, Pařil P, Pešić V, Tziortzis I, Verdonschot RC (2018) Biomonitoring of intermittent rivers and ephemeral streams in Europe: current practice and priorities to enhance ecological status assessments. Sci Total Environ 618:1096–1113. https://doi.org/10.1016/j.scitotenv.2017.09.137
5. Datry T, Singer G, Sauquet E et al (2017) Science and management of intermittent rivers and ephemeral streams (SMIRES). Res Ideas Outcomes 3:e21774. https://doi.org/10.3897/rio.3.e21774
6. Arthington AH, Bernardo JM, Ilheu M (2014) Temporary rivers: linking ecohydrology, ecological quality and reconciliation ecology. River Res Appl 30(10):1209–1215. https://doi.org/10.1002/rra.2831
7. Snelder TH, Datry T, Lamouroux N, Larned ST, Sauquet E, Pella H, Catalogne C (2013) Regionalization of patterns of flow intermittence from gauging station records. Hydrol Earth Syst Sci 17(7):2685–2699. https://doi.org/10.5194/hess-17-2685-2013
8. Datry T, Pella H, Leigh C, Bonada N, Hugueny B (2016) A landscape approach to advance intermittent river ecology. Freshw Biol 61(8):1200–1213. https://doi.org/10.1111/fwb.12645
9. Sekulić G, Radulović M (2020) The hydrology and hydrogeology of Montenegro. In: Pešić V, Paunović M, Kostianoy A (eds) The rivers of Montenegro. The handbook of environmental chemistry. Springer, Cham
10. Kostianoy AG, Soloviev DM, Kostianaia EA, Đurović B, Pestorić B (2017) Satellite remote sensing of the Boka Kotorska Bay. In: Joksimovich A, Djurovic M, Semenov AV, Zonn IS, Kostianoy AG (eds) The Boka Kotorska Bay environment. Springer, Cham, pp 495–520
11. Kostianoy AG, Soloviev DM, Kostianaia EA, Sirota AM (2018) Satellite remote sensing of Lake Skadar/Shkodra. In: Pešić V, Karaman G, Kostianoy AG (eds) The Skadar/Shkodra Lake environment. Springer, Cham, pp 89–120. https://doi.org/10.1007/698_2018_308
12. González-Ferreras AM, Barquín J (2017) Mapping the temporary and perennial character of whole river networks. Water Resour Res 53(8):6709–6724
13. Spence C, Mengistu S (2016) Deployment of an unmanned aerial system to assist in mapping an intermittent stream. Hydrol Process 30(3):493–500. https://doi.org/10.1002/hyp.10597
14. Zhou XW, Wang HL (2015) Application of Google Earth in modern river sedimentology research. J Geosci Environ Protect 3:1–8. https://doi.org/10.4236/gep.2015.38001
15. Akawwi E (2013) Geomorphology using geographic information system and global mapper. Am J Environ Sci 9(5):398–409
16. Goudie A (2013) Characterising the distribution and morphology of creeks and pans on salt marshes in England and Wales using Google Earth. Estuar Coast Shelf Sci 129:112–123
17. Boardman J (2016) The value of Google Earth™ for erosion mapping. Catena 143:123–127
18. Benda L, Miller D, Barquín J, McCleary R, Cai T, Ji Y (2015) Building virtual watersheds: a global opportunity to strengthen resource management and conservation. Environ Manag 57(3):722–739. https://doi.org/10.1007/s00267-015-0634-6

Do Molluscs Assemblages Reflect River Typology: A Case Study of Montenegro

Maja Raković, Momir Paunović, Jelena Tomović, Nataša Popović, Béla Csányi, Milica Jovanović, Peter Glöer, and Vladimir Pešić

Contents

Abstract Molluscs are diverse and abundant component of macroinvertebrate assemblages inhabiting different types of river ecosystems. According to the literature data, a total of 143 freshwater mollusc taxa have been recorded for freshwater ecosystems of Montenegro. Gastropoda was represented by 120 taxa from 14 families, while the Bivalvia was represented by 23 species from 4 families. The typology of river and lake water bodies of Montenegro has been carried out according to the EU Water Framework Directive (2000/60/EC). In this study we used a simple river typology of Montenegrin watercourses based on two descriptors, the river size and the altitude. Our preliminary result revealed that the obtained mollusc assemblage

M. Raković (✉), M. Paunović, J. Tomović, and N. Popović
Institute for Biological Research "Siniša Stanković", National Institute of Republic of Serbia, University of Belgrade, Belgrade, Serbia
e-mail: rakovic.maja@ibiss.bg.ac.rs; mpaunovi@ibiss.bg.ac.rs; jelena.tomovic@ibiss.bg.ac.rs; natasa.popovic@ibiss.bg.ac.rs

B. Csányi
Danube Research Institute, Göd, Hungary
e-mail: bela.csanyi@gmail.com

M. Jovanović and V. Pešić
Department of Biology, University of Montenegro, Podgorica, Montenegro
e-mail: vladopesic@gmail.com

P. Glöer
Biodiversity Research Lab, Hetlingen, Germany
e-mail: gloeer@malaco.de

Vladimir Pešić, Momir Paunović, and Andrey G. Kostianoy (eds.), *The Rivers of Montenegro*, Hdb Env Chem (2020) 93: 265–286, DOI 10.1007/698_2020_487,

clusters satisfactorily describe two main river types, the lowland portions of the large rivers in the Adriatic Sea Basin and the large- and medium-sized rivers at the mid-altitude. For a more confident water body typology based on mollusc assemblages, additional studies are highly needed.

Keywords Adriatic watershed, Bivalves, Black Sea watershed, Gastropods, Typology

1 Introduction

Non-marine molluscs belong to the second most diverse animal phylum in terms of number of described species [1]. For some taxonomic groups and geographical regions, diversity may have been overestimated, but for most groups such as molluscs and areas such as the Dinaric region, diversity is probably seriously underestimated. The number of taxa is increasing annually, as more research is carried out, especially in regions (and habitats) that are less well known. Non-marine molluscs are very diverse, not only in size and shape but also in their life cycle and habitat. Within Europe, there are major geographic differences in the level of endemism and species diversity of non-marine molluscs. There is an increase in the species diversity from Northern to Southern Europe, as it increases the number of (sub)species currently recognized in Southern and Eastern Europe [1].

The Western Balkan region is among the world hot spots of biodiversity [2]. Especially the Balkan freshwater fauna has long been recognized as highly diverse with remarkable degrees of endemism in practically all major taxa [3].

Previous reviews suggested that mollusc may have high diversity across the Balkan region, especially in its Mediterranean part. The Balkan repeatedly served as a glacial refugium [4]. Presumably this led to speciation and high level of endemism as the results of a lack or limited gene flow of the populations that survived in these refugia [4].

Several studies have shown that abiotic factors, such as pH, electrical conductivity and concentration of calcium, can influence mollusc assemblages diversity at both the local and regional scale [5, 6]. In addition to these factors, a number of other drivers such the habitat type and climate can shape diversity and abundance of mollusc assemblages [5–7].

Montenegro belongs hydrographically to the Black Sea and Adriatic Basins. The total surface of the Black Sea Basin in Montenegro is 7,260 km^2 or 52.5% of state territory. The most important Montenegrin rivers of the Black Sea Basin are the Tara (Fig. 1), the Piva, the Lim and the Ćehotina Rivers as part of the Drina River Basin (flowing into the Sava River), while the Ibar River is flowing into the Zapadna Morava. The Danube watershed placed in Montenegro is the southernmost part of the Black Sea drainage. On the other hand, the total surface of the Adriatic watershed

Fig. 1 Photographs of selected rivers (**a**) the Bijela, (**b**) the Rijeka Crnojevića, (**c**) the Tara, (**d**) Ravnjak, (**e**) the lower part of Morača River near Vukovci and (**f**) the Zeta River. Photo: (**a–c**) M. Raković; (**d–f**) V. Pešić

in Montenegro is 6,560 km^2 or 47.8% of state territory. The Zeta, the Morača (Fig. 1) and the Bojana Rivers together with the Skadar Lake form the main part of the Adriatic River Basin, flowing into the Adriatic Sea.

Small mountain streams and medium mid-altitude rivers dominate in the hydrological network of Montenegro (Fig. 2). Average length of river water body is 13.8 km, with the shortest one 1.3 km and longest 68.4 km.

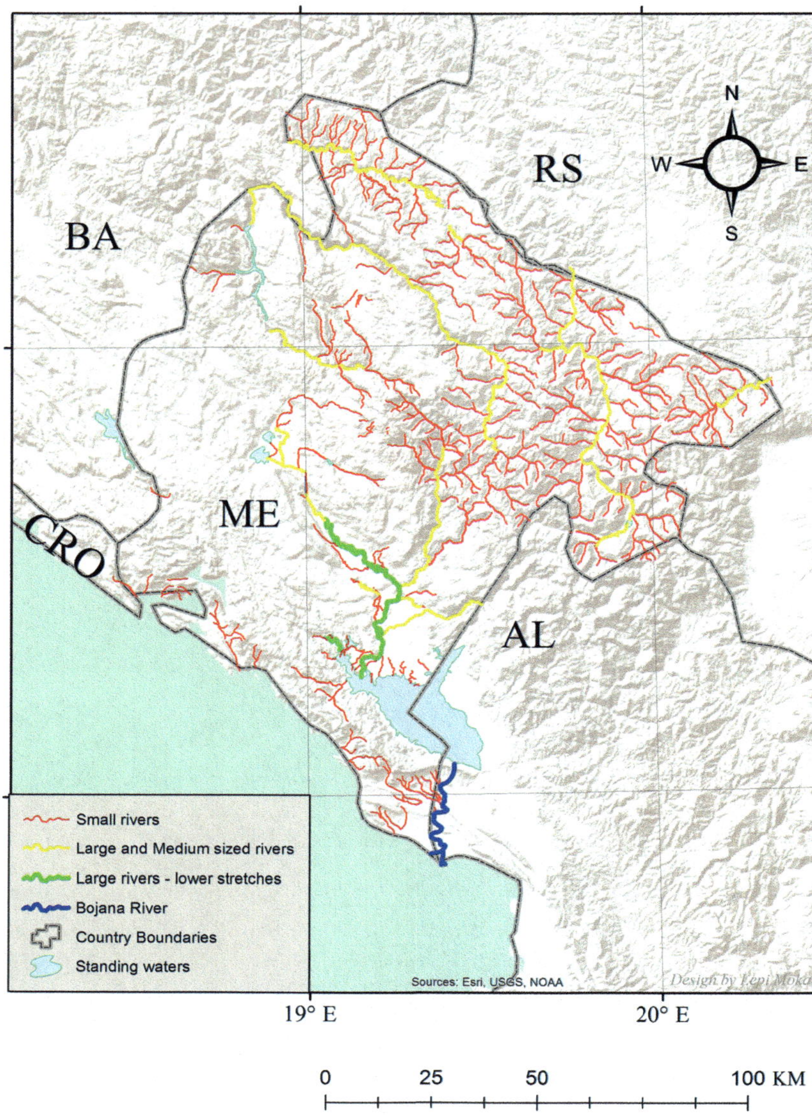

Fig. 2 Typology of river water bodies in Montenegro (Sources: Esri, USGS, NOAA)

The geographic range of Montenegro includes three freshwater ecoregions (defined by the Freshwater Ecoregions of the World project [8]): Dniester-Lower Danube, Dalmatia and Southeast Adriatic Drainages (Fig. 3). Recently Pešić and colleagues [9] redefined the FEOW ecoregional boundaries in Montenegro and moved the eastern border of the Dalmatian Ecoregion west of the River Sutorina.

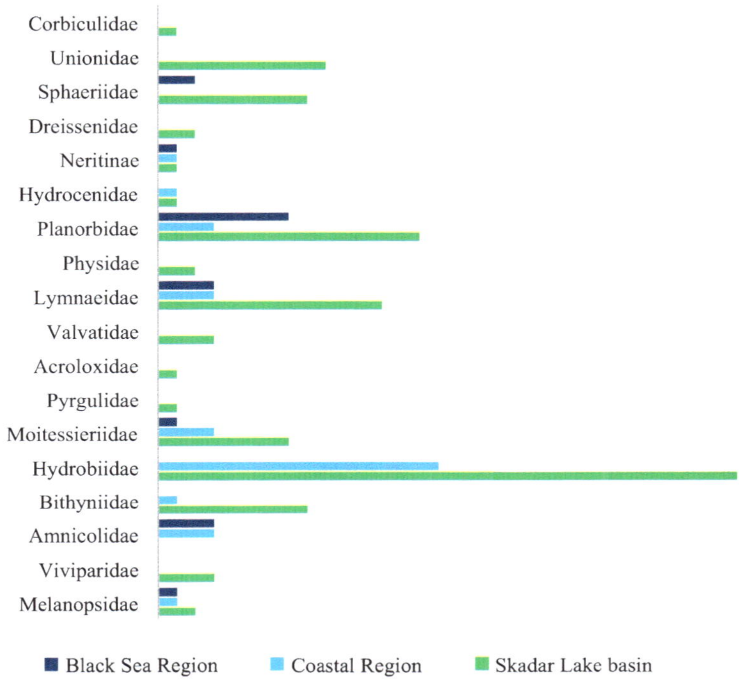

Fig. 3 Percentage share families of freshwater molluscs fauna from the three regions in the Montenegro

It is known that small (local) river basins can harbour this high regional diversity due to complex geological history and diverse habitats on local scales. For example recent studies revealed that the diversity of non-marine gastropods of the Lake Skadar Basin belongs to the richest in Europe [10]. Long-term permanency of a habitat is believed to be a precondition for the spreading of many taxa after in situ evolution [11].

2 Freshwater Molluscs of Montenegro

The overview of freshwater molluscs of Montenegro has been carried out based on the last checklist given for freshwater gastropods by Božana Karaman [12] and papers on published thereafter [12–30], as well as our recent investigation in 2019. A total of 143 freshwater mollusc taxa belonging to 18 families were recorded (Table 1). Gastropoda was represented by 120 species from 14 families, while Bivalves represented by 23 species from 4 families.

The Skadar Lake Basin has a higher number of mollusc taxa (69% of the total number) in comparison with the coastal region (20%) and the Black Sea Region

Table 1 List of freshwater molluscs of Montenegro

Kingdom*:* Animalia	*I*	*II*	*III*	*A*	*B*	*C*	*D*
Phylum*:* Mollusca							
Class*:* Gastropoda							
Family: Melanopsidae							
Amphimelania holandrii (Pfeiffer, 1828)	+	+	+	+	+	+	
Esperiana esperi (Férussac, 1823)							
Microcolpia daudebartii acicularis (Férussac, 1823)	+				+		
Family: Viviparidae							
Viviparus acerosus Bourguignat, 1862	+			+			
Viviparus contectus (Millet, 1813)	+			+			
Viviparus mamillatus (Küster, 1852)	+			+	+	+	
Viviparus viviparus (Linnæus, 1758)							
Family: Amnicolidae							
*_Bythinella dispersa_ Radoman, 1973			+			+	
*_Bythinella luteola_ Radoman, 1973			+			+	
*_Bythinella taraensis_ Glöer & Pešić, 2010			+			+	
*_Emmericia expansilabris_ Bourguignat, 1880		+				+	
*_Emmericia patula_ (Brumati, 1938)		+				+	
Emmericia ventricosa Brusina, 1870		+				+	
Family: Bithyniidae							
Bithynia hambergerae A. Reischütz, N. Reischütz & P. L. Reischütz, 2008	+				+		
Bithynia leachii (Sheppard, 1823)							
Bithynia majewskyi Frauenfeld, 1862	+					+	
Bithynia mostarensis Möllendorf, 1873	+				+		
Bithynia montenegrina (Wohlberedt, 1901)	+				+		
Bithynia radomani Glöer & Pešić, 2007	+			+	+	+	
Bithynia skadarskii Glöer & Pešić, 2007	+			+		+	
Bithynia tentaculata (Linnaeus, 1758)	+			+	+		
Bithynia zeta Glöer & Pešić, 2007	+	+		+		+	
Family: Hydrobiidae							
Adriohydrobia gagatinella (Küster, 1852)	+	+			+	+	
*_Antibaria notata_ (Frauenfeld, 1865)	+	+				+	
*_Belgrandiella bumasta_ Schüt, 1960							
**_Bracenica spiridoni_ Radoman, 1973	+					+	+
**_Bracenica vitojaensis_ Glöer, Grego, Erőss & Fehér, 2015	+					+	+
**_Hauffenia plana_ Bole, 1961		+				+	
Hydrobia cattaroensis (Westerlund, 1886)		+				+	
Hydrobia declinata (Frauenfeld, 1863)	+	+			+	+	
Hydrobia ventrosa (Montagu, 1803)	+					+	
*_Litthabitella chilodia_ (Westerlund, 1886)	+	+				+	+
* _Karucia sublacustrina_ Glöer & Pešić, 2013	+					+	
**_Plagigeyeria feheri_ Grego & Glöer, 2019		+					+
**_Plagigeyeria gladilini gladilini_ Kuščer, 1936	+					+	+

(continued)

Table 1 (continued)

Kingdom: Animalia	*I*	*II*	*III*	*A*	*B*	*C*	*D*
**Plagigeyeria montenigrina* Bole, 1961	+					+	+
**Plagigeyeria lukai* Glöer & Pešić, 2014	+					+	+
**Plagigeyeria mostarensis mostarensis* Kuščer, 1933	+					+	
**Plagigeyeria robusta robusta* Schütt, 1959							
**Plagigeyeria tribunicae* Schütt, 1963							
**Plagigeyeria zetaprotogona zetaprotogona* Scütt, 1960	+	+				+	+
**Plagigeyeria zetaprotogona pageti* Schütt, 1961	+	+				+	+
**Plagigeyeria zetaprotogona vitoja* A.Reischütz & P.Reischütz, 2008	+					+	+
**Plagigeyeria zetaprotogona zetadidyma* Schütt, 1960	+					+	+
**Plagigeyeria zetaprotogona zetatridyma* Schütt, 1960	+					+	+
Pseudamnicola conovula (Frauenfeld, 1863)		+				+	
Pseudamnicola orsinii (Küster, 1852)		+				+	
**Pseudamnicola troglobia* Bole, 1961							
Radomaniola curta curta (Küster, 1852)	+	+			+	+	+
**Radomaniola curta anagastica* (Radoman, 1973)	+					+	
Radomaniola curta pivensis (Radoman, 1973)			+		+	+	
**Radomaniola elongata* (Radoman, 1973)	+			+		+	
Radomaniola lacustris (Radoman, 1983)	+			+		+	
* *Radomaniola montana* (Radoman, 1973)	+	+				+	
**Stygobium hercegnoviensis* Grego & Glöer, 2019		+				+	+
**Vinodolia (Anagastina) gluhodolica* (Radoman, 1973)	+					+	+
**Vinodolia (Anagastina) hadouphylax* (Schütt, 1959)							
**Vinodolia (Anagastina) matjasici* (Bole, 1961)	+					+	+
Vinodolia (Anagastina) scutarica (Radoman, 1973)	+			+		+	
* *Vinodolia (Anagastina) vidrovani* (Radoman, 1973)	+					+	
Vinodolia (Anagastina) zetaevallis (Radoman, 1973)	+				+	+	
**Islamia montenegrina* Glöer, Grego, Erőss & Fehér, 2015	+					+	
**Iverakia hausdorfi* Glöer & Pešić, 2013	+					+	+
**Bythiospeum demattiai* Glöer & Pešić, 2013	+					+	+
**Arganiella tabanensis* Boeters, Glöer & Pešić, 2014	+					+	+
**Zeteana ljiljanae* Glöer & *Pešić*, 2014	+					+	+
Family: Moitessieriidae							
** *Bosnidilhia vitojaensis* Grego & Glöer, 2019	+					+	+
**Iglica absoloni* (A.J. Wagner, 1914)	+					+	+
**Montenegrospeum bogici* Pešić & Glöer, 2012	+					+	+
**Paladilhiopsis cattaroensis* Grego & Glöer, 2019		+				+	+
**Paladilhiopsis tarae* Bole & Velkovrh, 1987			+			+	+
**Paladilhiopsis (Lanzaia) matejkoi* Grego & Glöer, 2019		+				+	+
**Paladilhiopsis (Lanzaia) pesici* Glöer, Grego, Erőss & Fehér, 2015	+					+	+
**Paladilhiopsis (Lanzaia) vjetrenicae vjetrenicae* Kuščer, 1933	+					+	+
**Paladilhiopsis (Saxurinator) hadzii* (Bole, 1961)	+					+	+

(continued)

Table 1 (continued)

Kingdom: Animalia	*I*	*II*	*III*	*A*	*B*	*C*	*D*
*******Paladilhiopsis (Saxurinator) montenegrinus* Schütt, 1959		+					+
*******Paladilhiopsis (Saxurinator) schlickumi* Schütt, 1960							
*******Paladilhiopsis (Saxurinator) orthodoxus* Schütt, 1960	+					+	+
Family: Pyrgulidae							
Pyrgula annulata annulata (Linnaeus, 1758)	+			+		+	
Family: Acroloxidae							
Acroloxus lacustris (Linnaeus, 1758)	+			+			
Family: Valvatidae							
Valvata montenegrina Glöer & Pešić, 2007	+			+	+	+	
Valvata piscinalis (Müller, 1774)	+			+	+		
Valvata cristata Müller, 1774	+			+	+		
Family: Lymnaeidae							
Galba truncatula (Müller, 1774)	+	+	+	+	+	+	
Lymnaea raphidia (Bourguignat, 1859)	+			+			
Lymnaea stagnalis (Linnaeus, 1758)	+	+		+	+		
Omphiscola glabra (O.F. Müller, 1774)	+			+			
Radix ampla (Hartmann, 1821)	+			+	+		
Radix auricularia (Linnaeus, 1758)	+		+	+	+		
Radix balthica (Linnæus, 1758)	+			+	+		
Radix labiata (Rossmässler, 1835)	+	+	+	+	+		
Radix lagotis (Schrank, 1803)							
Radix skutaris Glöer & Pešić, 2007	+			+			
Stagnicola corvus (Gmelin, 1791)	+			+			
Stagnicola montenegrinus Glöer & Pešić, 2009	+			+	+		
Stagnicola palustris (Müller, 1774)	+			+	+	+	
Family: Physidae							
Haitia acuta (Draparnaud, 1805)	+			+			
Physa fontinalis (Linnæus, 1758)	+			+			
Family: Planorbidae							
Ancylus fluviatilis Müller, 1774	+		+	+	+	+	
Ancylus recurvus Martens, 1873	+		+		+	+	
Anisus spirorbis (Linnaeus, 1758)			+	+			
Anisus vortex (Linnæus, 1758)	+	+		+	+		
Bathyomphalus contortus (Linnæus, 1758)			+		+		
Ferrissia wautieri (Mirolli 1960)	+			+			
Gyraulus crista (Linnæus, 1758)			+	+			
Gyraulus albus (Müller, 1774)	+		+	+			
Gyraulus ioanis Glöer & Pešić, 2008	+			+			
Gyraulus meierbrooki Glöer & Pešić, 2007	+			+			
Gyraulus piscinarum (Bourguignat, 1852)	+			+			
Gyraulus shasi Glöer & Pešić, 2009	+			+			
Gyraulus laevis (Alder, 1838)							

(continued)

Table 1 (continued)

Kingdom*:* Animalia	*I*	*II*	*III*	*A*	*B*	*C*	*D*
Hippeutis complanatus Linnaeus, 1758		+	+	+	+		
Planorbarius corneus (Linnæus, 1758)	+			+	+		
Planorbis carinatus Müller, 1774	+			+	+		
Planorbis planorbis (Linnæus, 1758)	+	+		+	+		
Planorbis vitojensis Glöer & Pešić, 2010	+			+			
Segmentina nitida (Müller, 1774)	+			+			
Family: Hydrocenidae							
Hydrocena cattaroensis (Pfeiffer, 1841)	+	+				+	
Family: Neritinae							
Theodoxus danubialis (Pfeiffer, 1828)							
Theodoxus fluviatilis (Linnæus, 1758)	+	+	+	+	+	+	
Class: Bivalvia							
Family: Dreissenidae							
Dreissena polymorpha (Pallas, 1771)	+			+			
Dreissena carinata (Dunker, 1853)	+			+	+		
Dreissena blanci Westerlund, 1890	+			+			
Family: Sphaeriidae							
Pisidium casertanum (Poli, 1791)	+		+		+	+	
Pisidium moitessierianum Paladilhe, 1866	+				+		
Pisidium amnicum (Müller, 1774)	+				+		
Pisidium henslowanum (Sheppard, 1823)	+				+		
Pisidium personatum Malm, 1855			+			+	
Pisidium subtruncatum Malm, 1855	+				+		
Pisidium nitidum Jenyns, 1832	+				+		
Sphaerium (*Musculium*) *lacustre* (Müller, 1774)	+			+	+		
Sphaerium corneum (Linnaeus, 1758)	+						
Sphaerium rivicola (Lamarck, 1818)	+			+	+		
Family: Unionidae							
Sinanodonta woodiana (Lea, 1834)	+						
Anodonta cygnea (Linnaeus, 1758)	+			+	+		
Anodonta anatina (Linnaeus, 1758)	+			+	+		
Microcondylaea bonellii (Ferussac, 1827)	+			+	+		
Pseudanodonta complanata Rossmossler, 1835	+			+	+		
Unio tumidus Retzius, 1788	+						
Unio crassus Philipsson, 1788	+						
Unio pictorum (Linnaeus, 1758)	+			+	+		
Unio mancus Lamarck, 1819	+			+	+		
Family: Corbiculidae							
Corbicula fluminea (Müller, 1774)	+			+	+		

Crenobiotic taxa are marked with one asterisk, troglobiotic taxa with two asterisk (I – Skadar Lake Basin, II – coastal region, III – Black Sea Region, A – standing waters (lakes and wetland), B – rivers, C– springs and D – underground waters)

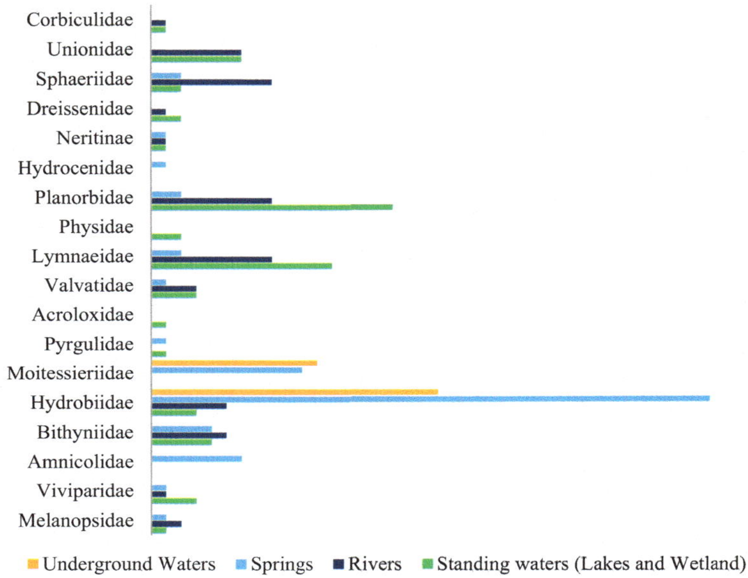

Fig. 4 Percentage share of freshwater mollusc diversity per different habitats in Montenegro

which harbour just 11% of the total number of freshwater molluscs known from Montenegro. The Skadar Lake Basin includes a large variety of water bodies, including lotic waters, pools and springs that has been for a long time in the malacological focus. An overview of the mollusc diversity of the latter basin recently has been given by Pešić and Glöer [10].

An analysis of diversity and endemism of molluscs inhabiting freshwater ecosystems of Montenegro revealed that hydrobiid snails dominate in richness (Fig. 3). The highest diversity and endemism was recorded in the Skadar Lake Basin. Most of the species of this family inhabit springs or underground waters [26]. Access to troglobitic species that inhabit underground layer is often difficult and is most often found in springs where they were washed and brought together. Around 82% of the hydrobioid species that inhabit the Lake Skadar catchment area are known only from type localities, while 33% are known only from empty shells [26].

An analysis of mollusc diversity per habitat revealed that highest number of recorded species inhabits springs (69 spp.) and standing (58 spp.) waters, followed by rivers (49 spp.). The lower number inhabits underground waters. The highest diversity and endemism was recorded in the Skadar Lake Basin. As it can be seen from Fig. 4 – the hydrobiid snails (family Hydrobiidae) is most diverse mollusc family in the springs and underground habitats. The family Moitessieriidae which includes troglobiotic species was reported only from the latter two habitats.

The most diverse mollusc families in standing water habitats (lakes and wetlands) are Planorbidea and Lymnaeidae. These two families are also frequent in rivers, especially in the lowland rivers (the Zeta and the Bojana/Buna) and the lower part of

Fig. 5 (**a**) Native *Dreissena carinata* attached on nonindigenous invasive species *Sinanodonta woodiana* (Lea, 1834) – specimens collected in the Skadar Lake, Montenegro. Photo: M. Raković; (**b**) *Dreissena polymorpha* in Šasko Lake. Photo: D. Molloy

the rivers near Skadar Lake (Morača River and the Skadar Lake tributaries such as Rijeka Crnojevića, Plavnica and others).

Molluscs from the Sphaeriidae family inhabit all types of stagnant and flowing waters and are among the most common representatives of freshwater invertebrates in Montenegro [31]. Such a widespread occurrence is probably due to their extreme ability to adapt to different habitat conditions but also to a specific life cycle.

Aquatic biotopes are ecosystems most predisposed to bioinvasions. Increasing colonization by allochthonous organisms has been observed during the last few decades due to their unique features [32].

Some species of freshwater mussels, such as *Sinanodonta woodiana* (Lea, 1834), *Dreissena polymorpha* (Pallas, 1771) and *Corbicula fluminea* (Müller, 1774), were found to be among the most prominent invaders of freshwater ecosystems [32–34]. During recent investigation conducted in 2019, the presence of these nonindigenous invasive species in the waters of Montenegro was confirmed (Fig. 5). The invasive *D. polymorpha* is only known from Lake Šasko and Bojana/Buna River [10, 26]. Pešić and Glöer [10] stated that there is no sign of the further expansion of *D. polymorpha* in the area populated by the native *D. carinata*, but monitoring of this highly invasive species is recommended.

Montenegro, especially the Skadar Lake Basin, can be considered as a biodiversity hotspot of worldwide importance as confirmed by recent studies that even very small springs or overlooked water outlets could host interesting, so far unrecognized, species [18, 19, 27]. The number of endemic species from the Lake Skadar Basin has already been included on the IUCN Red List of Threatened Species [1]. In total 34 gastropod species are listed by IUCN, and 7 of which are protected in Montenegro by national legislation [19].

During the recent investigation, the presence of *Microcondylaea bonellii* (Ferussac, 1827) was also detected. Nowadays this mussel became extremely rare and lives only in rivers along the Adriatic coast (Fig. 6).

Protection of these endemic and/or rare species and appropriate action plans should be based on reliable studies. Often the lack of recent finding results that

Fig. 6 *Microcondylaea bonellii* (Ferussac, 1827) mussels found in the Zeta River. Photo by B. Csányi

many species have been assumed to be critically endangered or even extinct (such as *Bracenica spiridoni* was thought to be extinct [35], but recently is rediscovered [20]).

Recent studies suggest that most, if not all, existing groups of the aquatic snails in the Balkan region show cryptic speciation (e.g. [16, 17]). Pešić and Glöer [10] stressed distinctiveness of many species that inhabits the Skadar Lake Basin require using molecular techniques.

3 Mollusc Assemblages as Indicator of Rivers Typology of Montenegro

Our investigation was performed from June to August 2019 with the aim to examine community composition and spatial distribution of freshwater molluscs in Montenegrin watercourses. Samples were collected at 15 sites in June, July and August 2019 (Table 2). Samples were taken by using FBA hand net (mesh size 250–500 μm; Kick and Sweep technique). Material are collected by the sixth (M.J.) and the last author (V.P). For preservation of biological samples, 96% ethanol was used. The collected molluscs were identified to the species level by Vladimir Pešić.

Table 2 Qualitative composition of freshwater molluscs at sampling sites in Montenegro from June to August 2019

Taxa/sites	R. Zeta – Brezovik (Nikšić)	R.Zeta – Danilovgrad	R.Zeta-Vranjske Njive	R. Zeta-Spuž	R-Morača_ Vukovci	R. Morača_ Bioče	R. Bojana_ Fraskanjel	R. Piva	Rijeka Crnojevića_ upper	Rijeka Crnojeviča_ middle	R. Cijevna_ Trgaja	R. Ćehotina_ Pljevlja	R.Tara_ Trebaljevo	R. Ibar_Bać	R. Mala rijeka
Family: Valvatidae															
Valvata montenegrina			+		+					+					
Family: Melanopsidae															
Amphimelania holandrii			+	+			+				+				
Family: Bithyniidae															
Bithynia radomani	+	+	+	+	+		+			+					
Family: Viviparidae															
Viviparus mamilatus		+	+							+					
Family: Neritinae															
Theodoxus fluviatilis		+	+	+	+	+	+		+	+	+			+	
Family: Planorbidae															
Planorbarius corneus										+					
Ancylus recurvus	+				+			+	+	+		+			+
Family: Lymnaeidae															
Stagnicola palustris										+					
Galba truncatula							+	+							
Radix ampla	+		+	+											
Radix labiata													+		
Radix auricularia/ scutaris							+	+		+					

(continued)

Table 2 (continued)

Taxa/sites	R. Zeta – Brezovik (Nikšić)	R.Zeta – Danilovgrad	R.Zeta-Vranjske Njive	R. Zeta-Spuž	R-Morača_ Vukovci	R. Morača_ Bioče	R. Bojana_ Fraskanjel	R. Piva	Rijeka Crnojevića_ upper	Rijeka Crnojeviča_ middle	R. Cijevna_ Trgaja	R. Ćehotina_ Pljevlja	R.Tara_ Trebaljevo	R. Ibar_Bać	R. Mala rijeka
Family: Physidae															
Haitia acuta							+								
Family: Hydrobiidae															
Vinodolia scutarica							+								
Radomaniola curta curta		+	+				+		+						
Family: Unionidae															
Unio mancus			+	+											
Family: Dreissenidae															
Dreissena carinata				+											
Dreissena polymorpha							+								
Family: Sphaeriidae															
Pisidium sp.			+				+								

A total of 19 freshwater molluscs taxa were recorded (Table 2). The Neritidae and Planorbidae species which had the highest occurrence frequencies in the investigated rivers of Montenegro were *Theodoxus fluviatilis* ($F = 0.70$) and *Ancylus recurvus* ($F = 0.47$), while the Hydrobiid snails – *Radomaniola curta* – was found to be the most dominant component of the molluscs community with 1893 individuals.

The smallest number of species (only one species per site) was recorded at six sites – *R. labiata* (Tara River near Trebaljevo), *Th. fluviatilis* (Ibar River near Bać and Morača River near Bioče) and *A. recurvus* (Ćehotina River after Pljevlja, Vezišnica near Pljevlja and Mala rijeka River). The Zeta River near Podgorica was the site with the highest taxa diversity (nine species – 53%).

Relationships between the analysed sites, derived from the obtained distance matrix using the single linkage clustering method for freshwater molluscs, are shown in Fig. 7. The UPGMA cluster diagram for freshwater molluscs of Montenegro revealed the occurrence of multiple clusters.

The first cluster was formed by site – Bojana River, near Fraskanjel. The second one included other 14 sites, which are split into several smaller clusters. Within the second cluster, there are a groupings rivers of the Black Sea Basin, which are characterized by a lower diversity of mollusc species (the Ćehotina, the Tara, the Piva and the Ibar Rivers), followed by grouping of middle and lower parts of large rivers such as the Rijeka Crnojevića, the Zeta and the Morača Rivers.

Recent 2019 study confirmed typology of water bodies in Montenegro based on molluscs assemblages. Thus, large lowland rivers of the Adriatic Basin (the Bojana River and the most downstream stretch of the Morača River) are characterized by diverse and abundant freshwater mussels' assemblages, together with typical snail communities – families Bithyniidae (genus *Bithynia*), Viviparidae (*Viviparus mamillatus*) and Valvatidae (*Valvata montenegrina*) which are not found in the rivers belonging the Black Sea Basin. The large- and medium-sized rivers at the mid-altitude are characterized by more uniform and less diverse mollusc assemblages. *Ancylus recurvus* and *Th. fluviatilis* (Fig. 8d–e) were the most frequently collected species. The typological complexity of Montenegrin freshwater mollusc fauna is additionally illustrated by the distribution of frequent species, e.g. *Th. fluviatilis*, but also by distribution of endemic taxa, e.g. endemic species of the genus *Bithynia*.

4 Discussion

In this Chapter we explored spatial patterns of mollusc assemblages of selected rivers and their basins in Montenegro aimed to determine main drivers that shape its diversity at local and regional scale. We use a simple river typology of Montenegrin watercourses based on two descriptors, size and altitude. Our preliminary result revealed that the obtained mollusc assemblage clusters well describe two main river types, the lowland portions of the large rivers in the Adriatic Sea Basin and the large- and medium-sized rivers at the mid-altitude (Fig. 7).

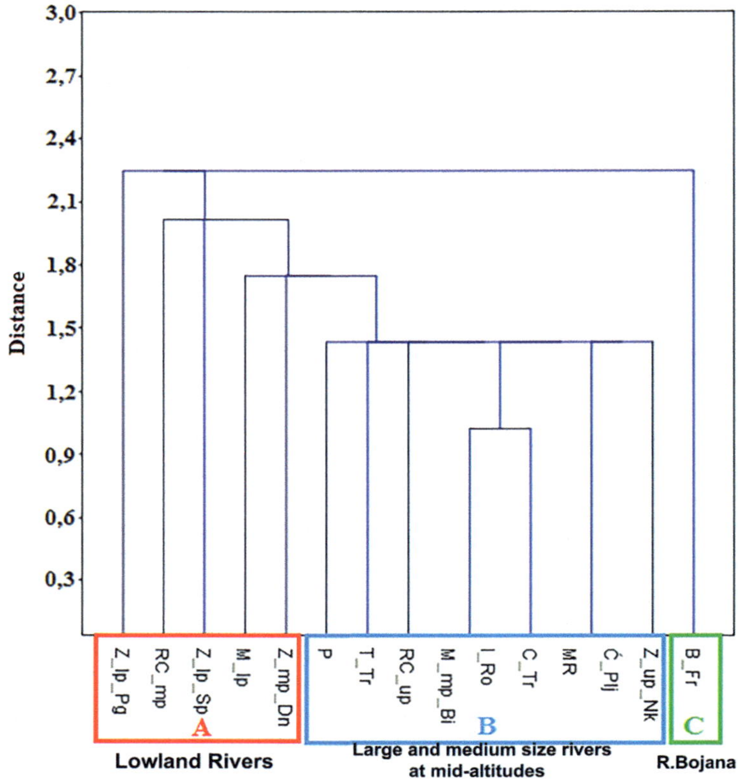

Fig. 7 UPGMA cluster diagram of 15 sites of Montenegro in the period from June to August 2019 based on Single Linkage Euclidean distances for freshwater molluscs. Sites abbreviations: *Z_mp_Dn*. Zeta River, middle part, near Danilovgrad, *Z_lp_Pg* Zeta River, lower part, Vranjske Njive, *Z_up_Nk* Zeta River, upper part, near Brezovik (Nikšić), *Z_lp_Sp* Zeta River, lower part, near Spuž, *B_Fr* Bojana River near Fraskanjel, *C_Tr* Cijevna River, near Trgaja, *RC_up* Rijeka Crnojevića River, upper part, *RC_mp* Rijeka Crnojevića River, middle part, *Ć_Plj* Ćehotina River, after Pljevlja, *T_Tr* Tara River, near Trebaljevo, *I_Ro* Ibar River, Bać, *M_mp_Bi* Morača River, middle part, near Bioče, *M_lp* Morača River, lower part, *MR* Mala Rijeka River, *P_br* Piva River

Moreover the results of our study revealed difference between the Adriatic and Black Sea Basins, both in terms of species composition and species richness. Based on a limited number of investigated sites, the estimated local mollusc diversity for Black Sea Basin was one to two species per site (in total 19 species), and it was lower than for sites across the Adriatic Sea Basin which varies from two to eight species per site (in total 31 species).

Our results suggest that climate might have substantial effects on mollusc assemblages. In the continental region, molluscs occur at low densities and low species richness. Moreover, our data confirm that endemism of mollusc and bio-diversity within the more continental Black Sea Basin is considerably lower than within the

Fig. 8 Photographs of some characteristic representatives of the mollusc assemblages of studies rivers: (**a**) *Viviparus mamillatus* (Küster, 1852), (**b**) *Bithynia radomani* Glöer & Pešić, 2007, (**c**) *Valvata montenegrina* Glöer & Pešić, 2007, (**d**) *Ancylus recurvus* Martens, 1873, (**e**) *Theodoxus fluviatilis* (Linnæus, 1758), (**f**) *Radix labiata* (Rossmässler, 1835). Photographs: P. Glöer

Mediterranean and sub-Mediterranean region of the country that belongs to Adriatic Sea Basin (Table 1).

The studied rivers in the Black Sea Basin are characterized by a relatively uniform fauna with the presence of a small number of species. *A. recurvus* was the most commonly found species. The ecology of this species, as well as other species in the Black Sea Basin, has been poorly researched and lacks ecological studies of abundance. The monthly monitoring of the Tara River conducted in 2019 at three localities, i.e. the mouth of the Drcka River, Tara near Mateševo and Tara near Trebaljevo, revealed the finding of *R. labiata* at the latter site. *R. labiata* occurred here from July to October, reaching maximum abundance of 275 ind/m^2 in August.

The results of our study revealed the lack of some species (and families) of characteristic rivers in the Adriatic Basin. This is especially true of the species *V. mamillatus*, as well as species of the genus *Bithynia* that are absent in the Black Sea Basin of Montenegro. Moreover, Hydrobiidae and Moitessieriidae are found to be the principal components of springs and groundwater habitats (Fig. 6) in Adriatic Sea basin. The families Planorbidae, Lymnaeidae and Physidae, as well as freshwater mussels (Unionidae), are typical inhabitants of standing waters and larger rivers in the region (Fig. 9).

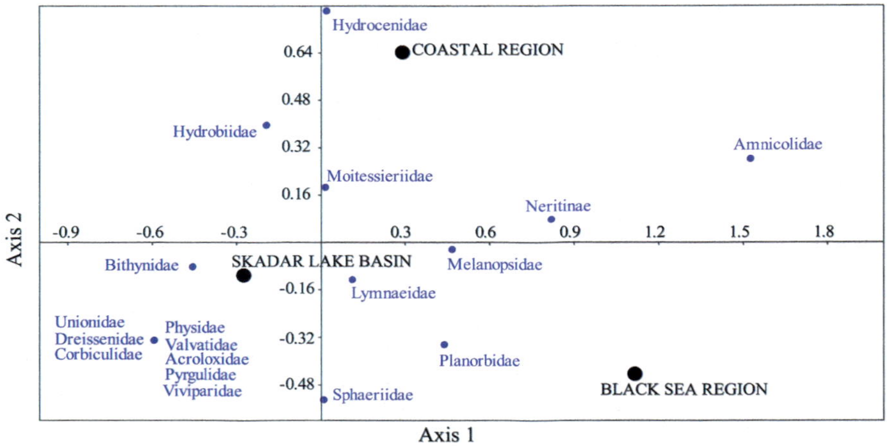

Fig. 9 CA plot of three regions and families in the space of the first two correspondence axes (Axis 1 and Axis 2) for freshwater molluscs of Montenegro

The results revealed that the Zeta and the Crnojevića Rijeka Rivers have the richest mollusc assemblage. The both rivers are lowland watercourses characterized by a large number of micro-habitats with domination of different bottom substrate – ranging from clay, mud and fine sand to gravel, pebbles and stone of different size. It is of particular interest that the middle and lower reaches of the Zeta River are dominated by the hydrobiid genus *Radomaniola* with *R. curta*. This species inhabits adjacent springs as well as small tributaries of the Zeta River, including those characterized by intermittent conditions (e.g. Rimanić stream; see [36]). This species seems to be sensitive to organic pollution. For example, at a microlocality located directly below the sewage outlet into the Zeta River (near Danilovgrad), the abundance of *Radomaniola* decreases while the abundance of filter-feeding *V. mamillatus* increases. Apart from *R. curta*, the Zeta River harbours rich populations of *Bithynia radomani* and *V. mamillatus*, especially in lentic parts of the river on the muddy substratum. In the upper part of the Crnojevića Rijeka River, *R. curta* and *Th. fluviatilis* dominated, while in the lower part of the river, filter-feeding *Viviparus* and lymaneids dominate.

A rich gastropod community is present in the lower part of the Morača River, below the Podgorica City. In the lotic part, *A. recurvus* (distribution of these species in Montenegro is unclear as in the past it is often confused with *A. fluviatilis*; thus all records of *A. fluviatilis* need confirmation) predominated and a less numerous *Th. fluviatilis*. The lentic part (see Fig. 1e) is dominated by *B. radomani*, with the significant presence of *Th. fluviatilis*.

The mollusc richness on the sites along the rivers in the region of the Montenegrin Coast (with exception of the Bojana/Buna River) was much lower than in the Skadar Lake Basin, or even Black Sea Basin. Likely this is due to intermittent nature of many of the watercourses (see [36] for discussion) which dry up in the summer. The only exception is the Bojana/Buna River which harbours a rich mollusc fauna.

In conclusion, our still preliminary study revealed that mollusc assemblages are an effective typological descriptor for analyses on regional level. This approach was previously confirmed for long sections of large lowland rivers, e.g. the Danube [37]. However, to provide more confident water body typology based on Molluscs, additional data is needed.

Acknowledgements The part of the work is supported by the Ministry of Education, Science and Technological Development of the Republic of Serbia. We thank Daniel Molloy (USA) for permission to use his photograph of *Dreissena polymorpha* from the Šasko Lake.

References

1. Cuttelod A, Seddon M, Neubert E (2011) European red list of non-marine molluscs. Publications Office of the European Union, Luxembourg, p 97
2. Griffith HI, Kryštufek B, Reed JM (2004) Balkan biodiversity. Kluwer Academic Publishers, Dordrecht, p 358
3. Bănărescu PM (2004) Distribution pattern of the aquatic fauna of the Balkan Peninsula. In: Griffith HI et al (eds) Balkan biodiversity. Kluwer Academic Publishers, Dordrecht, pp 203–217
4. Hewitt GM (1996) Some genetic consequences of ice ages, and their role in divergence and speciation. Biol J Linn Soc 58:247–276
5. Horsák M, Hájek M, Díte D, Tichý L (2007) Modern distribution patterns of snails and plants in the western Carpathian spring fens: is it a result of historical development? J Molluscan Stud 73 (1):53–60
6. Sturm R (2007) Freshwater molluscs in mountain lakes in eastern Alps (Austria): relationship between environmental variables and lake colonization. J Limnol 66(2):160–169
7. Heino J, Muotka T (2006) Landscape position, local environmental factors, and the structure of molluscan assemblages of lakes. Landsc Ecol 21(4):499–507
8. FEOW (2008) Freshwater ecoregions of the world-conservation science program. The Nature Conservancy and World Wildlife Fund, Washington
9. Pešić V, Grabowski M, Hadžiablahović S, Marić D, Paunović M (2020) The biodiversity and biogeographical characteristics of the river basins of Montenegro. In: Pešić V, Paunović M, Kostianoy A (eds) The rivers of Montenegro. The handbook of environmental chemistry. Springer, Cham
10. Pešić V, Glöer P (2018) The diversity and conservation status of the molluscs of lake Skadar/Shkodra. In: Pešić V, Karaman G, Kostianoy A (eds) The Skadar/Shkodra lake environment. The handbook of environmental chemistry, vol 80. Springer, Cham, pp 295–310
11. Glöer P, Albrecht C, Wilke T (2010) Enigmatic distribution patterns of the Bithyniidae in the Balkan region (Gastropoda: Rissooidea). J Conchol 40(2):179–189
12. Karaman B (2014) Catalogue: Fauna of land and freshwater snails gastropoda (Mollusca) of Montenegro. Catalogues 9, The section of natural science, vol 8. Montenegrin Academy of Science and Arts, pp 1–406
13. Glöer P, Pešić V (2009) *Stagnicola montenegrinus* n. sp., a new species of Montenegro (Gastropoda: Lymnaeidae). Mollusca 27:53–56
14. Glöer P, Pešić V (2010) The Planorbis species of the Balkans with the description of *Planorbis vitojensis* n. sp. (Gastropoda: Planorbidae). J Conchol 40:249–257

15. Gloer P, Pešić V (2010) The freshwater snails of the genus *Bythinella* Moquin-Tandon (Gastropoda: Rissooidea: Hydrobiidae) from Montenegro. Arch Biol Sci Belgrade 62 (2):441–447
16. Falniowski A, Szarowska M, Glöer P, Pešić V, Georgiev D, Horsák M, Sirbu I (2012) Radiation in *Bythinella* Moquin-Tandon, 1856 (Mollusca: Gastropoda: Rissooidea) in the Balkans. Folia Malacol 20(1):1–10
17. Falniowski A, Szarowska M, Glöer P, Pešić V (2012) Molecules vs Morphology in the taxonomy of the Radomaniola/Grossuana group of Balkan Rissooidea (Mollusca, Caenogastropoda). J Conchol 41(1):19–36
18. Pešić V, Glöer P (2012) A new species of Bythiospeum Bourguignat, 1882 (Hydrobiidae, Gastropoda) from Montenegro. Biol Nyssana 3(1):17–20
19. Pešić V, Glöer P (2013) *Montenegrospeum*, a new genus of Hydrobiid snails (Gastropoda: Risooidea) from Montenegro. Acta Zool Bulg 64(4):565–566
20. Pešić V, Glöer P (2013) A new freshwater snail genus (Hydrobiidae, Gastropoda) from Montenegro, with a discussion on gastropod diversity and endemism in Skadar Lake. ZooKeys 281:69–90
21. Glöer P, Pešić V (2014) New subterranean freshwater gastropods of Montenegro (Mollusca: Gastropoda: Hydrobiidae). Ecol Monten 1(2):82–88
22. Glöer P, Pešić V (2014) Two new subterranean freshwater gastropods of Montenegro (Mollusca: Gastropoda: Hydrobiidae). Ecol Mont 1(4):244–248
23. Boeters H, Glöer P, Pešić V (2014) *Arganiella tabanensis* n. sp. from Montenegro (Mollusca: Gastropoda: Hydrobiidae). Ecol Mont 1(3):131–139
24. Glöer P, Pešić V (2015) The morphological plasticity of *Theodoxus fluviatilis* (Linnaeus, 1758) (Mollusca: Gastropoda: Neritidae). Ecol Mont 2(2):88–92
25. Glöer P, Grego J, Erőss ZP, Fehér Z (2015) New records of subterranean and spring molluscs (Gastropoda: Hydrobiidae) from Montenegro and Albania with the description of five new species. Ecol Mont 4:70–82
26. Pešić V, Gadawski P, Gligorović B, Glöer P, Grabowski M, Kovács T, Murányi D, Płóciennik M (2018) The diversity of the zoobenthos communities of the lake Skadar/Shkodra basin. In: Pešić V, Karaman G, Kostianoy A (eds) The Skadar/Shkodra lake environment. The handbook of environmental chemistry, vol 80. Springer, Cham, pp 255–293
27. Grego J, Glöer P, Falniowski A, Hofman S, Osikowski A (2019) New subterranean freshwater gastropod species from Montenegro (Mollusca, Gastropoda, Moitessieriidae, and Hydrobiidae). Ecol Mont 20:71–90
28. Pešić V, Hofman S, Rysiewska A, Osikowski A, Falniowski A (2019) Species distinctness of *Bithynia cettinensis* Clessin, 1887 and *B. zeta* Glöer et Pešić, 2007 (Caenogastropoda: Truncatelloidea). Folia Malacol 27:111–118
29. Rysiewska A, Hofman S, Osikowski A, Beran L, Pešić V, Falniowski A (2019) *Viviparus mamillatus* (Küster, 1852), and partial congruence between the morphology-, allozyme- and dna-based phylogeny in european viviparidae (Caenogastropoda: Architaenioglossa). Folia Malacol 27:43–51
30. Cvijetić-Burzanović K (2003) A contribution to the knowledge of the Bivalvia (Mollusca) of Bjelasica mt. in Montenegro, Serbia and Montenegro. Nat Monten 2:57–62
31. Pešić V (2007) On some Pisidium species (Bivalvia, Sphaeriidae) from river Zeta near Vranjske Njive (Podgorica, Montenegro). Glas Republ Zavoda Zašt Prirode Podgorica 29–30:171–173
32. Tomovic J, Zoric K, Simic V, Kostic M, Kljajic Z, Lajtner J, Paunovic M (2013) The first record of the Chinese pond mussel *Sinanodonta woodiana* (Lea, 1834) in Montenegro. Arch Biol Sci Belgrade 65(4):1525–1531
33. Paunović M, Csány B, Simić V, Stojanović B, Cakić P (2006) Distribution of *Anodonta* (Sinanodonta) *woodiana* (Lea, 1834) in inland waters of Serbia. Aquat Invasions 1:154–160
34. Lajtner J, Crnčan P (2011) Distribution of the invasive bivalve *Sinanodonta woodiana* (Lea, 1834) in Croatia. Aquat Invasions 6(1):119–124

35. Regnier C, Fontaine B, Bouchet P (2009) Not knowing, not recording, not listing: numerous unrecognized mollusk extinctions. Conserv Biol 23:1214–1221
36. Pešić V, Pavićević A, Savić A, Hadžiablahović S (2020) The intermittent rivers of South Montenegro: ecology and biomonitoring. In: Pešić V, Paunović M, Kostianoy A (eds) The rivers of Montenegro. The handbook of environmental chemistry. Springer, Cham
37. Tomović J, Paunović M, Atanacković A, Marković V, Gačić Z, Csányi B, Simić V (2014) Biotic typology of the Danube River based on distribution of mollusc fauna as revealed by the Second Joint Danube Survey (2007). Acta Zool Bulg 66:527–537

The Rivers of Montenegro: From Conflicts to Science-Based Management

Vladimir Pešić, Momir Paunović, Andrey G. Kostianoy, and Vesna Vukašinović-Pešić

Contents

Abstract This concluding chapter briefly describes the main conclusions presented in the book *The Rivers of Montenegro* which has 12 chapters aimed at providing an extensive overview of the various hydrological and environmental aspects of the rivers and their basins in Montenegro. The book emphasizes the current gaps, recent advances, and future prospects of our knowledge of aquatic ecosystems in Montenegro which in the last decade has become the focus of ecological, social, and economic changes. This has led to an increase in potential conflicts over the

---

V. Pešić (✉)
Faculty of Sciences, University of Montenegro, Podgorica, Montenegro
e-mail: vladopesic@gmail.com

M. Paunović
Institute for Biological Research "Siniša Stanković" - National Institute of Republic of Serbia, University of Belgrade, Belgrade, Serbia
e-mail: mpaunovi@ibiss.bg.ac.rs

A. G. Kostianoy
P.P. Shirshov Institute of Oceanology, Russian Academy of Sciences, Moscow, Russia

S.Yu. Witte Moscow University, Moscow, Russia
e-mail: kostianoy@gmail.com

V. Vukašinović-Pešić
Faculty of Metallurgy and Technology, University of Montenegro, Podgorica, Montenegro
e-mail: vesnav@ucg.ac.me

Vladimir Pešić, Momir Paunović, and Andrey G. Kostianoy (eds.), *The Rivers of Montenegro*, Hdb Env Chem (2020) 93: 287–302, DOI 10.1007/698_2020_480,

management of water resources. Conflict between biodiversity conservation on the one hand and general human activity on the other is most commonly recognized in the public. However, recent emerging lessons from the development of small hydropower plants in Montenegro revealed that conflicts are much more complex and can severely affect both socioeconomic and ecological parameters.

This book gives an ecologist's perspective on these conflicts through a consensus approach that could expedite water resource protection taking into account the interests of all involved parties including the aims of governments and local communities, as well as environmental and flood control concerns.

Keywords Environmental conflicts, Montenegro, River basin, Rivers, Water management, Water resources

1 Rivers as Water Resources of Montenegro

The Montenegro is extremely rich in water resources. Speaking in numbers – an average perennial specific module of runoff is estimated at around 43 L s^{-1} km^{-2}, while the total mean annual runoff from the territory of Montenegro is estimated at around 595 m^3 s^{-1} (with the total runoff volume of 18.75×10^9 m^3) [1]. The contribution of transit waters is estimated at 29 m^3 s^{-1} [2], meaning that waters formed on the territory of the country make up 95% of the total runoff from Montenegro [1].

The waters of Montenegro flow into the Danube River (and the Black Sea) or into the Adriatic Sea, with approximately 45.4% (6,268 km^2) of its surface belonging to the Adriatic and 54.6% (7,545 km^2) to the Black Sea (Danube) basin [1]. The rivers of the Black Sea basin (Fig. 1) are generally longer and characterized by more water than those of the Adriatic Sea basin. However, despite that it is larger, the Black Sea basin produces only about two thirds (7.855×10^6 m^3) of the total runoff volume of the Adriatic Sea basin (11.814×10^6 m^3) [1].

The spatial and temporal distribution of water resources is highly uneven and mainly determined by the morphological and hydrogeological characteristics of the karst that dominates the Montenegrin landscape [3]. The spatial arrangements of the mountain ranges in Montenegro that extend in the north-northwest direction are the major barriers to the surface runoff [3]. The water stability in the karstic terrain is mainly determined by the underground water drainage [3]. As the consequence of the karst flow regime, the contribution of the surface runoff to the river basin discharge varies greatly, both spatially and temporally, as well as between and within the river basins.

The Adriatic Sea basin, although richer in water, has a less developed river network. Western part of Montenegro, which is interestingly one of the areas

Fig. 1 The river Tara. The canyon of the river is protected as a part of the Durmitor National Park, and since 1977 it has been a UNESCO World Heritage Site. Photo by S. Popović

with the highest rainfall in Europe [1], almost lack surface water. Most of the atmospheric water here disappears into the permeable ground and flows through complex underground structures. Around one third of the Adriatic Sea basin is drained by subterranean routes, which has not yet been reliably investigated [1]. In the dry season, the discharges are more than a thousand times lower than the discharges in the rainy period, resulting in the most rivers and streams in the Adriatic Sea basin dry up in the summer months [1].

2 Hydropower Potential as a Source of Possible Conflicts

Benefits of hydropower, which gaining increasing importance in Montenegro as a developing country, are reflected in the fact that they provide low carbon, renewable energy and in the same time increase water storage capacity. Montenegro is officially committed to meet the target of the EU Renewable Energy Directive (2009/28/EC) to increase the use of renewable energy sources to 20% of total energy consumption by 2020. The approximate requirements for electricity supply in Montenegro vary, and according to the National Energy Strategy (NES), the energy consumption in 2021 is estimated at around 6.9 TWh per year [4].

Montenegro has an enormous amount of hydroelectric potential that is only partially utilized. It is estimated that the total amount of hydropower potential in Montenegro that could be utilized is within the range 5.6–6.1 TWh per year [5]. This estimate does not include small hydropower plants (SHPPs) whose potential is estimated to be about 0.45 TWh per year [6].

Recent studies (e.g., [5, 6]) have shown that there are significant differences in hydropower between and within river basins. The largest hydro-potentials have the Tara River (2,225 GWh/year), followed by the Zeta (2,007 GWh/year), the Morača (1,469 GWh/year), the Lim (1,438 GWh/year), and the Piva (1,361 GWh/year), while the potential of other watercourses is lower [5]. The existing hydropower facilities generate 1.6 TWh per year, and the installed power generation is currently set to 658 MW, which are only 25–27% of the real potential that could be utilized [5]. Montenegro has two large hydropower plants (LHP), i.e., HPP "Perućica" on the Zeta River and HPP "Piva" on the Piva River, and a larger number of SHPPs with the limits of the installed capacity set up to 10 MW. In Montenegro, 53 small hydropower plants are currently planned or under construction, while 18 SHPPs are active, of which 13 SHPPs are owned by 7 private hydropower operators and 5 SHPPs are owned by the national EPCG (Elektroprivreda Crne Gore) energy company [7].

The average production capacities of large- and medium-scale hydropower plants in Montenegro are estimated to be about 5,650 GWh per year [5]. On the other hand, the production of SHPPs is about 14 times less (about 417 GWh per year) indicating that LHP employment contributes significantly more to energy production, questioning the energy effectiveness of further SHP capacity development [5].

In the recent decades, the construction of SHPPs has been systematically stimulated by the government as a "desirable" alternative aimed to tap the renewable hydropower potential. Intensive small hydropower (SHP) development is justified by the reduced impacts of these facilities on the environment due to their smaller size, as well as less demanding land and infrastructure requirements. Last but not least, SHP development was stimulated by the state financial subsidies (fixed feed-in tariffs and feed-in premiums) to hydropower operators, which consumers paid through electricity bills, which was the source of a number of controversies in the public.

In recent years, it has become clear that massive deployment of small-scale hydropower projects can cause important environmental impacts (see, e.g., [3, 8]). The main environmental problem recognized with SHP development is that the most of small rivers and streams in Montenegro where plants have been built do not have enough water in the summer months to meet the prescribed minimum of ecological flow [3]. Most planned SHPPs work on the principle of reroute water through pipelines and tunnels bringing water directly to turbines, leaving often long stretches of river with disturbed natural flow (or even a dry). The river distance that left disturbed depends on the SHP plant size (Fig. 2) and the hydro-morphological characteristics of the site on which they were built. For example, the Šekular SHP plant uses a 4-km-long tunnel along the river.

An analysis of 15 watercourses out of a total of 16 that were the subject of the First Tender of the Ministry of Economy of Montenegro for the construction of SHPPs has established that sections with disturbed (reduced) flow have a total length of 60 km [9]. This sum is obtained as the sum of the distances between the water catch and release points for each hydropower project. This study also revealed that 43% of this length has guaranteed less than 60 L s^{-1} of water [9].

In the last years, SHP development has been accompanied by an increasing number of conflicts with the local community over these facilities. This has been recognized in the public as a consequence of the lack of an established set of equitable sharing principles among the contending parties [3]. The result was that the government faced difficult decisions to balance between the SHP

Fig. 2 Photo of SHPP "Majstorovina" on the Bistrica stream in the Bijelo Polje Municipality. Photo by System-mne company

development and environmental and social concerns. In October 2019, the government is deciding to terminate the contract for the construction of seven SHPPs, including two planned to be built on the Bukovica River, which was a source of intense conflict with the local community. Such complex but also costly decisions indicate the need for up-front consideration during the planning phase of the construction of hydropower facilities.

3 The Lack of Scientific Data Affects the Protection of Aquatic Biota

Montenegro is characterized by an extraordinary rich and diverse aquatic biota that inhabits various types of running waters. On the global scale, the region of Western Balkan (Dinaric area) to which Montenegro belongs is recognized as one of the most important hotspots of biodiversity and endemism. This is the result of a turbulent geological history of the area over the last 30 million years, during which an enormous and dynamic landscape change occurred, resulting in a great habitat complexity [10]. The karst had a decisive influence on the hydrographic isolation that has led to allopatric speciation and vicariant events resulting in a high endemism of local biota (Fig. 3) [10].

Isolated habitats such as canyons which often dominate the highly modified karstic landscapes of Montenegro are important habitats for shaping biodiversity. Karadžić and colleagues [11] analyzing biodiversity in mesic forests showed that the components of α-diversity (species richness and evenness) are greater in canyons than in non-ravine habitats. Higher plant diversity in canyons is attributable to heterogeneous microclimatic conditions along strong thermal and moisture gradients [11].

Fig. 3 Photo of the Vlado's shrimp (*Atyaephyra vladoi*), recently described species that inhabit the Skadar lake and adjacent springs and rivers, illustrates the outstanding biodiversity of Montenegrin aquatic ecosystems. Photo by B. Sket

Biogeographically, the territory of Montenegro is located within two regions, the Mediterranean and Alpine biogeographical regions, whose borders run approximately parallel to the shore of the Adriatic Sea, coinciding with the boundary between the Adriatic and Black Sea basins of the country [12]. According to the FEOW (Freshwater Ecoregions of the World) ecoregion system, the geographic range of Montenegro includes three freshwater ecoregions, i.e., the Dniester-Lower Danube, the Dalmatian, and the Southeastern Adriatic [13]. The biogeographical and ecoregional boundaries are not always clear, and sometimes they require revision. Contrary to the accepted FEOW global freshwater ecoregion system, which places the eastern border of the Dalmatian Ecoregion near the town of Budva [13], Pešić and colleagues in this book [10] moved the eastern border of the latter ecoregion west of the river Sutorina, located near the administrative border of Montenegro. According to this concept, the northwestern boundaries of the Southeastern Adriatic Ecoregion overlap with the distribution border of the fish *Pelasgus minutus* [10].

In this book (see [10]), Pešić and colleagues gave a brief overview of the diversity and ecology issues related to aquatic biota of the main river basins in Montenegro. They highlighted the large gaps in knowledge of some well-known groups of aquatic biota in Montenegro, especially those that have been proven to be good biogeographical indicators such as fish [10]. At the same time, the authors [10] emphasized the biogeographical significance of other groups of organisms such as underground fauna as potentially good biogeographic indicators, especially in karst areas where surface running waters are scarce and almost all water drains underground. Regardless of the uncertain paleogeography of the Montenegrin river network, recent studies (see [14] for an overview) demonstrated the significance and uniqueness of the aquatic biota of Montenegrin river basins proving their usefulness as a model system for future biogeography and evolution studies.

4 The Global Climate Change May Accelerate the Negative Impact on Biodiversity

The biodiversity of aquatic ecosystems is threatened by a number of factors that have not yet been sufficiently studied which may undermine proposing and implementing appropriate conservation plans. Hydrological regime alteration, climate change, pollution, and invasive species are generally recognized as the main threats for all types of aquatic and wetland habitats [15]. The abovementioned threats are also among the most important factors that shape the integrity of the rivers and their basins in Montenegro affecting structural and functional connectivity of their habitat patches.

Climate change is perceived as one of the most serious threats to the planet's biodiversity [16]. The change in drought (and less precipitations) and flooding (and rising precipitations) regime affects both individual species and the entire aquatic assemblage. Generalist species, such as most aquatic plants, are assumed to be relatively tolerant of a wide range of environmental factors. On the other hand, diversity and abundance of specialist species are often related to the relative habitat quality making them more susceptible to disturbance events.

Climate change has increased markedly since the early 1990s which certainly has an impact on aquatic ecosystems. The available data showed that the 2001–2010 was the warmest decade since records began, with temperature increasing by +1.40°C in the northern mountainous region and +1.30°C in the coastal region [17]. The rise of ambient air temperature leads to changes in water temperature what consequently led to changes in biological processes in the rivers. Vukašinović-Pešić and colleagues [18] proved an increase in the frequency of anomalous and extreme events of water temperature in the Black Sea basin over the studied 10-year period.

Generally, the available but still scarce long-term data shows that variability in precipitation become more extreme [17]. The changes are reflected in the considerable interannual variability in the flow between and within the river basins [1]. This is likely to have an effect on the physicochemical equilibrium of the rivers [18] that in turn affect their living world.

The hydrological regime changes are particularly visible in the reduced flow in the summer but also in the increased proportion of intermittent rivers and ephemeral streams (IRES) in the hydrological network [3, 19]. Intermittent rivers (IRs) that periodically cease to flow are particularly common in the southern part of Montenegro and form a high proportion of river networks in the Mediterranean and sub-Mediterranean parts of the country. These habitats were often neglected by ongoing biomonitoring programs [20]. However, recent studies have shown that these habitats provide essential ecosystem services, for example, in contributing to nutrient and carbon cycling [21].

Further climate change will be expected to lead to an increased prevalence of intermittent rivers in the local and regional hydrological network [19]. At the same time, it can be expected that natural intermittent rivers will also be affected by changing their natural regime from near-perennial to episodic flow [19]. In this book [19], Pešić and colleagues gave an overview of the species and communities that inhabit intermittent rivers in the southern part of Montenegro. They stressed the importance of further research on these habitats and the need for their inclusion in ongoing monitoring programs ensuring that the assessment of their ecological status encompassed both phases, the flowing and the dry one (Fig. 4), and the biota that inhabits them [19].

Climate change significantly contributes to spreading invasive species. The lack of historical data for the most invasive species in aquatic ecosystems of Montenegro and the human impact that accelerates their spread hampers our knowledge. Although the lists and detailed studies are still missing, the number of invasive species has been recorded in many aquatic biotic groups, especially fish and mollusks. A total of 13 freshwater fish species have been introduced in aquatic ecosystems of Montenegro, while several fish (such as the grayling that entered into the Morača and the Skadar's chub that entered into the

Fig. 4 Photo of the Sitnica (Matica) River near Podgorica. Intermittent sites can harbor rich biodiversity as a result that different taxa colonize the same location over a temporal gradient, e.g., lotic taxa colonize stream riverbed when the flow resumes, while terrestrial taxa colonize dry sediment. Photo by V. Pešić

Piva basin) have been translocated from one to another basin [22]. Recent studies (e.g., [22]) showed that the Adriatic Sea basin contains higher number of allochthonous (introduced + translocated) species (13 spp.) than the Black Sea basin (7 spp.).

The most invasive mollusk in the Montenegrin aquatic ecosystems is *Dreissena polymorpha*, a species recognized by the IUCN I as 1 of 100 in the World's Worst Invasive Alien Species [23]. In Montenegro, the presence of this species was found in the lake Šasko and the Bojana/Buna River, and at the moment *D. polymorpha* does not show evidence of spreading to other water bodies in Montenegro [24].

5 Water Quality Monitoring Program as a Tool Toward Science-Based Management

As is the case with other countries in the region, Montenegro is formally committed to implementing the European Union Water Framework Directive 2000/60/EC (WFD) [25]. The ultimate goal of WFD implementation is to achieve the "good ecological status" for all watercourses on state territory. The first step in fulfilling this goal is identifying factors that may lead to deterioration of water quality. This is achieved through the implementation of the National Surface Water

Monitoring Programme of Montenegro, which monitors the physical, chemical, and microbiological indicators. The monitoring was conducted by the Institute for Hydrometeorology and Seismology of Montenegro [18].

The results of monitoring have recognized municipal sewage as the main source of surface water pollution in Montenegro [18]. Despite the consistent improvement of wastewater management in the country, municipal sewage treatment is still insufficient. In addition, a significant amount of raw sewage is directly released to streams and rivers. This is especially common during heavy rains as a result of the low capacity of channels that are used at the same time to transport sewage as well as to transport rainwater from street inlets to the sewage treatment plant.

Pollution of agricultural origin as well as chemical pollution generated by industry is ranked among the more dangerous threats [18]. The excess nutrient load originating from the fertilizers in the adjacent fields is a cause of increased aquatic ecosystem eutrophication [26], especially in lentic habitats.

The hydrological alteration may cause deterioration of water quality in rivers by influencing sources, migration, and transformation of pollutants [18]. Vukašinović-Pešić and colleagues [18] studied the interannual variability of the frequency of occurrence of extreme values of the chemical parameters of water quality over a 10-year period (2009–2018). They found that the frequency of anomalies in the Adriatic Sea basin decreased from an average of 20.8 in 2009 to 10.4 anomalous events in 2018 [18]. On the other hand, the Black Sea basin exhibited higher fluctuations in anomalies over the studied period, with a weak decreasing trend in the frequency of anomalous events and an increasing trend for extreme events over the studied 10-year period [18].

Notwithstanding the significant threats described above, the water quality index (WQI) over a 10-year period, however, revealed an increase in water quality of rivers in Montenegro since 2012 [18]. This was confirmed also by the analysis of microbiological parameters in the same period, which showed that the overall microbiological water quality of rivers in Montenegro is quite good [27]. Kolarević and colleagues [27] showed that about 80% of investigated sites of Montenegrin rivers have bathing water quality.

The National Surface Water Monitoring Programme of Montenegro recognized certain localities and watercourse as hotspots of pollution. Based on the WQI results, the worst water quality had the Ćehotina River and its tributary the river Vezišnica, as well as the Morača River below the effluent outlet from the sewage treatment plant in Podgorica and the river Ibar below Rožaje [18]. Among the studied rivers, fecal pollution was the largest at the sites (i.e., river Ćehotina downstream of Pljevlja, river Vezišnica, river Lim downstream of Bijelo Polje, and Morača River in Podgorica near the outlet of wastewater treatment plant in Podgorica) that were exposed to municipal wastewater pollution. The highest level of *E. coli* and intestinal enterococci indicator bacteria was found in the vicinity of wastewater outlet in Podgorica [27], indicating the presence of a high risk of poor implementation of wastewater treatment in Montenegro.

Among the most striking examples of polluted rivers in Montenegro are the Ćehotina River and its tributary the river Vezišnica. For decades, they received pollutants (chemicals but also ash and slag) from the Pljevlja thermal power plant and a coalmine [18]. Over that time there were a number of episodes of the illegal release of chemicals into the water of the Ćehotina River, often with fatal consequences for the aquatic biota [10].

Recent studies (see [28] for an overview) have shown that deterioration of the water quality of river ecosystems as a result of chemical and/or fecal pollution may affect their ecotoxicological status. This finds confirmation in the results of some bioassay tests that have revealed high toxicity at sites exposed to chemical and/or fecal contamination. For example, Kračun-Kolarević and colleagues [28] found the highest responses of biomarkers indicating embryotoxic, genotoxic, and phytotoxic effects in zebra fish embryo test and in roots of Allium vaccine in samples from Ćehotina and Tara Rivers downstream of urban centers (Pljevlja and Mojkovac towns, respectively). On other hand, genotoxic endpoints in zebra fish bioassay test recognized sites on the Morača and Lim rivers which are considered to be more exposed to fecal pollution [28].

In addition to pollution, the intensive regulatory works along rivers are recognized as one of the main pressure threats to the integrity of the rivers in Montenegro. The "river regulation" has been frequently performed in recent decades under the pretext of controlling floods. In reality, this was often an excuse for the sand and gravel extraction activities (e.g., in the lower part of the Morača River, see Fig. 5a) but also for large infrastructural development. The most recent example is the impact of the Bar-Boljare highway construction project on the UNESCO protected Tara River (Fig. 5b).

The urban streams and rivers that in most cases flow through concreted canals likely are examples of aquatic ecosystems that are most heavily transformed and disturbed. In Montenegro, even there have been attempts to concrete the riverbed of some urban streams, as was the case with the river Ribnica in the center of Podgorica. It is known that river regulation has negative impacts on the riverine biota by reducing floodplain areas which are well-known as spawning grounds and rearing sites for juveniles of fishes and other organisms [29, 30].

6 Further Perspectives

The protection of water and aquatic ecosystems is increasingly becoming a central concern at the institutional, scientific, and social levels. Increased demand for water (projected to increase some 55% by 2050, [31]) and energy projected to increase in the coming decades (by 30% by 2040 [32]) is likely to increase the pressure to maximize the use of water resources. In recent years, the rivers in Montenegro have become a point of contention that involved a number of contending parties, from environmentalists and private businesses to governmental and nongovernmental institutions.

Fig. 5 (**a**) The sand and gravel extraction in the lower part of the Morača River. Photo by V. Pešić. (**b**) The regulation of the Tara River near Mateševo during the construction of the Bar-Boljare highway. Photo by D. Marić

In countries that have been in transition for more than a few decades, such as Montenegro, the conflict has meant choosing one of two sides. In the public, especially the one that is environmentally friendly, the conflict is recognized as a suppression of the project aimed at environmental protection in relation to the projects aiming at energetic (and economic) valorization of hydropower potential. The fair estimates of available water resources and projections for their future use were often undermined by political decisions, which are best seen in the case of small hydropower (SHP) development in Montenegro.

Available studies (e.g., [5]) have shown that the energy production capacity of SHP is as much as 14 times smaller than the large hydropower plants (LHP). Furthermore, other indicators (see [33]) such as the number of dams recognized as obstacles to flow velocity and the habitat loss measured as river length with disturbed natural flow per kWh produced were also much higher for SHP than for LHP. Concluding, in addition to the fact that SHPPs are much less efficient in terms of energy production, they also more likely to generate high cumulative impact.

An emerging lesson from the recent experience with SHP development in Montenegro has shown that only implementing studies that analyze the cumulative implications of such development on river ecosystems from an aggregated perspective would allow planning sustainable energy-environment systems. Ensuring collaboration between all of the contending parties from scientists and decision-makers to the representatives of a local community is the main factor that can prevent potential conflicts and guarantee sustainable planning of the water resources.

As a step toward this goal is the book *The Rivers of Montenegro* which is addressed to different types of specialists, from environmental chemists, hydrologists, and water resource managers to those working in the various fields of fundamental and applied ecology. As mentioned in the introductory chapter [3], this book is aimed at stimulating toward a science-based management approach in the use of water resources in Montenegro.

Acknowledgments We are indebted to all the authors of chapters for their enthusiasm, commitment, and insights that have resulted in this book. Moreover, we thank the many peoples who helped us with discussion and ideas for this book. Thanks to Saša Popović (Podgorica), Ljubiša Perović (System-mne d.o.o, Podgorica), Boris Sket (Ljubljana), and Drago Marić (Podgorica) for their permission to use their photos for this chapter.

References

1. Sekulić G, Radulović M (2020) The hydrology and hydrogeology of Montenegro. In: Pešić V, Paunović M, Kostianoy A (eds) The rivers of Montenegro. The handbook of environmental chemistry. Springer, Cham
2. IJC (Institute Jaroslav Cerni) (2001) Water master plan of Montenegro. Government of Montenegro, Podgorica. (in Serbian)
3. Pešić V, Paunović M, Kostianoy V (2020) The rivers of Montenegro – introductory remarks. In: Pešić V, Paunović M, Kostianoy A (eds) The rivers of Montenegro. The handbook of environmental chemistry. Springer, Cham
4. Energy Strategy of the Republic of Montenegro to 2025 (2006) Book D: development plan of the power system of the Republic of Montenegro – master plan. Institute for research in energy, ecology and technology, Ljubljana & Energy Institute "Hrvoje Pozar", Zagreb. (In Montenegrin)
5. Sekulić G (2020) The utilization of the hydropower potential of rivers in Montenegro. In: Pešić V, Paunović M, Kostianoy A (eds) The rivers of Montenegro. The handbook of environmental chemistry. Springer, Cham. https://doi.org/10.1007/698_2019_412

6. Đorđević B, Šaranović M (2007) Hydropower potential of Montenegro. Montenegrin Academy of Science and Arts, Special Issues (Monographs and Studies), Book 55, Podgorica (in Serbian)
7. COTEE (Montenegrin Electricity Market Operator) (2019) Annual energy report of the RES and VEC service for 2018. http://www.cotee.me/index.php/2012-10-16-12-20-18 (In Montenegrin)
8. Fu X, Tang T, Jiang W, Li F, Zhou S, Wu N, Cai Q (2008) Impacts of small hydropower plants on macroinvertebrate communities. Acta Ecol Sin 28:45–52
9. Bajković I. (2013) Minimal (ecological) flows in Montenegro. Diploma Thesis, University of Montenegro (In Montenegrin)
10. Pešić V, Grabowski M, Hadžiablahović S, Marić D, Paunović M (2020) The biodiversity and biogeographical characteristics of the River basins of Montenegro. In: Pešić V, Paunović M, Kostianoy A (eds) The rivers of Montenegro. The handbook of environmental chemistry. Springer, Cham
11. Karadžić B, Bulić Z, Jarić S, Mitrović M, Pavlović P (2020) Vegetation in ravine habitats of Montenegro. In: Pešić V, Paunović M, Kostianoy A (eds) The rivers of Montenegro. The handbook of environmental chemistry. Springer, Cham
12. EEA (European Environment Agency) (2016) Biogeographic regions of Europe. https://www.eea.europa.eu/data-and-maps/figures/biogeographical-regions-in-europe-2
13. FEOW (2008) Freshwater ecoregions of the world-conservation science program. The Nature Conservancy and World Wildlife Fund, Washington DC
14. Grabowski M, Jabłońska A, Wysocka A, Pešić V (2018) The obscure history of the Lake Skadar and its biota: a perspective for future research. In: Pešić V, Karaman G, Kostianoy A (eds) The Skadar/Shkodra lake environment. The handbook of environmental chemistry, vol 80. Springer, Cham, pp 47–61
15. Janssen JAM, Rodwell JS, Garcia Criado M, Gubbay S, Haynes T, Nieto A, Sanders N, Landucci F, Loidi J, Ssymank A, Tahvanainen T, Valderrabano M, Acosta A, Aronsson M, Arts G, Attorre F, Bergmeier E, Bijlsma R-J, Bioret F, Bita-Nicole C, Biurrun I, Calix M, Capelo J, Carni M, Chytry M, Dengler J, Dimopoulos P, Essl F, Gardjfell H, Gigante D, Giusso del Galdo G, Hajek M, Jansen F, Kapfer J, Mickolajczak A, Molina JA, Molnár Z, Paternoster D, Piernik A, Poulin B, Renaux B, Schaminee JHJ, Sumberova K, Toivonen H, Tonteri T, Tsiripidis I, Tzonev R, Valachovic M (2016) European red list of habitats, part 2: terrestrial and freshwater habitats. EU, Brussels, 44 pp
16. Malcolm JR, Liu C, Neilson RP, Hansen L, Hannah LEE (2006) Global warming and extinctions of endemic species from biodiversity hotspots. Conserv Biol 20:538–548
17. The Second National Communication on Climate Change (2015) The Second National Communication on Climate Change of Montenegro to the United Nations Framework Convention on Climate Change (UNFCCC). Ministry of Sustainable Development and Tourism, Podgorica. http://www.mrt.gov.me/ResourceManager/FileDownload.aspx?ridĽ205823&rTypeĽ2. Accessed 1 May 2019
18. Vukašinović-Pešić V, Blagojević N, Savić A, Tomić N, Pešić V (2020) The change in the water chemistry of the rivers of Montenegro over a 10-year period. In: Pešić V, Paunović M, Kostianoy A (eds) The rivers of Montenegro. The handbook of environmental chemistry. Springer, Cham. https://doi.org/10.1007/698_2019_417
19. Pešić V, Pavićević A, Savić A, Hadžiablahović S (2020) The intermittent rivers of South Montenegro: ecology and biomonitoring. In: Pešić V, Paunović M, Kostianoy A (eds) The rivers of Montenegro. The handbook of environmental chemistry. Springer, Cham
20. Stubbington R, Chadd R, Cid N, Csabai Z, Miliša M, Morais M, Munné A, Pařil P, Pešić V, Tziortzis I, Verdonschot RC (2018) Biomonitoring of intermittent rivers and ephemeral streams in Europe: current practice and priorities to enhance ecological status assessments. Sci Total Environ 618:1096–1113. https://doi.org/10.1016/j.scitotenv.2017.09.137
21. Datry T, Foulquier A, Corti R, von Schiller D, Tockner K, Mendoza-Lera C et al (2018) A global analysis of terrestrial plant litter dynamics in non-perennial waterways. Nat Geosci 11:497–503. https://doi.org/10.1038/s41561-018-0134-4

22. Marić D (2019) Fauna of freshwater fish (Osteichthyes) of Montenegro. Montenegrin Academy of Science and Arts, Special edition, Book 149, Department of Natural Sciences, vol 48 (in Serbian)
23. Lowe SJ, Browne M, Boudjelas S (2000) 100 of the world's worst invasive alien species. IUCN/SSC Invasive Species Specialist Group (ISSG), Auckland. http://www.issg.org/pdf/publications/worst_100/english_100_worst.pdf
24. Pešić V, Glöer P (2018) The diversity and conservation status of the Molluscs of Lake Skadar/Shkodra. In: Pešić V, Karaman GS, Kostianoy AG (eds) The Skadar/Shkodra lake environment. The handbook of environmental chemistry, vol 80. Springer, Cham
25. European Commission (2000) Directive 2000/60/EC, establishing a framework for community action in the field of water policy. Off J Eur Comm Brussels L 327:1–71
26. Pešić V, Karaman GS, Kostianoy AG, Vukašinović-Pešić V (2018) Conclusions: recent advances and the future prospects of the Lake Skadar/Shkodra environment. In: Pešić V, Karaman GS, Kostianoy AG (eds) The Skadar/Shkodra lake environment. The handbook of environmental chemistry, vol 80. Springer, Cham
27. Kolarević S, Kračun-Kolarević M, Jovanović J, Ilić M, Paunović M, Kostić-Vuković J, Martinović R, Jokanović S, Joksimović D, Pešić V, Kirschner A, Linke R, Ixenmaier S, Farnleitner A, Savio D, Reischer G, Tomić N, Vuković-Gačić B (2020) Microbiological water quality of rivers in Montenegro. In: Pešić V, Paunović M, Kostianoy A (eds) The rivers of Montenegro. The handbook of environmental chemistry. Springer, Cham
28. Kračun-Kolarević M, Kolarević S, Jovanović J, Đorđević J, Ilić M, Sunjog K, Kostić-Vuković J, Divac Rankov A, Ilić B, Pešić V, Vuković-Gačić B, Paunović M (2020) Impact of pollution on rivers in Montenegro – ecotoxicological perspective. In: Pešić V, Paunović M, Kostianoy A (eds) The rivers of Montenegro. The handbook of environmental chemistry. Springer, Cham
29. Penczak T, Zięba G, Koszaliński H, Kruk A (2003) The importance of oxbow lakes for fish recruitment in a river system. Arch Hydrobiol 158:267–281
30. Ivanchev VP, Ivancheva EY, Tereshchenko VG (2011) Significance of a floodplain in the formation of fish communities in small rivers of Ryazan oblast. J Ichthyol 51:618–632
31. OECD (2012) OECD environmental outlook to 2050. Organization for Economic Co-operation and Development, Paris
32. IEA (2016) World energy outlook 2016. Paris, International Energy Agency
33. Mayor B, Rodríguez-Muñoz I, Villarroya F, Montero E, López-Gunn E (2017) The role of large and small scale hydropower for energy and water security in the Spanish Duero Basin. Sustainability 9:1807

Printed by Printforce, the Netherlands